24th International Conference of the Center for Chemical Process Safety 2009

Topical Conference at the 2009 AIChE Spring National Meeting

Tampa, Florida
26 – 30 April 2009

ISBN: 978-1-61567-255-4

Printed from e-media with permission by:

Curran Associates, Inc.
57 Morehouse Lane
Red Hook, NY 12571

CURRAN ASSOCIATES INC. proceedings.com

Some format issues inherent in the e-media version may also appear in this print version.

Printed by Curran Associates, Inc. (2009)

For permission requests, please contact AIChE
at the address below.

AIChE
3 Park Avenue
New York, NY 10016-5991

Phone: (203) 702-7660
Fax: (203) 775-5177

www.aiche.org

Additional copies of this publication are available from:

Curran Associates, Inc.
57 Morehouse Lane
Red Hook, NY 12571 USA
Phone: 845-758-0400
Fax: 845-758-2634
Email: curran@proceedings.com
Web: www.proceedings.com

TABLE OF CONTENTS

Welcome to the 2009 Center for Chemical Process Safety's 24th Annual International Conference: Process Safety-The Next Generation.

The last several years has brought considerable change and challenges. A weak global economy, attrition of process safety resources, increasing regulatory activities, mergers and acquisitions, and expanding technical concerns are just some of the issues shaping the future. Process safety must continually adapt to meet these challenges to ensure the long-term viability and survivability of our facilities, business, and industries.

To speak to these issues, this year's conference brings together many professionals from different countries, sectors, and industries to share ideas, methods, and tools for addressing the challenges which the future holds. The topics and sessions are designed to provide opportunities for you to learn new techniques, sharpen your existing skills, and learn what other professionals are developing as they lead their organizations into the next generation of process safety.

Thank you for your participation in process safety and your leadership in advancing our discipline.

Eric Freiburger/NOVA Chemicals, Inc.
2009 CCPS Conference Chair

On behalf of the CCPS Conference Committee and Session Chairs:

Scott Berger/CCPS
Roxy Schneider/CCPS
Karen Tancredi/DuPont
John Wincek/Croda Incorporated
Donald J. Connolley/BP Products North America
Cheryl Grounds/BP Exploration & Production
Christie Arseneau/AcuTech Consulting Group
Robert F. Wasileski/NOVA Chemicals, Inc.
Jatin Shah/Baker Engineering and Risk Consultants
Shakeel Kadri/Air Products & Chemicals Inc.
William Bridges/Process Improvement Institute, Inc.

2009 Global Congress on Process Safety

Plenary and Luncheon Speakers

2009 Global Congress Opening and Keynote Address, April 27, 8:30AM-9:30AM

- The Global Congress Keynote Address by Invited Speaker, **John Bresland**, Chairman and CEO of the U.S. Chemical Safety Board.
- *Ballroom A, Tampa Convention Center*

5th Global Congress on Process Safety Keynote Luncheon, April 27 11:45AM-1:15PM

- Luncheon Keynote will be given by **Laura Itle,** Technical Lead for the Office of SAFETY Act Implementation. Dr. Itle will give a presentation titled "SAFETY Act and You".
- *Ballroom C, Tampa Convention Center*

5th Global Congress on Process Safety Luncheon Talk, April 29, 12:00PM-1:15PM

- **Lisa Morrison**, PSM Audit Manager of BP International, will discuss BP's PSM Audit Program.
- *Ballroom C, Tampa Convention Center*

MONDAY, APRIL 27

	Session #16 *Joint Plenary Session (CCPS/LPS/PPSS)*
8:00 am	Introductions & Welcome, **John Murphy** and **Scott Berger**
8:10 am	Introductory Remarks – **Jean-Paul Lacoursiere**, LPS Chair – **John Champion**, PPSS Chair – **Eric Freiburger**, CCPS Chair
8:20 am	William H. Doyle LPS Best Paper Award
8:30 am	Keynote Address: **Invited Speaker, Mr. John Bresland**, Chairman and CEO, U.S. Chemical Safety & Hazard Investigation Board
9:20 am	**Coffee and Networking Break in Exhibit Hall**

	43rd Annual Loss Prevention Symposium	11th Process Plant Safety Symposiun	24th CCPS International Conference
	Session #18: **Fire, Explosion and Reactive Chemicals** Location: Room 24/25 **Session Co-Chairs: Stanley S. Grossel & John A. Noronha**	***Session #19:*** **Applications of LOPA** Location: Room 22/23 **Session Co-Chairs: Jack Chosnack & James Gary**	***Session #20:*** **Auditing Process Safety** Location: Ballroom D **Session Chair: Christie Arseneau**
10:00 am	Approach for the Development of a More Efficient and Safer Process in the Pharmaceutical Industry *(L. R. Saenz)*	Key Issues with Implementing LOPA (Layer of Protection Analysis) – Perspective from One of the Originators of LOPA *(Bill Bridges)*	Can Auditing Be Improved to Drive PSM Performance Gains? *(James A. Klein)*
10:30 am	Predicting Explosibility Properties of Chemicals from Quantitative Structure — Property Relationships *(Patricia Rotureau)*	Experience of a Very Large Layer of Protection Analysis with More than 2000 'Cause – Consequence Pairs' Covering a Large Gas Terminal in Europe *(Richard Gowland)*	Auditing in Four Levels to Assure Completeness *(Adrian L. Sepeda)*
11:00 am	Minimum Ignition Energy of Dust Cloud. The Effect of the Amount of Inert Dust *(Luca Marmo)*	Improved Integration of LOPA with Hazop Analyses *(Dick Baum)*	Auditing Global Logistics Operations – A Process Safety Focus *(Michael R. Green)*
11:45 am	Luncheon With Speaker: **Dr. Laura Itle**, Officer of Safety ACT Implementation		
	Session #24: **Consequence Modeling** Location: Room 24/25 **Session Co-Chairs: David G. Clark & Brian R. Dunbobbin**	***Session #25:*** **Safety Culture and Operating Discipline** Location: Room 22/23 **Session Co-Chairs: James R. Thompson & James A. Klein**	***Session #23:*** **Implementing PSM Globally** Location: Ballroom D **Session Chair: Shakeel Kadri**
1:30 pm	Development of Algorithms for Predicting Ignition Probabilities and Explosion Frequencies *(Mike Moosemiller)*	The Impact of the BP Baker Report *(Stephanie Payne)*	Training Tools to Support a Global Process Safety Culture *(Dan Reinke)*
2:00 pm	On the Use of Models for Accident Investigations *(Kees van Wingerden)*	Identifying "Culture Issues" Using Root Cause Analysis *(Mark Paradies)*	Implementing and Sustaining a Global Process Safety Program *(Christie A. Arseneau)*
2:30 pm	A Versatile Approach to Explosion Vent Design for Enclosures Handling Flammable Vapors *(David Herrmann)*	Implementing a Culture of Safety and Operations Excellence at Newport Chemical Demilitarization Facility (NECDF) *(Tulanda Brown)*	Implementing PSM Globally *(David Cummings)*
3:00 pm	**Coffee and Networking Break in Exhibit Hall**		

MONDAY, APRIL 27 Continued

	Session #24: **Consequence Modeling (Continued)** Location: Room 24/25 **Session Co-Chairs: David G. Clark & Brian R. Dunbobbin**	***Session #25:*** **Safety Culture and Operating Discipline (Continued)** Location: Room 22/23 **Session Co-Chairs: James R. Thompson & James A. Klein**	***Session #41:*** **Tools to ID Hazards** Location: Ballroom D **Session Chair: Robert F. Wasileski**
3:30 pm	Consequence Modeling Lack of Accuracy and its Effects on Quantitative Risk Analysis Results *(Mohammed J. Ayyoubi)*	From Training to Human Reliability: Lessons Learned During the Implementation of a Human Reliability Function in the Pharmaceutical Industry *(Ginette Collazo-Ramos)*	Accident Investigation Using Process Control Event Diagrams *(D. R. Morrison)*
4:00 pm	Advanced CFD Modeling on VCE and Vapor Dispersion *(Anna Qiao)*	Improving Operating Discipline Through the Successful Implementation of a Behavior-Based Safety Program *(B. Vaughen)*	The Retro-Synthesis of Risk Assessment Tools for Hazard Evaluation and Screening Analysis *(Fred Henselwood)*
4:30 pm	LNG Modeling for Facility Siting *(David Moore)*	Safety Culture Implementation Programs in a Developing Country *(Robin Pitlado)*	A Practical Approach to Hazard Identification for Operations and Maintenance Workers *(Christy Franklyn)*

TUESDAY, APRIL 28

	43rd Annual Loss Prevention Symposium	**11th Process Plant Safety Symposiun**	**24th CCPS International Conference**
	Session #46: **Hazard Evaluation and LOPA** Location: Room 24/25 **Session Co-Chairs: Robert W. Johnson & Mike Moosemiller**	***Session #47:*** **Mechanical Integrity — State of the Art** Location: Room 22/23 **Session Co-Chairs: Steven D. Emerson & James Willis**	***Session #49:*** **PSM Metrics and Culture** Location: Ballroom D **Session Chair: John Wincek**
8:00 am	Oversights and Omissions in Process Hazard Analyses: Lessons Learned from CSB Investigations *(Mark Kaszniak)*	New Statistical Insights into Reasons for Initial Failures in Pressure Relief Valves *(Julia Bukowski)*	Cheddar or Swiss? How Strong are Your Barriers? *(Michael P. Broadribb)*
8:30 am	Optimizing Qualitative Hazard Evaluation *(William (Bill) Bridges)*	Validation of Spring Operated Pressure Relief Valve Time-to-Failure and the Importance of Statistically Supported Maintenance Intervals *(Robert E. Gross and Stephen P. Harris)*	Implementation of Process Safety Management Into Diverse Corporate Cultures *(C. A. Soczek)*
9:00 am	LOPA as Practiced at Pfizer's Global Manufacturing API Facility in Kalamazoo, MI *(G.E Evenson)*	Lessons Learned from Corrosion Under Insulation *(Tim Overton)*	Process Safety Leadership – A Systematic Approach *(C. D. Shelton)*
9:30 am	**Coffee and Networking Break in Exhibit Hall**		

(Tuesday, April 28 continued on next page)

TUESDAY, APRIL 28 Continued

	Session #46: **Hazard Evaluation and LOPA (Continued)** Location: Room 24/25 **Session Co-Chairs: Robert W. Johnson & Mike Moosemiller**	***Session #47:*** **Mechanical Integrity — State of the Art (Continued)** Location: Room 22/23 **Session Co-Chairs: Steven D. Emerson & James Willis**	***Session #49:*** **PSM Metrics and Culture (Continued)** Location: Ballroom D **Session Chair: John Wincek**
10:15 am	The Buncefield (U.K.) Fire and Explosion: Improving Layer of Protection Analysis Practise to Determine the Required Degree of Protection to Meet Regulator Requirements *(Richard Gowland)*	Implementing Mechanical Integrity at Solutia *(F. Russ Davis)*	Safety Culture: From Black Art to Real Improvements *(Cinzia De Santis)*
10:45 am	The New Approach of Risk Analysis in France for Seveso Establishments: Strengths and Weaknesses *(Jerome Taveau)*	Adoption of Mechanical Integrity Principles at Air Liquide Large Industries US *(Danny Ledbetter)*	OECD Guidance on Safety Performance Indicators *(Kim Jennings and Fran Schulberg)*
11:15 am	Hazard Evaluations and Layer of Protection Analysis: Independent Protection Layers, A New CCPS Guideline Book *(John Murphy)*	Implementing Risk-Based Inspection on a Global Scale *(Andrew Morey)*	CCPS Process Safety Metrics Status Report *(Dorothy Kellog)*
12:00 NOON	**Networking Luncheon**		
	Session #62: **Food and Consumer Products Process Hazards** Location: Room 24/25 **Session Co-Chairs: John E. Going & Nir Keren**	***Session #63:*** **Process Safety – From the Top Down to Bottom Up** Location: Room 22/23 **Session Co-Chairs: Philip M. Myers & Donald Abrahamson**	***Session #61:*** **Transportation Risk** Location: Ballroom D **Session Chair: Cheryl Grounds**
1:30 pm	Preventing Combustible Dust Explosions in the Food and Consumer Products Industries *(John Bresland)*	The Process Safety Guerilla: Giving Your Company a Future, Even if it Doesn't Want One *(Scott Berger)*	Guidelines for Chemical Transportation, Safety, Security and Risk Management *(Brad Fuller)*
2:00 pm	From the Top Down – Hazard Identification and Risk Evaluation in Food Processing – Toxic Releases, Dust Explosions and Flammable Excursions *(Catherine Kollhoff)*	Using Playbooks – A Unique Strategy in Technical Oversight of Highly Hazardous Operations *(James Gary)*	A Model to Evaluate Pipeline Failure Frequencies based on Design and Operating Conditions *(Vania De Stefani)*
2:30 pm	Nothing to Sneeze at – Combustible Dust Hazard Process Safety *(Guy R. Colonna)*	An Impetus for Management Involvement in Process Safety *(Jack Chosnek)*	Analysis of the Explosion Linked to the Formation of an Explosive Atmosphere In a Shop Carrying Fuel Products *(Sébastien Evanno)*
3:00 pm	**Coffee and Networking Break in Exhibit Hall**		
	Session #62: **Food and Consumer Products Process Hazards (Continued)** Location: Room 24/25 **Session Co-Chairs: John E. Going & Nir Keren**	***Session #63:*** **Process Safety – From the Top Down to Bottom Up (Continued)** Location: Room 22/23 **Session Co-Chairs: Philip M. Myers & Donald Abrahamson**	***Session #77:*** **Fires, Explosions, and Quantitative Analysis** Location: Ballroom D **Session Chair: Jatin Shah**
3:30 pm	Promoting the Safe Use of Ammonia for Industrial Refrigeration *(Bruce Badger)*	What Does Management Need to Know To Lead Process Safety Improvement? *(Mark Paradies)*	Protection of Permanent Buildings at an Existing Upstream Oil and Gas Facility *(Mark Aspengren)*
4:00 pm	Process Hazards in the Distilled Spirits Industry, Birkmire *(John C. Birkmire)*	Process Safety – Perspectives from the Top *(Tom Archibald)*	Basic Tools for a QRA of an Oilfield Cold Vent *(Pablo Zunana)*
4:30 pm	Replacement of HCFC-22 Refrigerants with Ammonia: a Challenge for Hockey Arenas in Quebec *(Jean Paul Lacoursiere)*		Blast Response of Soft-Skinned Shelters and Occupant Vulnerability *(D.R. Whitehouse)*

WEDNESDAY, APRIL 29

	43rd Annual Loss Prevention Symposium	11th Process Plant Safety Symposiun	24th CCPS International Conference
	Session #82: **Audits, Inspection and Regulation** Location: Room 24/25 **Session Co-Chairs: Lisa Long & Frank H. Gurry**	*Session #83:* **Applications of Risk Analysis Techniques** Location: Room 22/23 **Session Co-Chairs: Alvin Waller & Douglas J. Ferguson**	*Session #79:* **Managing Organizational Change** Location: Ballroom D **Session Chair: Donald J. Connolley**
8:00 am	Auditing Your Process Safety Culture *(David M. Heller)*	Procter & Gamble Global Risk Ranking *(Frank H. Gurry)*	Identifying Key Safety Roles during Organizational Change *(P.A. Davidson)*
8:30 am	Preliminary Findings from OSHA's Refinery National Emphasis Program *(Lisa Long)*	Overfill Protection Systems – Complex Analysis, Simple Solution *(Angela Summers)*	Risk Assessment Process for Organizational Changes *(John C. Wincek)*
9:00 am	Overview of the Chemical Facility Anti-Terrorism Standards *(D. R. Deziel)*	Pressurization of Fixed Roof Storage Tanks Due to External Fire *(Fabien Fouillen)*	An Overview of the CCPS Guideline on Process Safety Consideration During the Merger and Acquisition Process *(G. Melham)*
9:30 pm	**Coffee and Networking Break in Exhibit Hall**		
	Session #82: **Audits, Inspection and Regulation (Continued)** Location: Room 24/25 **Session Co-Chairs: Lisa Long & Frank H. Gurry**	*Session #83:* **Applications of Risk Analysis Techniques (Continued)** Location: Room 22/23 **Session Co-Chairs: Alvin Waller & Douglas J. Ferguson**	*Session #94:* **Risk Criteria** Location: Ballroom D **Session Chair: Karen Tancredi**
10:15 am	Dust Explosion Prevention: Regulations, Standards, and Mitigation Techniques *(Thimothy J. Myers)*	Quantifying the Effectiveness of Explosion Protection Measures *(Peter Moore)*	Single Scenario Risk Criteria: Why Would I Want Them and How do I get Them? *(J. Wayne Chastain)*
10:45 am	Lessons Learned in Auditing Automated Systems for PSM Compliance *(Angela E. Summers)*	The Failure to Communicate: Risk Assessment Challenges Outside the Chemical Process Industries *(Catherine Kollhoff)*	The Application of Peak Risk Concepts within a Corporate Risk Program *(Fred Henselwood)*
11:15 am	A Practical Approach to Implementing a Process Safety Management System Upgrade in Response to an OSHA Refinery PSM NEP Audit *(Robert Walter)*	Are Our Facilities Safe?... And Now?... And Now? *(Melina Alpert)*	Choosing Appropriate Quantitative Safety Risk Criteria: Applications from the New CCPS Guidelines *(Dave Jones)*
12:00 Noon	Luncheon With Speaker: **Lisa Morrison**, BP International		

(Wednesday, April 29 continued on next page)

WEDNESDAY, APRIL 29 Continued

	Session #96: **Case Histories — Joint (CCPS/LPS/PPSS) Session** Location: Ballroom D **Session Chairs: Henry L. Febo, Cheryl Grounds, William Bridges, & Kathy Pearson**
1:30 pm	Lessons Learned from the 2006 Facility Explosion in Danvers, MA *(Scott Davis)*
2:00 pm	Root Cause Analysis of a Cryogenic Refrigeration System Explosion *(Delmar Morrison)*
2:30 pm	Valero – McKee Refinery Propane Release and Fire *(Jim Lay)*
3:00 pm	**Coffee and Networking Break in Exhibit Hall**
3:15 pm	An Oven Explosion Incident – Lessons Learned on PSI, SIS and MOC *(Michael D. Cazabon)*
3:45 pm	Polymerization Reactor Explosion at the XYZ Chemical Company *(Peter Howell)*
4:15 pm	Tank 15 – The Perfect Storm *(Jeffrey Oland)*
4:45 pm	Global Congress Closing Remarks

THURSDAY, APRIL 30

	All papers presented Thursday belong to the 43rd Annual Loss Prevention Symposium (LPS) — Session #111
	43rd Annual Loss Prevention Symposium ***Session #111:*** **Explosion Prevention and Mitigation — LPS Session** Location: Room 24/25
	Session Chairs: Dr. Erdem A. Ural & Brian Kelly
8:00 am	The Dynamics of Vented Gas Explosions *(C. Proust)*
8:30 am	Safe Thickness for Combustible Dust Layers *(Erdem Ural)*
9:00 am	Explosion Characteristics of Hydrogen-Air Mixtures in a Confined Space *(Dan Crowl)*
9:30 am	**Coffee and Networking Break in Exhibit Hall**
10:15 am	Prevention and Mitigation of Polyethylene and Hydrocarbon/Polyethylene Explosions *(P.R. Amyotte)*
10:45 am	Flammable Gas Venting/Treatment Lessons Learned and Best Practices *(James VanOmmeren)*
11:15 am	Process Management Lessons Learned from a Petroleum Refinery Spent Caustic Explosion *(K.P. Bloch)*
11:45 am	Flammability Characteristics of Low GWP Refrigerant HFO-1234yf *(Barbara Minor)*
12:15 pm	**Session Closing Remarks**

PSM AUDITING – PREPARING FOR OSHA'S CHEMICAL NEP'S

This presentation will review the 33 citations which OSHA has issued to date against refineries as a part of its National Emphasis Program ("NEP") and will focus on the citation patterns, OSHA's interpretations of the PSM Standard and its use of industry standards in the enforcement process. Lastly, the presentation will outline a strategy for incorporating this data into your PSM audit programs to improve workplace safety and to prepare for OSHA's Chemical NEP.

Mark S. Dreux
Arent Fox LLP
1050 Connecticut Avenue, N.W.
Washington, D.C. 20036
(202) 857-6405 (Direct)
(202) 857-6395 (fax)

Auditing (Process Safety) in Four Levels To Assure Completeness

Adrian L. Sepeda, PE—President, A. L. Sepeda Consulting Inc.
ADRIAN_L@swbell.net

"It's not what we don't know that gets us in trouble.
It's what we don't know that we don't know that gets us in trouble."
Author unknown

ABSTRACT

An audit must have a defined structure, an understood purpose, and enough evidentiary details to assure completeness. It should be divided into discrete levels with specific topics so that the auditor can fully identify and understand the degree to which each facility or company complies with the requirements. Each of these levels should also be designed to analyze different types of information, implementation actions, and infrastructure. This paper proposes and discusses a four-level approach to conducting a comprehensive and complete audit.

The four levels are:

1. *Recognition and acknowledgment of the need to comply with a specific set of rules or regulations—senior-level sponsored documents and actions;*
2. *An implementation strategy and infrastructure—procedures, practices, guidance documents and responsibilities necessary to frame and support the compliance efforts being measured;*
3. *Evidence of compliance—verifiable documentation, training records, and interviews that confirm necessary and required actions are institutionalized into the normal work practices; and*
4. *Quality assessments—confirmation that appropriate systems are in place and working as intended plus a mechanism for identifying and implementing improvement opportunities.*

The content and purpose of each level as well as the interaction between levels is discussed and explained in this paper.

Auditing (Process Safety) in Four Levels To Assure Completeness

Adrian L. Sepeda, PE—President, A. L. Sepeda Consulting Inc.
ADRIAN_L@swbell.net

INTRODUCTION AND BACKGROUND

An audit is a process used to gather sufficient facts and information, including statistical information, to verify or refute compliance with standards or requirements. Audits are conducted for a variety of reasons. These reasons range from trying to confirm that unlawful actions have taken place to validating conformance to a particular commitment, standard or law—and include almost everything in between. The typical evidence for conformance is verifiable proof and/or documentation that a particular action has been taken or that a specific physical item is in place.

Process safety auditing conforms to this same model.[1] A process safety audit confirms or refutes claims that a particular process safety-related activity or action has taken place or regularly takes place as scheduled. From a regulatory standpoint, the base model for requirements is the OSHA PSM regulation 29CFR 1910.119[2] and the OSHA Auditor's Guidance Document[3]. Buried in these documents are requirements at a variety of administrative and functional levels.

The OSHA PSM standard, 29CFR 1910.119(o)[4] requires that "Employers (covered by the standard) shall certify that they have evaluated compliance with the provisions of this section at least every three years to verify that the procedures and practices developed under the standard are adequate and are being followed." In general, industry covered by the OSHA PSM standard has implemented the provisions thereof, including the auditing requirement.

This paper addresses how the process safety auditing requirement should be completed to ensure not only that the appropriate actions are taking place, but that they are sustainable and instutionalized within the organization.

FOUR LEVELS OF AUDITING

OVERVIEW

The approach proposed in this paper involves four discrete levels of activities to be audited all the while considering the interaction between the different levels. In essence, these four levels start out as four questions that must be addressed:

1. Is there a sustainable commitment to do it?
2. Do you know how to do it?
3. Are you doing it?
4. How well are you doing it?

The OSHA auditor's Compliance Guidelines and Enforcement Procedures[3] do not specifically define these four questions but essentially asks them via a variety of questions on different topics throughout the investigative process.

Changing the questions into definable actions allows a more systematic and comprehensive auditing approach. By doing so, the four levels of auditing can then be shown as Figure 1 and expressed as:

1. Is there a commitment to do it? Becomes Recognition and acknowledgment of the need to comply with a specific set of rules or regulations
   - Typically this includes senior level sponsored documents and actions demonstrating support and expectations.
   - It also includes sustainability commitments (i.e. compliance on an ongoing basis).

2. Do you know how to do it? Becomes An implementation strategy and operational infrastructure
   - This involves the generation and application/enforcement of procedures, practices, guidance documents, responsibilities, and staffing/expertise necessary to understand, scope, and support the compliance efforts being measured.

3. Are you doing it? Becomes Evidence of compliance
   - Verifiable documentation, training records, and interviews that confirm necessary and required actions are taking place,[5] are institutionalized into the normal work practices, and that staffing and expertise is adequate.[6]

4. How well are you doing it? Becomes Quality control
   - Confirmation that appropriate systems are in place, sustainable and adequately resourced, working as intended,[5] plus a mechanism for identifying and implementing improvement opportunities is in place.

Figure 1—Four Levels of Auditing

DETAILED DISCUSSION

Each of these four levels contains discrete yet complementary and supporting information, actions and commitments necessary for full and sustainable implementation of the other levels. For example,

1. **Recognition and acknowledgment of the need to comply with a specific set of rules or regulations** generally involves actions from senior management and executives of the company. They must openly commit to meet the obligation(s) and make compliance a company philosophy, priority, and requirement. In so doing, they will typically generate documents and demonstrate such actions to provide sustainability even during adverse times. Typical documentation and actions are policy statements, mission statements, supporting and commitment speeches, staff meeting agenda items for updates on specific activities or projects, and support of membership in allied and supporting organizations (such as the American Chemistry Council, The Center for Chemical Process Safety, Mary Kay O'Connor Process Safety Center, etc.). These actions are visible indicators to company employees of senior level commitment to compliance. They help ensure that the commitment is appropriately prioritized and resourced. Further, they are the foundation for all other actions in support of compliance.

2. **An implementation strategy and infrastructure** is comprised of procedures, practices, guidance documents, and responsibilities necessary to frame and support the compliance efforts being measured. It ensures that the appropriate

training and expertise is in place and being used correctly. In reality, this mid-level management and supervision team converts the senior level commitment into implementable actions with "how-to" instructions and specific responsibilities, schedules, and timetables. This level is also usually responsible for developing and maintaining the update, revision, and management of change documentation and schedules.

3. **Evidence of compliance** gets down to the "show, tell and prove it" part of an audit. It involves retrieving and reviewing documentation that proves policies, procedures and practices were followed, that appropriate inspections reviews and analyses were conducted on schedule and recorded, and that appropriate decisions were made commensurate with the data obtained. In other words, a complete or statistically representative review of all pertinent records, decisions and actions is required to fulfill this section of the audit. The auditor must also seek and find information (and documentation) that confirms that the process of staying in compliance is instutionalized into the normal work practices and that staffing and expertise is adequate. Most of this information is in the form of written documentation, but a portion should be gathered from interviews with a representative sampling of employees performing the various functions and jobs.

4. **Quality assessment** is the process of determining just how well they are doing the actions that are required. The previous step evaluated if they were doing what they were supposed to do; this step judges the quality of that effort. The auditor should confirm that the systems in place are sustainable, working at a high level of efficiency and accuracy, and that there is a mechanism in place for identifying and implementing improvement opportunities. Examples of deficiencies found and corrected as well as implemented improvement opportunities can be used as documentation of the presence of such efforts.

INTERDEPENDENCY BETWEEN LEVELS

Knowing that there are four identifiable levels to be audited and what they address is not enough, however. The auditor must also recognize and plan the audit to understand and identify the interdependencies between the different levels. Further, the auditor must also recognize that there is a priority on which the interdependencies are built. Much like a pyramid, the foundation must be laid and then each successive layer is put in place depending on the layer beneath it and supporting the layer above it.

For auditing, the actions which address the first level, "Recognition and acknowledgment of the need to comply with a specific set of rules or regulations," become the foundation and support all of the other actionable items above it. It ensures that those above it do not fail. It provides the assurance that today's finding in the field will be the same findings tomorrow. Without it being rock solid, any findings in the other three audit levels represent simply a point in time that could change drastically from day to day and from person to person, often depending on individual effort and expertise. In this step, senior management recognizes that compliance goes away when held in the hands of one or a few but lives forever if it is the company philosophy and mantra.

And so it is with each of the four audit levels. The second level, "Implementation strategy and infrastructure," gets its support from the "Recognition and acknowledgment of the need to comply with a specific set of rules or regulations" level which provides the resources to develop appropriate procedures, practices, and guidance documents. These provide the "what is expected" and the "how to do it" instructions. It is interdependent with the level below it and the level above it. See Figure 2.

The third level, "Evidence of compliance," is where the novice auditor tends to spend most of their time believing that conformance here indicates a compliant program at the facility. Unfortunately, that belief may not be true. Findings here may exist for only fleeting moments in time if they are not fully supported and demanded by the previous level, "Implementation strategy and infrastructure." Heroic efforts of one or a few are not enough for sustainability and today's findings could be significantly different from tomorrow's findings. This level is very interdependent on the foundation on which it is built, the level below it, and the level above it, for which it provides the foundation. See Figure 2.

The fourth audit level, "Quality assessment,"[7] interacts with all levels below it. It evaluates just how well the actions met the intent and need and provides a mechanism for identifying and implementing improvement opportunities. This is where sustainability throughout the process could be evaluated holistically. Because it involves all three levels below it and because it affects all four levels including its self, it is interdependent with all other levels. See Figure 2.

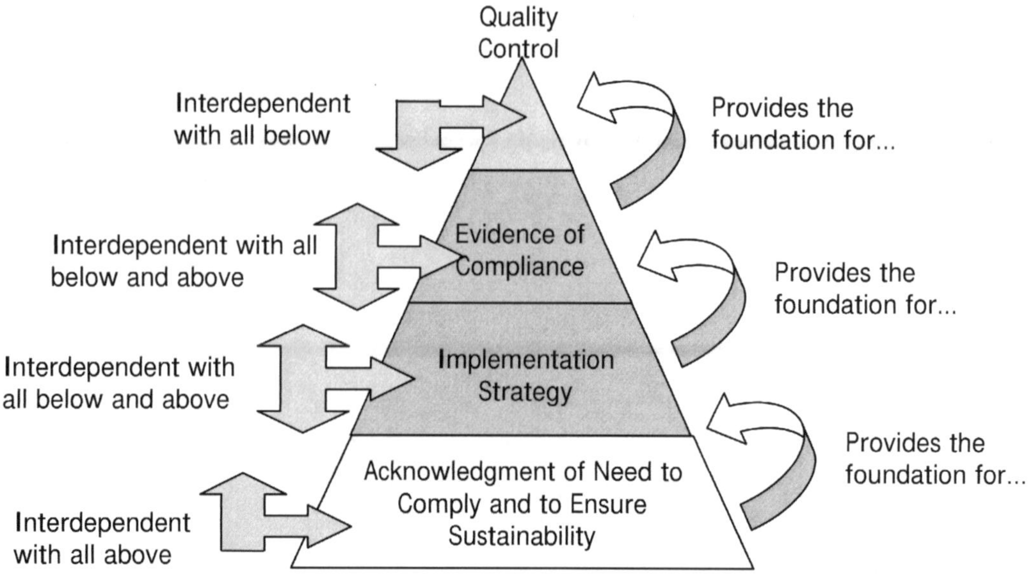

Figure 2—Foundations and Interdependencies

PATH FORWARD

For an audit to be successful in determining compliance and sustainability (i.e. compliance on an ongoing basis), it must consider not only the immediate findings but also evaluate the sustainability of all the compliance efforts and programs.[7] Heroic efforts of one or a few are not enough for sustainability. Today's findings could be significantly different from tomorrow's findings without sustainability. Sustainability is provided from the top of the company, not on the shop floor. The company senior executives must support a mechanism for ensuring that sustainability exists today and continues into tomorrow.

Auditors must recognize the four levels of auditing, how each provides a foundation for those above it, and the interdependencies of each level. They should organize their audit approach accordingly. That usually means that they have to start at the bottom and work their way up the pyramid. Deficiencies found in one level can manifest themselves as a multitude of symptoms and non-conformances (i.e. "findings") in levels above.[6] It is useless and frustrating to the auditor and to those audited if all of these are detailed out when the corrective action really lies in the preceding level. It is also misleading if a dozen or more findings are recorded in the upper levels and only one corrective action is needed in a lower level. Accept the fact that one deficiency in a level will cause a multitude of "findings" in the levels above, and move on.

If sufficient resources and auditors are available to audit more than one level at a time, then particular care should be taken to not fall in the trap described in the above paragraph and that time and resources are not wasted. As soon as a "finding" is discovered in any level above the first level, "Recognition and acknowledgment of the need to comply with a specific set of rules or regulations," that auditor should inform the auditors of the preceding levels. Together, they should properly position the "Finding Cause" by determining if the "finding" is valid and specific for the level in which it was found or, systemic throughout the process because the infrastructure and support is not there from the preceding levels. They can do this by continuously asking the question—Is this "finding" caused by actions or inactions specific to and controlled within this level, or is it caused by insufficient direction and support from the previous audit level(s)?

If the "finding" exists because of a fault in the preceding level(s), the preceding level(s) should record the nonconformance. Identifying its source at the lowest level will correct numerous "findings" at higher levels and keep corrective efforts focused and cost efficient. See Figure 3 for an example of positioning the cause of the "finding."

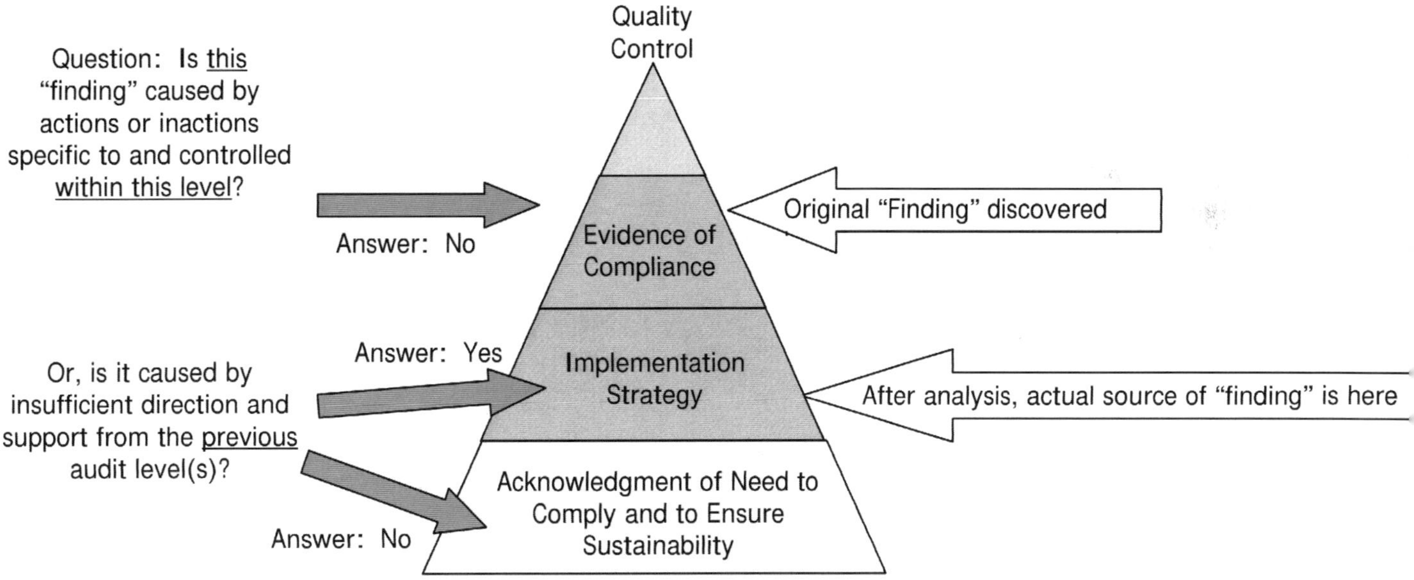

Figure 3—Example of Positioning "Finding" Cause

REFERENCES:

[1] Process Safety Management Guidelines for Compliance, U.S. Department of Labor Robert B. Reich, Secretary, Occupational Safety and Health Administration Joseph A. Dear, Assistant Secretary, OSHA 3133 1994 (Reprinted), Compliance audits

[2] OSHA Regulations (Standards-29 CFR) Process Safety Management of Highly Hazardous Chemicals-1910.119

[3] OSHA CPL 2-2.45A (REVISED) - Process Safety Management of Highly Hazardous Chemicals -- Compliance Guidelines and Enforcement Procedures

[4] 29CFR 1910.119 (o) Compliance Audits

[5] *Guidelines for Implementing Process Safety Management Systems*, Chapter 4, Center for Chemical Process Safety, AIChE, New York, 1994

[6] *Guidelines for Risk Based Process Safety*, Chapter 21, Center for Chemical Process Safety, AIChE, New York, 2007

[7] "The Use of a Ranking System and Protocol for Process Safety Management Audits," David A. Moore, PE, CSP

Auditing Global Logistics Operations - A Process Safety Focus

Michael R. Green, BP, Geneva, IL
Rick Lide, BP, League City, TX

Getting products to customers safely. This simple statement reflects a desired outcome of any logistics movement. This paper will discuss an approach for auditing logistics operations using the principles encompassed within BP's Safety and Operations (S&O) Audit Program and will feature an emphasis on process safety. The paper concludes with three hypothetical case studies to help illustrate portions of the process.

INTRODUCTION

Within BP our goals are simply stated - no accidents, no harm to people and no damage to the environment. BP is one of the world's largest energy companies, providing our customers with fuel for transportation, energy for heat and light, retail services and petrochemicals products for everyday use. BP has operations in more than 100 countries across six continents. As of December 2007 BP was active in exploration in 29 countries and had production activities in 22, held an interest in 17 refineries and had daily production of 2.4 million barrels of oil and natural gas liquids and 8.1 billion cubic feet of natural gas (1.4 million barrels of oil equivalent). BP's daily refined product sales were 5.6 million barrels.

BACKGROUND

Safety & Operations Audit was formed in 2006 and now includes more than 50 dedicated BP auditors with experience in operations, process safety, and regulatory compliance. The auditors come from a variety of industries. Audit teams are often joined by one or more non-BP auditors who are Subject Matter Experts and who generally have industry experience in line with the entity being audited and who have been trained in the S&O Audit process.

S&O Audit conducts audits of BP facilities to confirm that proper management programs are in place to effectively manage operational risk and assure ongoing compliance with internal and external requirements. S&O Audit also conducts Process Safety Management (PSM) audits at BP sites in the US.

TECHNICAL APPROACH

Successful auditing of logistics operations requires a high level of understanding of the key parties, transportation modes, hazard classes of dangerous goods moved and applicable requirements. Key parties include suppliers, customers and logistics service providers such as carriers and terminals. Within BP principal transportation modes include pipeline, marine, rail and road. Principal hazard classes include flammable gases, flammable liquids and combustible liquids and to a much lesser extent toxic, corrosive and explosive materials. Mandatory, mutually-agreed and voluntary requirements/specifications include government regulations, industry association codes of practice, company standards, and manufacturer's recommendations. Detailed knowledge of the applicable requirements within the sites/units in scope is required for effective auditing. This is particularly challenging when covering logistics operations that span multiple countries with differing legal jurisdictions. Additional complexity can be introduced via commercial agreements which specify variable points of custody and risk transfer for the goods and hence influence when various requirements become applicable to the parties involved in the logistics movement.[1]

A key factor in auditing logistics is to understand that the focal point is the logistics network. The logistics network can be thought of as the collection of "ship from" locations, "ship to" locations and connecting routes. The network would consist of all modes of transport used (e.g., road, rail, pipeline, marine, air) as well as all logistics service providers (e.g., carriers, terminals, warehouses, freight forwarders, brokers, etc.) supporting the movement of goods within the network. Within the network a single material movement can be thought of as a "connection" and a site can be thought of as a "node". Many audit programs are site centric and, as a result, may test only portions of a logistics network. A site dominated perspective within an audit program may not fully evaluate risk variants in logistics networks that if left uncontrolled may introduce additional risk into site based logistics operations.

Consider a material transfer operation by tank truck between a manufacturer and customer. At first glance from a site centric perspective this may be viewed as a simple operation. Tank trucks come in, load and depart; however, there are many possible commercial variants and if the focus is only on the manufacturer's site, the audit is examining just one part of the movement (i.e., the route(s) and "ship to" location(s) are not examined). If the business has many customers being served from this one manufacturing plant, the resulting logistics network originating from this single "ship from" location could be very complex.

In addition, consider the commercial variations possible for a single material movement by tank truck from a "ship from" to a "ship to" location. The truck and/or the tank trailer can be owned by the manufacturer, the customer, or a distributer, or could be supplied by a carrier which is contracted either directly or indirectly (via a broker) to the manufacturer, the customer or the distributer. The driver of the truck may be an employee of the manufacturer, the distributer, the carrier, or the customer. The truck, trailer, and/or driver may be dedicated to the movement or the movement may employ different trucks, trailers and drivers. The loading and unloading operation may be performed by the driver, or by an employee or contractor at the loading/unloading site. The transfer hose may be supplied by the loading/unloading site or by the carrier. The loading site (i.e., "ship from" location) may be the manufacturer's plant, a terminal contracted to the manufacturer, or a competing manufacturer (i.e., a swap). The unloading site (i.e., the "ship to" location) may be the customer's plant or a terminal contracted to the customer. The point of custody and risk transfer of the material can vary from resident in the "ship from" location's loading tank to exiting the "ship to" location's unloading tank. The operation could begin in one country's legal jurisdiction and end in another country's legal jurisdiction. The route from the loading point to the unloading point may be specified or may be left to the discretion of the driver.

UNDERSTANDING HOW THE AUDITED ENTITY OPERATES

In order to effectively develop a plan to audit a logistics network, the planning auditor needs to understand the number and types of logistics movements (i.e., modes, service providers, "ship from" locations, "ship to" locations and routes) to determine which operations are higher risk and should be included during the audit. The audit team member assigned to audit the logistics operation(s) needs to understand the movement(s) being audited in order to determine the stakeholders and their respective applicable requirements. The auditor also needs to understand the self assurance program of the audited entity (i.e., internal reviews conducted to ensure the procedures to identify risk and manage operations are being followed). The planning auditor will need to work with the audited entity to define the logistics network and to understand what the relative risks are for each network component.

Logistics often involves sites and operations not under direct control of the audited entity. The decision to include such sites within the scope of an audit should be undertaken in light of the contractual agreement(s) in place with the other operating company(ies) to ensure clarity of the audit requirement scope as well as the findings agreement and corrective action resolution processes.

From a process safety perspective, the US PSM regulations apply to covered processes within fixed sites. US DOT regulations apply to persons and functions involved in the transportation of hazardous materials in commerce.[2] At some fixed sites, both regulations may apply.[3] Aspects of loading or unloading operations may be subject to US PSM regulations, if they are covered processes or part of covered processes. As with many logistics operations there are several variants. Consider a material that is loaded at one site and unloaded at a different site. The loading and unloading sites may both be covered by PSM, both not be covered, or one covered and one not covered. In the case where both loading and unloading operations are covered by PSM, the logistics movement may be interfacing with two different PSM systems. The logistics movement must meet the applicable requirements of each system. Although these requirements may be similar, there will likely also be differences depending on the exact details of the movement and the procedures implementing the process safety regulations at each site.

Within the US many logistics operations are not covered by PSM. Nevertheless it is important to understand the key risks presented by these movements and how they relate to process safety so that the risks can be effectively managed.

Considering a broad definition of process safety we can outline several key points to help manage risk for logistics operations: 1) ensure operators are competent 2) prevent transportation accidents 3) keep the material contained and 4) if hazardous materials are released, minimize their impact to people and the environment. Applicable regulations provide the regulatory framework, and managing regulatory compliance should be a key aspect of any logistics operation.

SAMPLING STRATEGY

To facilitate planning a logistics audit, the planning auditor should generate a sampling strategy. The sampling strategy documents what programs (e.g., PSM regulations, DOT regulations, internal company requirements, etc.) will be audited and where (e.g., at what "ship from" locations, at what "ship to" locations and for which routes) the program elements will be audited. The sampling strategy should include the higher risk segments of the network, but should also be constructed in such a manner that the conclusions drawn from the sample can be reasonably applied across the full logistics network and not just specifically to the network segments reviewed by the audit team. Once the sampling strategy has been defined and agreed, the planning auditor should work with the lead auditor to generate Terms of Reference for the audit.

SAMPLING AND TESTING PLAN

The planning auditor in conjunction with the lead auditor (and other auditors as appropriate) should develop a sampling and testing plan (S&T Plan) for the audit. The S&T Plan should consist of a defined set of activities to be completed during the audit. Activities might include record reviews, interviews, field checks and field interviews. Each activity should be designed to test an audit population against a set of requirements. Activities are different than protocol questions. Whereas the latter directly follow from the applicable requirements (e.g., regulatory, contract, company, etc.) and provide "what" to audit, they often do not specify "how" to audit. Activities should be constructed to provide the practical instructions needed to gather the evidence required to make the determination on whether there is a finding.

The sample population for each activity should be determined in a defined manner from the activity's full population. Times to complete the activity per sample unit should be estimated. This will allow time estimates per activity to be computed. The sum of the activity times will provide an estimate of the resource requirements for the audit and thus the audit can be appropriately staffed with auditors.

Activities should then be assigned to auditors. Auditors should not be assigned an activity unless they have been determined to be competent in auditing that portion of the program.

DAILY AUDIT SCHEDULE

Using the sampling strategy and the S&T Plan, the planning auditor should develop a daily audit schedule for the audit. This schedule should provide the mid level detail for the audit by providing a timeline that specifies when and where a program will be audited by a member of the audit team. If the activity involves a formal interview it should specify the individual who will be interviewed.

Schedules can be difficult to manage for large audits. Data checks during schedule development should be done to ensure auditors have sufficient time to complete their scheduled activities (i.e., to ensure an auditor is not scheduled to complete 105 hours of auditing in 80 hours) and that the times scheduled for activities within the daily audit schedule match those calculated in the S&T plan (e.g., if the S&T plan calls for 20 hours for activity A, then the daily audit schedule should show 20 hours for activity A vs. only 12 hours for activity A).

ON SITE ACTIVITIES

Field auditing should assess the management components that are in place for each program audited, and additionally, test the effectiveness of programme implementation. Protocols should be available for each program audited. Since logistics networks may span different legal jurisdictions, several country specific protocols may be required.

Program assessments may be included which would provide evaluations of the overall program management system. Program assessments should address key components of a management system, which can include: 1) Processes and Procedures, 2) Clarity of Responsibility and Accountability and 3) Self Assurance. The implementation of the program management system may also be addressed.

The audited entity should be advised of the progress of the audit and the audit team's findings, including program assessments, findings and corrective actions with agreed dates for closure. The final audit report, when provided to the audited entity and/or other relevant parties, should also include that information.

Findings should be written at the management system level. Corrective actions should be written to remedy deficiencies in the evidence listed in the finding and also to address the management system deficiency. Management system deficiencies may require the undertaking of action(s) that address segments of the network that were not examined during the audit, but those actions may be necessary in order to correct underlying system deficiencies. There should be a process in place that provides for the tracking and verification of actions as they are completed.

Figure 1 graphically represents the process described above.

Figure 1. General Process Overview

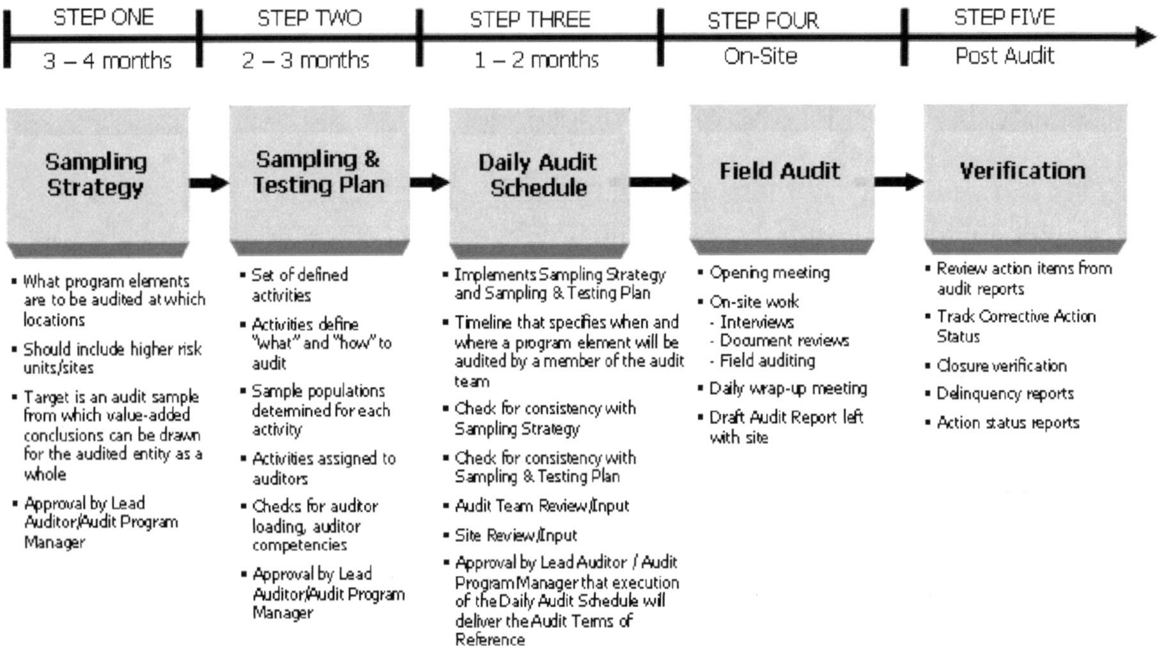

The hypothetical case studies that follow illustrate portions of the process described above with an emphasis on findings and corrective actions. [4]

HYPOTHETICAL CASE STUDIES

I. PSM Audit

Audited Entity (FACILITY 1):	A facility for storage, distribution, and processing of compressed flammable gases.
Key Logistics Operations:	Rail car unloading and truck loading.
Employees/Contractors:	< 25
Sampling Strategy:	Sample all PSM elements at each covered process.

As part of the review of process safety information for electrical classification, the sampling and testing plan called for all electrical classification drawings to be reviewed. There was one electrical classification drawing for FACILITY 1. It showed that all of the six loading racks were within electrically classified areas. The auditor performed a record of the electrical classification drawing to ensure consistency with FACILITY 1 procedure requirements and also with recognized and generally accepted good engineering practices (RAGAGEP). Field checks were made of all six loading racks to determine whether equipment and/or processes that were present in classified areas were consistent with the classification.

Based on the information collected a finding was raised.

Finding

FACILITY 1 did not maintain all equipment present in classified areas consistent with the required classification as documented on the electrical classification drawing.

Evidence

- Loading racks #4, #2, and #5 were classified as Class I, Division 2, Groups C&D on drawing ECD-0403, ECD-0204, and ECD-0503.
    - The operator shed at loading rack #4 had a portable radio plugged into an electrical receptacle. The radio was not labeled indicating suitability for use in a classified area.
    - The operator shed at loading rack #2 had a desktop refrigerator installed. The refrigerator was not labeled indicating suitability for use in a classified area.

- The operator shed on loading rack #5 had an air conditioner installed. The air conditioner was not labeled indicating suitability for use in a classified area.

Corrective Actions

1) Determine whether the radio, refrigerator and air conditioner listed in the evidence of the finding are suitable for use in the classified area in which they are presently installed. If so, label them accordingly. If not, remove them from service.

2) Define and document a self-assurance process to verify the continued suitability of electrical equipment for its designated area classification.

3) Conduct and document the first self-assurance audit as per the process defined in Action 2.

II. S&O audit - PSM regulations not applicable

Audited Entity (Business 1):	Multi-site distributed business
Key Logistics Operations:	Storage tanks, truck loading/unloading
Employees/Contractors:	< 300 among all sites; < 30 per site.
Sampling Strategy:	To audit conformance to Business 1 Logistics Network Management Procedure test 40% of the "ship from" locations and test 4% of the "ship to" locations. ***Special Instructions:*** Perform a 3 hour observation of truck inspections at Site 2.

Four of Business 1's ten "ship from" locations and twenty-four of Business 1's 600 "ship to" locations were visited by the audit team. In conjunction with visits to the "ship to" locations, auditors accompanied drivers to perform field checks on routes. The business utilized tank trucks that were carrier owned and operated. The carriers were both dedicated (i.e., tank truck/driver hauled exclusively for Business 1) and not dedicated (i.e., tank truck/driver did not haul exclusively for Business 1).

The auditors reviewed the Business 1 Logistics Network Management procedure, the list of approved carriers and the dates of previous audits for the carriers to be visited by the audit team. The auditors examined the protocols used by Business 1 to conduct audits of the logistics network, the audit reports and the status of actions originating from the audits. Guided by the audit sampling strategy and the audit sampling and testing plan, an auditor observed three hours of truck inspections for carriers coming to load at Site 2 conducted by Site 2 operators. Site 2 operators performed the inspections using the inspection checklist contained in the Business 1 Logistics Network Management Procedure. The observation was performed at Site 2 because this site was recently acquired by Business 1 and the carriers were not dedicated. Eighteen truck inspections were observed. The Site 2 Transportation Manager did not allow any truck failing inspection to load until the deficiency was remedied.

Finding

Business 1 did not provide evidence that it had implemented the self assurance process per procedure requirement to ensure that it met the requirements specified in the Business 1 Logistics Network Management Procedure. Business 1 was not meeting all of the requirements specified in the Business 1 Logistics Network Management Procedure.

Evidence

- Business 1 did not provide any evidence of past self assurance audits of the Business 1 Logistics Network Management Procedure.

- Twenty-four "ship to" locations were examined. Two of the twenty-four locations (8.3%) were not on the approved "ship to" location list. The two locations were Ship to Location #1 and Ship to Location #2. Shipment to each site began one month prior to the audit.

- At Business 1 Site 1 four of fifteen carriers (26.7%) observed loading had not had an initial health, safety, security and environment (HSSE) audit performed by Business 1. The four carriers were Carrier #1, Carrier #2, Carrier #3, and Carrier #4.

- At Business 1 Site 2 seven of eighteen carrier inspections (38.9%) observed found deficiencies. The deficiencies are summarized below.

    - Trucks with tires with insufficient tread
        - Truck 64, Carrier #5, front driver-side tire.
        - Truck 43, Carrier #6, rear passenger-side external tire.

    - Trucks with non-functioning lights
        - Truck 24, Carrier #7, passenger-side headlight did not illuminate.
        - Truck 235, Carrier #8, rear brake lights did not illuminate.
        - Truck 89, Carrier #8, right rear turn signal did not illuminate.
        - Truck 52, Carrier #9, reverse lights did not illuminate.

    - Trucks with tank inspection deficiencies
        - Truck 24, Carrier #7, tank inspection was overdue.
        - Truck 64, Carrier #5, tank inspection was overdue.

    - Drivers
        - Truck 24, Carrier #7, Driver, attempted to begin loading without wearing safety glasses.
        - Truck 33, Carrier #10, Driver, driver's log was not current.

- At Site 3 one of twelve carriers (8.3%) reviewed was not on the approved carrier list. The carrier, Carrier #11, had recently been removed from the approved list. A record of communication of this removal to the sites was not provided. Additionally, two of the twelve carriers (16.7%) reviewed were overdue for a periodic HSSE audit, Carrier #12 and Carrier #13. The two carriers were dedicated carriers.

Corrective Action

1) Halt shipments per procedural requirement to "Ship to" Location #1 and "Ship to" Location #2 until each is qualified for loading under the Business 1 Logistics Network Management Procedure.

2) Halt loading operations per procedural requirement for Carrier #1, Carrier #2, Carrier #3, and Carrier #4, respectively, until each is qualified for loading under the Business 1 Logistics Network Management Procedure.

3) Communicate to carriers loading at Site 2 that the requirements specified in Business 1 Logistics Network Management Procedure and agreed to in carrier contracts will be enforced effective immediately. Document the communications.

4) Communicate to all sites that Carrier #11 has been removed from the approved carrier list. Document the communications.

5) Halt loading operations per procedural requirement for Carrier #12 and Carrier #13, respectively, until each has successfully completed the periodic HSSE audit required by Business 1 Logistics Network Management Procedure.

6) Complete and document the results of the first self assurance audit against the Business 1 Logistics Network Management Procedure.

7) Develop and implement a corrective action plan with due dates and responsible parties to address all issues identified in the self assurance audit in action 6.

III. Combined S&O/PSM audit

Audited Entity (SITE A):	Large multiunit site
Key Logistics Operations:	Tank truck and rail car loading/unloading
Employees/Contractors:	> 500
Sampling Strategy:	The full sampling strategy is too large to provide the full details for this paper, hence a summary is provided. For logistics operations the PSM elements sampled included training, process safety information and mechanical integrity. Sampling included 50% of the loading racks for truck and rail loading/unloading.

For loading racks, records reviews, field checks and field interviews were performed. Specific activities performed by the auditors in select areas are listed below.

Training

The training element of PSM was reviewed. The auditor noted that several employees recently retired and were replaced with experienced hires. In conjunction, several other employees rotated in job assignments. The auditor conducted record checks and field interviews. During record checks the auditor examined training records for a sample of operators to confirm that they had received the required training. The auditor examined whether the operators were trained in an overview of the process and on the specific health and safety hazards associated with the job tasks. The auditor also examined if the operators had been trained on safe work practices and on all applicable operating procedures including emergency operations and emergency shutdown. During the record review the auditor checked the records to determine whether they documented employee identity, the date of the training, and the means used to verify understanding of the training material. Additionally, the auditor reviewed whether operators had received refresher training at least every three years.

During field checks the auditor observed the operators to determine whether they adhered to the current operating procedures. During field interviews the auditor questioned the operators on their understanding of the current operating procedures, their understanding of safety systems and their functions, and on whether the operators were consulted on the frequency of refresher training.

On the basis of the record reviews, field checks and field interviews, a finding was raised.

Finding

SITE A did not ensure that all employees were trained in the applicable operating procedures before being involved in operating the process.

Evidence

- Three of three newly hired employees had not been trained in the operating procedures for the processes they were operating.

- Two of four employees whose roles were recently changed had not been trained in the operating procedures for the processes they were operating.

Corrective Actions

1) Remove from active operator status the operators who have not been trained in the operating procedures for the processes to which they are assigned.

2) Train the operators discussed in the evidence of the finding.

3) Revise SITE A training procedure (TP-01) to specify that newly hired employees must be trained in the operating procedures to which they are assigned prior to being assigned to active operator status.

4) Issue the revised TP-01 procedure listed in action 3 following the applicable MOC process for procedural change. Procedure implementation is to include but not be limited to the following steps: identify personnel affected by the changes to the procedure, communicate the changes to all affected personnel, record communications.

Process Safety Information - P&IDs

During a review of process safety information, piping and instrumentation diagrams (P&IDs) were reviewed. A high level review was performed to see whether lines and equipment were labeled, connecting lines were labeled, and relief devices included set points. The high level review was a record review to determine whether overall the P&IDs met SITE A standards as defined by the site's internal procedure (SITE A Procedure 12). The review also made a determination of whether the P&IDs conformed with recognized and generally accepted good engineering practices (RAGAGEP). A more thorough field check was made of several P&IDs. The lines chosen by the auditor for the more thorough review were lines for which SITE A indicated recent changes had been made. The lines were walked down in both the modified areas and the non-modified areas by the auditor who compared what was observed in the field to what was on the current P&IDs provided by SITE A.

Based on the information collected a finding was raised.

Finding

SITE A did not document relief valve settings on the P&IDs for the loading racks and did not document some field changes on the current revision of the P&IDs.

Evidence

- Three of five P&IDs reviewed did not list the relief setting of the relief valves. P&ID 10-04 showed two relief valves without set points (PSV-210 and PSV-216). P&ID 02-08 showed three relief valves without set points (PSV-326, PSV-458 and PSV-501). P&ID 03-10 showed one relief valve without a set point (PSV-101). There were ten P&IDs covering the loading racks.
- Two of five P&IDs reviewed did not accurately reflect field status of equipment. P&ID 04-04 did not show a sample valve that had been installed on the line. P&ID 03-10 did not show a relief valve that was added as a result of a recommendation from the last PHA.
- SITE A P&ID Procedure 12 did not specify when to update P&IDs following changes to process equipment.

Corrective Actions

1) Update P&IDs 10-04, 02-08 and 03-10 to show the relief setting for each relief valve.

2) Update P&IDs 04-04 and 03-10 to reflect field status of equipment as outlined in the evidence.

3) Review P&IDs 01-03, 02-08, 08-05, 09-05, and 10-04 vs. field status of equipment. Update, as needed, so that P&IDs accurately reflect field status of equipment. (Reader Note: These P&IDs are the remaining P&IDs covering the loading racks that were not examined in the audit sample).

4) Revise SITE A P&ID Procedure 12 to specify the timeframe during which P&IDs updates following changes to process equipment are to be completed.

5) Issue the procedure revised in action 4 following the applicable MOC process for procedural change. Procedure implementation is to include but not be

limited to the following steps: identify personnel affected by the changes to the procedure, communicate the changes to all affected personnel, record communications.

Mechanical Integrity - Inspection & Testing

For mechanical integrity inspection and test records an auditor reviewed the inspection and testing procedures for loading rack equipment. The auditor also reviewed a sample of recent equipment inspection/test records for loading rack equipment (i.e., piping systems & components, relief & vent systems/devices, emergency shutdown systems, controls (including monitoring devices & sensors, alarms & interlocks), pumps, and other equipment types if important to safety). For the equipment test/inspection records chosen, the auditor reviewed that all requirements were met and that the record documented:

- the inspection/test date (checking to verify whether inspections/tests had been performed within the required frequency),
- the name of the person who performed inspection/test,
- the unique identifier of the equipment,
- the description of test performed,
- the inspection/test results, and
- whether identified equipment deficiencies were corrected before further use or otherwise addressed in a safe and timely manner (i.e., safety measures had been put in place to reduce the risk associated with the hazard to an acceptable level).

Based on the results of the review a finding was raised.

Finding

SITE A had not fully implemented the requirements of the SITE A procedure for Inspection and Testing of Loading Assemblies.

Evidence

- Multiple inspection tags/labels were present on the loading arm assemblies and hoses at two of two loading racks examined. The tags/labels displayed conflicting information on what the last test date was for each loading arm and hose.

- The inspection records presented recorded the loading line number for each loading assembly but did not record the hose engraving number.

- Evidence of sign-off by Operations accepting the loading assemblies following the last inspection was not presented. This evidence is required by the SITE A procedure for Inspection and Testing of Loading Assemblies.

Corrective actions

1) Review and update the procedure for Inspection and Testing of Loading Assemblies to include recording of hose engraving numbers in inspection/test records.

2) Issue the procedure reviewed in Action 1 following the MOC process for procedural change. Procedure issuance to include but not be limited to the following steps: identification of personnel affected by the change to the procedure, identification of the requirements for competency of personnel affected by the procedure, actions required to close any gaps in competency identified, communication of the change to all affected personnel, and record of communications.

3) Investigate, determine and document the extent of the gaps in the loading assemblies vs. the requirements of the updated procedure for Inspection and Testing of Loading Assemblies for each Loading Rack at SITE A.

4) Develop a corrective action plan with due dates and responsible parties to address any issues based on the investigation detailed in Action 3.

5) Implement and close out the corrective action plan to address any issues based on the investigation detailed in Action 3.

6) Conduct the first self-assurance audit on the implementation of the procedure updated in Action 1.

References:

1. For an excellent discussion of Transportation Risk Management see Guidelines for Chemical Transportation Safety, Security and Risk Management, 2nd edition, Center for Chemical Process Safety, Wiley, 2008.

2. See 49 CFR Part 171 Section 1, Applicability of Hazardous Materials Regulations (HMR) to persons and functions.

3. As taken from OSHA Standard Interpretations 09/09/1993 - HHCs as it applies to Hydrogen Peroxide delivered from a tank truck.

   "When visiting a work site at which there is a covered process, cargo tank motor vehicles (CTMV's) are covered by the PSM standard. These CTMV's would be covered by OSHA standards to the extent they are not covered otherwise by the Hazardous Material Regulations of the Department of Transportation (DOT) (See 49 CFR Subchapter C and particularly, Part 177 - Carriage by Public Highway). These DOT regulations cover the cargo tank design, construction, maintenance (including repairs) and operations of CTMV's. Operations regulated by DOT include loading and unloading, connected cargo transfer hoses whether provided from the CTMV or not, attendance by a qualified person during loading and unloading and associated training.

   When the Department of Transportation (DOT) has jurisdiction of the unloading of a CTMV, an attendant who is a qualified person must be present, in compliance with the Hazardous Material Regulations, 49 CFR Subchapter C. Furthermore, DOT's jurisdiction ends and OSHA's jurisdiction begins at the connection between the hose from the CTMV and the process when unloading hazardous material as described above."

4) The authors gratefully acknowledge the contributions of the S&O Audit team in developing the S&O Audit process and also the management of the S&O Function for the opportunity to share this work within the process safety technical community.

Training Tools to Support a Global Process Safety Culture

Dan Reinke, P.E., CSP, CPEA
Environmental Resources Management
190 E 5th Street, Suite 255
St. Paul, MN 55101
+1-651-225-4554

Gurmukh Bhatia
The Sherwin-Williams Company
101 Prospect Avenue
Cleveland, Ohio 44115
+1-216-566-2160

David Thaman
PPG Industries
4325 Rosanna Drive
Allison Park, PA 15101
+1-412- 492-5267

Prepared for Presentation at
American Institute of Chemical Engineers
2009 Spring National Meeting
5th Global Congress on Process Safety
43rd Annual Loss Prevention Symposium
Tampa, Florida
April 26–30, 2009

UNPUBLISHED

SUMMARY

Many companies have discovered that the process safety culture at some locations is focused around complying with the letter of regulatory or corporate requirements and not on the broader goals of preventing low probability, high severity events. As a result these facilities may be going through the motions but could fall short of the expected performance. This is being seen across industry groups worldwide.

As a result, The Sherwin-Williams Company, PPG Industries and Environmental Resources Management (ERM) decided to form a working relationship and work with VGI Productions to develop a series of four professionally produced, interactive training programs to support each company's efforts to increase the understanding and attention to the fundamentals of process safety worldwide. In this paper we will share the process used to develop these high-end productions, the structure of the final products and how they are being used to drive process safety program enhancements in each company's global operations. In our presentation we will also share clips taken from the referenced video recordings.

The four modules use interviews, stories and footage from company employees at all levels, from operations to senior management, to relate the keys to effective process safety management and how they have benefited from their involvement in process safety activities. The series is titled "PS – It's A Whole new World." The first of these modules provides an introduction to process safety, the benefits of having a strong program and the keys to success. The second module, subtitled "Understand Your Process," focuses on key elements such as process safety information and hazard analyses. The third module, subtitled "Control and Maintain Your Operations," deals with topics such as operations, mechanical integrity and management of change. The fourth segment, subtitled "Building A Process Safety Culture," reinforces the importance of employee engagement and operations leadership.

BACKGROUND

For years many major corporations, including The Sherwin-Williams Company and PPG Industries, have implemented global process safety (PS) standards to reduce the likelihood of process incidents. For US-based corporations these standards have typically mirrored the US OSHA Process Safety Management (PSM) Standard, 29 CFR 1910.119.

Corporate expectations regarding implementation of the 14 PSM elements are sometimes not clearly communicated or understood at the plant level, resulting in varied program implementation and interpretations. Some safety professionals who do not come from a background in chemical processing do not always understand the overall PS objectives and thus treat the process safety management standard as steps that must be followed as efficiently as possible without disrupting the manufacturing process. These safety professionals struggle in engaging facility leadership, engineering, production, maintenance and other key functional areas and try to comply with standards through their own heroic efforts.

ERM and corporate process safety auditors are concerned when they see this lack of effective implementation of these global standards. A typical comment is that the facility personnel are focused on PSM instead of PS, spending their time without adequately understanding the objective of their efforts.

To improve the understanding and acceptance of the importance of process safety, project participants have developed training materials that now are being used in company operations worldwide. These training materials include a series of four professionally-produced videos that use company settings and personnel to drive home the keys to successful management of PS.

HISTORY

The program was conceptualized and initiated by corporate process safety professionals at The Sherwin-Williams Company. To aid in the development and financial costs of their vision they invited PPG Industries and Environmental Resources Management (ERM) to join them in this initiative. VGI Productions in Cleveland, Ohio was retained by the parties to turn the concepts into reality, producing the four videos and associated supporting materials.

Work progressed through months of meetings and calls to develop first an overall shared vision and outline. Many options on organization and presentation techniques were presented and discussed before settling on the final approach, mixing an actor serving the role of a plant manager, Andy, and personnel from all levels of the project participants telling their stories of process safety.

The training videos introduce Andy as a plant manager very supportive of employee safety, but unaware of the meaning, importance and value of process safety. Through a series of discussions and experiences presented in the four modules Andy becomes a champion for process safety in his plant and throughout the organization.

PROCESS

Once the general vision and outline were developed, VGI toured and filmed company operations and interviewed personnel from facilities around the world. Taping was timed to coincide with meetings that brought representatives from outside the US to local plants. VGI used prepared questions to get interviewees' personal stories about how they manage process safety in their daily activities. Interviews were filmed with a green screen backdrop to allow the use of process footage and backdrops. Interviewees were positioned so they would be seen on the screen as talking to Andy the plant manager, who is shown in the foreground.

Senior leadership representatives were also interviewed on tape to provide comments that were used as custom introductions for each of the project participants. Each party then is able to demonstrate and communicate the upper-most support and understanding of the importance of process safety in their organization.

With the tapes of employee interviews in hand, VGI and the project participants worked to develop scripts, interjecting messages for Andy to ask questions of the interviewees and share his personal journey of process safety understanding. VGI spent countless hours reviewing and selecting segments of the plant and studio interviews to piece into the overall storyline. Each script went through several iterations, with participants editing for technical accuracy and clarity of the messages for a global audience.

With the extensive up-front work, once rough cuts of the videos were produced the final editing process was able to focus on opportunities to enhance the appearance of some graphics and clarity of key messages. This step was also time consuming but led to final products that all participants see as valuable to their organizations.

VGI also aided in the preparation of supporting materials, including Powerpoint presentations, quizzes and instructor and student lesson kits for each module.

The objectives of each of the modules are as follows:

Module #1: Process Safety Is Good Business

Goals:

1. Counter the most common misperceptions about process safety:
    a. It is unnecessary when plant injury rates are improving;
    b. It is unnecessary when plants already have good behavioral or employee engagement program;
    c. It applies only to chemical companies and to some plant areas covered by regulations
    d. It is complex, unwieldy, impractical, and costly for most facilities.
2. Make the case that process safety is good manufacturing practice, good business, and essential to organizational sustainability.

Module #2: Building A Process Safety Culture

Goals:

1. Communicate practical aspects of/ideas for organizing for PSM success:
   a. Commitment: there has to be a leader, and it has to be you;
   b. Teamwork: most of team is already in your plant as are elements of a good program;
   c. Method: good managerial practice- setting goals; allocating people, time, resources, accountability;
   d. Additional resources.
2. Describe the foundation:
   a. Starting point: gap analysis or PSM audit, importance of establishing MOC;
   b. Process safety information;
   c. Process hazard analysis;
   d. Standard Operating Procedures.
3. Communicate that the goal is integrating an ongoing PSM culture, continuous improvement.

Module #3: Plant-Wide Process Safety

Goals:

1. Instill the team's broad logic for guiding PSM:
   a. Know your process;
   b. Run and maintain your process;
   c. Manage changes in your process.
2. The nuts & bolts of PSM in action, small to large applications.

Module #4: Process Safety – Where Do I Fit In?

Goals:

1. Provide managers with PSM employee engagement tools they can use to:
   a. Address production workers and supervisors directly;
   b. Communicate the PSM difference and its value for employees and the business.
2. Present examples of employee involvement, contributions.

RESULTS

While the last three modules were only recently completed, the first module was finished in summer 2008 and has been used by all participants for training and awareness building. While The Sherwin-Williams Company and PPG Industries are using this solely within their own organizations, their participation with ERM allows the materials to be used by ERM clients around the world that seek to enhance their process safety culture.

Next steps for the process are to translate the videos and materials into languages covering all participants' major operations. This work is planned for 2009.

Images from the videos are presented below.

corp psm director

consultant

Implementing and Sustaining a Global Process Safety Program

Christie A. Arseneau, David A. Moore, Michael J. Hazzan
AcuTech Consulting Group - Alexandria, VA
www.acutech-consulting.com

ABSTRACT

With the globalization of the chemical process industries and the recognition of the need to maintain high standards of process safety throughout an enterprise, process safety is now more widely practiced throughout the world. Process Safety incidents are not limited by geography and nor should the implementation of Global Process Safety programs be so limited. In the past, some may have viewed Process Safety as a compliance activity and held different regions to different standards based on regulatory requirements, but more companies are implementing global standards for Process Safety for reasons relating to corporate governance and process improvement in addition to improving their ability to compete on a global stage.

Balancing the facilities in countries that have been subject to regulation and are ready for more sophisticated programs with facilities in other regions or that are struggling to manage basic occupational safety requirements, for example, presents significant challenges. Local customs, value systems, and cultural differences make for significant obstacles in extending a common process safety platform and in expecting common performance.

This paper outlines a process for companies to use when implementing and sustaining a global process safety program. Included is consideration of these site-specific differences to allow flexibility, while maintaining strong philosophical fundamentals in process safety programs. Allowing for 'best practices' to emerge from participatory engagement of leaders throughout the enterprise, rather than top-down demands for compliance, has proven successful.

INTRODUCTION AND BACKGROUND

Flixborough, Seveso, Mexico City, Bhopal. Texas City, Danvers, Buncefield. Lessons learned from company-specific incidents as well as milestone industry events were incorporated in the development of various industry-wide guidelines, corporate engineering and safety standards, and national laws regarding Process Safety.

Incidents in recent years have heightened the standard for Process Safety on a global basis. The Baker Panel report[1] on the BP Texas City incident defined ten recommendations specific to BP, but that may also be applied throughout industry:

- Process Safety Leadership
- Integrated and Comprehensive PSM system
- Process Safety Knowledge and Expertise
- Process Safety Culture
- Clearly Defined Expectations and Accountability for Process Safety
- Support for Line Management
- Leading and Lagging Performance Indicators for Process Safety
- Process Safety Auditing
- Board Monitoring
- Industry Leadership

Published in 2007, the CCPS Guidelines for Risk Based Process Safety[2] established a new framework for process safety, building upon the original ideas published in the early 1990s. The framework includes four pillars upon which the process safety elements are established:

- Commit to Process Safety
    - Process Safety Culture
    - Standards, Codes, Regulations, and Laws
    - Process Safety Competency
    - Workforce Involvement
    - Stakeholder Outreach
- Understand Hazards & Risk
    - Process Knowledge Management
    - Hazard Identification and Risk Analysis
- Manage Risk
    - Operating Procedures
    - Safe Work Practices
    - Asset Integrity and Reliability
    - Contractor Management
    - Training and Performance Assurance
    - Management of Change
    - Operational Readiness
    - Conduct of Operations
    - Emergency Management
- Learn from Experience
    - Incident Investigation
    - Measurements and Metrics
    - Auditing
    - Management Review and Continuous Improvement

The Buncefield Incident Report[3], published in December 2008, identified three aspects that are key to addressing specific recommendations and preventing another similar incident from occurring:

- Leadership
- The role of standards and guidance
- The timetable for implementation of specific recommendations resulting from the incident investigation

These publications provide some elements of a common framework for developing, implementing, and sustaining a global process safety program. However, within these and other, established frameworks, there continue to be challenges faced by those charged with the task.

Many multi-national corporations have adopted internal Process Safety standards and applied them to global operations. Challenges facing corporations choosing to implement a single process safety standard globally, especially in countries where no formal regulation exists, are numerous. These challenges may include establishing a common Process Safety culture and managing the abilities of local human capital with the inherent risks of the process. In an ever-evolving world where markets, economies, and competition drive operating companies to sustain, expand, acquire, or build their global operations, process safety becomes an increasingly important and challenging need.

Companies that choose to implement a global Process Safety program utilizing a common framework will need to overcome these challenges. By following a strategic approach with a common philosophical baseline, companies will be establishing the foundation for implementing and sustaining a global Process Safety program.

Challenges to Establishing a Global Process Safety Culture

A common challenge to the chemical industry is to implement Process Safety when a functioning occupational safety management system or a strong safety culture is not in place. As identified in the Baker Panel Report and the CCPS *Guidelines for Risk Based Process Safety*, Process Safety Culture is a key component of a strong management system. Lack of a strong safety culture may be due to:

- Implementation of varying internal corporate cultures developed in various international affiliates, as compared to domestic business affiliates.
- Mergers and acquisition activity where, as the old saying goes, "old habits die hard".
- Poor implementation of an occupational safety program.
- Availability, sophistication, and education of human capital to understand hazards, risk, and process safety fundamentals.

Another challenge to a corporation implementing a global Process Safety program and establishing a strong Process Safety culture is when local culture and values conflict with the safety cultures and values being implemented in the program. For example, in one country, local drivers may drive with their ears, rather than with their eyes. The drivers listen for other car horns and proceed to drive ahead without slowing, stopping, or looking for oncoming traffic at an intersection. While this driving practice is embedded in the culture outside the plant fence line, it may create a challenge when reinforcing good fork truck driving practices within the fenceline, especially when handling highly hazardous chemicals and navigating through highly hazardous process areas.

Managing the implementation of Process Safety in countries or jurisdictions that are not currently subject to Process Safety regulations, while raising the standard for other facilities that have been subject to Process Safety regulations for some time, can create additional challenges. The problem is one of timing of the development of the industry and the regulatory framework on process safety. A multi-national corporation with facilities, for example, in Italy, UK, USA, Brazil, China, and Nigeria envisions implementing a common global Process Safety standard. The facilities in Italy, UK, and USA have been subject to Process Safety regulations for many years and have a well-established program. Certainly, there are opportunities for continuous improvement and also to raise the standard by implementing new programs such as Process Safety Metrics and LOPA. The corporation is faced with the challenge to support the facilities with well-established programs and also to support the facilities in Brazil, China, and Nigeria that are in the initial stages of understanding Process Safety, developing a strategy for implementation, and establishing a strong Process Safety Culture.

Challenges to Managing Local Human Capital

Management of local human capital is both a key component and an ongoing challenge to implementing and sustaining a Global Process Safety program. Local resources will be tasked with the accountability to implement and maintain the Process Safety program. The efforts undertaken to implement Process Safety are often undertaken in an open fashion with knowledge and interest in the progress at multiple levels in the organization. This heightened level of interest may lead to the local resources feeling a higher level of external pressure to deliver on and exceed expectations, an increased sense over being overwhelmed with the details, and perhaps a sense of isolation due to the program being developed without their input or imposed on them from external sources.

Whether intentional or not, international operating facilities may not always have had the support for their Process Safety efforts. Process facilities in domestic countries (i.e., in the same country as the corporate headquarters) may have had an advantage over global operations due to the availability and relative location of the corporate headquarters to support Process Safety activities. Facilities in countries or other jurisdictions that are subject to Process Safety-related regulations may have received a relatively high level of support for Process Safety activities due to compliance requirements. International operations that are many time zones away, not subject to regulatory activity, or perhaps even recently acquired, may feel disenfranchised or isolated because they may not have received past support for their process safety programs. With little or no support, the human capital in those locations did not evolve their Process Safety knowledge with the rest of the facilities that received support, thus there is a corporate knowledge and process safety talent gap at technical and management levels within these organizations.

Another challenge to managing human capital is hazard awareness and recognition. Personnel working in international operations may have a lesser understanding of process related hazards than their counterparts. This could be due to a number of reasons including the education level of the local workforce, high rates of turnover, and ineffective training. Facilities that do not understand and recognize the hazards of their processes are challenged to establish a sense of vulnerability to the existing hazards. While this challenge is not limited strictly to international operations, as it can also be a challenge faced by facilities with very mature Process Safety programs, it is important to recognize when a sense of vulnerability has been lost.

An obvious challenge to any business entity operating in another country is language and cultural customs. While the English language is often adopted as an official business language used in communications among facilities in multiple countries, there may not be local talent with the language skills to understand the technical aspects of Process Safety and effectively articulate the requirements in the local language. The impact challenge is highlighted during Corporate-delivered training and audits often performed by non-local personnel.

Implementation Strategy

While corporations seeking to implement and sustain Global Process Safety programs face many challenges, the classical management approach of "Plan-Do-Check-Act" can be expanded into the "Global Implementation Strategy" model displayed in Figure 1 to help implement such a program more effectively. This approach takes into account some of the additional challenges facing implementation of Global Process Safety program that are above and beyond the normal project planning cycle.

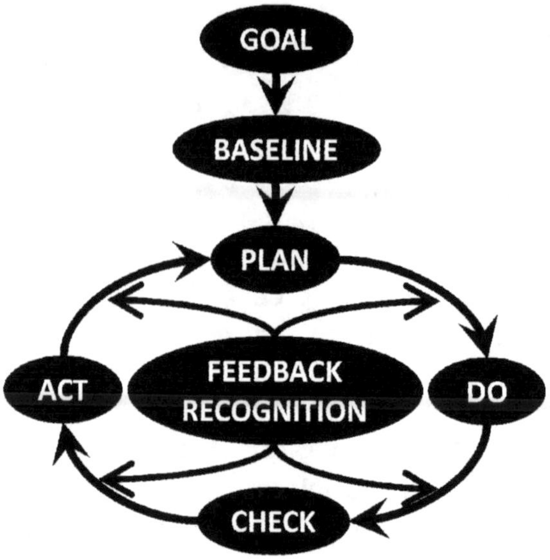

Figure 1. Global Implementation Strategy Model

APPROACH

When a corporation is initially engaging in or maintaining a global Process Safety program, an overarching goal should be established and committed to at the highest levels within the organization. The goal may be to have a standardized Process Safety program for all global facilities implemented within a defined number of years. Alternately, the goal may be to comply with local laws only. Whatever the goal is, the goal should be clearly defined and communicated throughout the organization.

When establishing goals for a global Process Safety program, additional aspects to consider at a Corporate level may include:

- Defining Standards and the applicability of the standards. Such definition will not only build a framework from which an integrated and comprehensive Process Safety system can be built, but it will allow international affiliates to understand where the expectations are being set. This framework will also allow for the establishment of a scope for the program so that affiliates may understand if these standards will apply to them.
- Committing to Process Safety by establishing Process Safety leadership at all levels of the organization. This may include a senior business manager who serves as a champion for Process Safety within the business and establishing a Corporate Process Safety subject matter expert as a central point of contact. Having these resources established will reinforce the corporate commitment to Process Safety and nurture a positive Process Safety Culture.

BASELINE

When implementing a global Process Safety program, it is important to fully appreciate the task ahead and understand the current state of the Process Safety program's development. Establishing a baseline of the status of the current Process Safety program can be completed through a number of means that may include:

- Asking facilities to complete a self-assessment regarding the current status of Process Safety related management systems such as training, operating procedures, process safety information, and mechanical integrity, as compared to the centralized standards.
- Reviewing available audit reports that may identify gaps against the company standards.
- Conducting risk assessments to identify process hazards and potential gaps in other elements of a Process Safety management system.

These methods can identify distinct areas for local improvement and identify opportunities to address in the balance of the implementation strategy. This baseline process can also establish an understanding of the current knowledge level regarding the topic and provide an opportunity to provide a "big picture" education to the personnel who will have direct accountability for the program. This education should reinforce that much of the Process Safety program can be built upon existing management systems already in place. For example, most facilities are likely to have an existing training program in place for operators. While the training program may not meet the established Corporate standards, minor modifications may be made to meet the intent of the program.

When establishing a baseline for a global Process Safety program, additional aspects to consider may include:

- Partnering with individual facilities or their central businesses (if a large number of facilities) to identify potential barriers to implementing a global Process Safety program. These could be cultural barriers, inadequate resources (people, equipment, or budget), language, etc. For example, one facility in a native-English speaking country may utilize a large population of immigrants who speak English as a second or third language. Local supervisors speak English and utilize a team leader "translator" as an intermediary between management and operators. This language barrier is important to recognize early on as part of the baseline. This aspect is often completed at a centralized or Corporate level.
- Understanding local requirements and how to comply with them. If a facility does not meet local requirements, ensure that resources are provided to bring them to a level where the requirements are met prior to implementing additional requirements. This approach can be modified if sufficient resources are available to manage implementation of local regulations and additional Corporate

requirements, or if the implementation of local regulations can be done with little effort. This aspect is often completed at a centralized or Corporate level.

- Utilizing the information gathered, Corporate can better understand the level of work that needs to be done and better understanding the resources available to assist.

PLAN

The Plan phase of the Implementation Strategy is focused on strategy development at both the Corporate level and at the level of the individual entity.

Corporate Planning

- Depending on the hazards associated with the processes, there may be varying levels of risk identified. As part of the planning phase, the centralized or Corporate level should plan for a risk based strategy for Process Safety that can be adjusted based on local conditions. This planning strategy may include such logistics as:
    - Engage stakeholders in the Process Safety programs at all levels by forming employee-based teams to spearhead the effort in planning and implementing the details of the program.
    - Determine whether or not there will be a phased-in implementation. Will some or all facilities implement a few topics first? Or, will some or all facilities implement all topics immediately? Consider using a research and development approach to implementing Process Safety by starting with experimental trials of select aspects at select facilities. Incrementally move towards a broader pilot of the process and finally scale it up towards full implementation.
    - Utilize the baseline to help determine which facilities are immediately ready to participate in the planning and implementation phase. Adjust the centralized or Corporate plans for specific facilities that need additional time or resources or that have less risk.
    - Identifying and developing templates for use. In rolling out a Global Process Safety program, especially for facilities that have common manufacturing processes with common hazards, the use of templates can provide increased efficiency. Utilize best practices from similar sister facilities that have established Process Safety programs as a starting point for templates. Ensure that the templates meet the minimum requirements established within the Corporate standards and are readily accessible, for example, through a Sharepoint site or company Intranet.
    - Develop a training strategy and secure a budget to implement. Training of international affiliates can be time consuming and expensive, especially if done in person. The Corporate group may consider providing intensive

Process Safety training to a small number of individuals in various regions. These persons can then be seen as Regional Process Safety Experts that can support the program and may be able to respond quicker. If this is not a practical approach due to resources available, another consideration would be the development of a Process Safety training series that may be done in shorter, more frequent sessions through computer based or webmeeting forums.

- Develop a communication strategy. Frequent communication is important to gauge facility progress, identify roadblocks to implementation, or to detect other issues that may arise. Frequent communication is also a means to provide a continuous flow of information to the field to continue building Process Safety knowledge and reinforcing a positive Process Safety culture.

Individual Facility Planning

Information used in the development and implementation of the Goal and Baseline established both the scope and the current state of the program in the field. Discrete tasks identified during the baseline can be fed into an individual facility's Gantt Chart (Figure 2), or other scorecard, with a timetable for implementation. These tools help the facility manage implementation of the program, allows the Corporate group to track their progress, and to plan local and Corporate resources accordingly.

When establishing a plan for a global Process Safety program, additional aspects to consider may include:

- The planning phase is an appropriate time to assign accountability and Process Safety responsibilities to local management, and local subject matter owners.
- Engage operators and other facility employees in this early stage of the implementation process.
- Consider recruiting and assigning Process Safety Mentors from more experienced plants to support the less experienced plants. These mentors can serve as an additional expert and resource to support the facility while sharing "best practices" from their individual locations.

Task – Operating Procedures	Year 1 Q1	Year 1 Q2	Year 1 Q3	Year 1 Q4	Year 2 Q1	Year 2 Q2	Year 2 Q3	Year 2 Q4
Identify list of Operating Procedure Needs	X							
Prioritize Operating Procedures		X						
Establish template for Operating Procedures		X						
Complete Priority 1 Procedures			X	X				
Complete Priority 2 Procedures					X	X		
Complete Priority 3 Procedures							X	X

Figure 2. Example Gantt Chart

DO

The Do phase of the implementation strategy will take significant effort. For Corporate resources utilized to support the implementation, it is important to recall the experience of implementing Process Safety in the past. The Seveso Directive was enacted and the AIChE Center for Chemical Process Safety was formed in the 1980's. The US OSHA Process Safety standard and US EPA Risk Management Plan regulation were promulgated in the 1990's. Often, there is an institutional memory of these periods of Process Safety implementation that can be utilized during a broader, global implementation of Process Safety. During this phase, facilities may need to be supported by Corporate resources.

CHECK

Provide opportunities for routine management and Corporate review of the implementation progress. The level of detail provided for review may vary depending on the level of management. For example, the operations or plant manager may seek a very detailed report of the implementation status. A Vice President of Operations may only want to know a higher level of detail; he or she may want to know what is on target, what is not on target, and what the plan is to get back on target. The Gantt chart can be updated and used as a scorecard to communicate progress.

Corporate audits or self-assessments may provide another opportunity to check the status of implementing the Process Safety program.

Over time, the Corporate group may implementing increasingly sophisticated leading and lagging indicators to track progress of program implementation and overall process safety performance.

ACT

Process Safety is an ongoing, iterative process that demands continuous attention. Just as facilities are never "done" with occupational safety programs, facilities implementing Process Safety will never truly be "done" either.

This continual process of following a Corporation's standards and guidelines for Process Safety requires significant resources, activities, and management attention. It also requires clear programmatic plans and guidance so that there is a consistent effort throughout the enterprise.

FEEDBACK RECOGNITION

Recognition and feedback are important aspects of implementing and sustaining a global Process Safety program. Figure 1 depicts Recognition and Feedback in the center of the traditional "plan-do-check-act" model, with spokes to each element. These are essential elements and should be integral to the program.

Recognizing those that reach milestones, as well as recognizing those that fall behind is a key component to sustaining the program. Providing constructive feedback on the work products or processes provides encouragement to those who have accountability to implement the program. Feedback is important to know where you stand.

Recognition and feedback can be formal or informal and come from any level in the organization. It is key that the feedback and recognition is regular and constructive.

CONCLUSION

For a Corporation to be successful in implementing and sustaining a Global Process Safety program, they must recognize and overcome numerous challenges while implementing a strategic plan of iterative improvements at both the corporate and local level. As the Corporation is starting down the path towards a global program, they may reference numerous frameworks that have been recently published. The BP Baker Panel Report, the Buncefiled Incident Investigation Report, and the CCPS Guidelines for Risk Based Process Safety all outline various aspects that can be considered when implementing a global Process Safety program.

Once the challenges are recognized and resolved, the next step is to develop the suggested Global Implementation Strategy. This modified "Plan-Do-Check-Act" model adds a global goal, establishment of a baseline, and recognition/feedback loops at regularly phases of the cycle.

Following this approach, corporations will be building a foundation upon which a flexible and risk based Process Safety program can be implemented and sustained.

References:

1. *The Report of the BP U.S. Refineries,* Independent Safety Review Panel, 2007

2. *Guidelines for Risk Based Process Safety*, Center for Chemical Process Safety, AIChE, New York, 2007

3. *The Buncefield Incident - 11 December 2005: The final report of the Major Incident Investigation Board, Volume 1,* Buncefield Major Incident Investigation Board, 2008

A Comprehensive Approach for Global Systems to Sustain and Improve Process Safety Management

David E. Cummings, MS, CIH

Global Process Safety Management Competency Leader
Safety, Health, Environment and Sustainable Growth Center
E. I. du Pont de Nemours and Company
Wilmington, DE, USA 19898
(302) 774-9558
david.e.cummings-1@usa.dupont.com

ABSTRACT

Process safety management implementation is complex because it crosses over many functional areas of an organization, including operations, engineering, technical, R&D, maintenance, sourcing, and construction. The development, implementation and continuous improvement of a standardized and sustainable global process safety management system is essential for business success, including the maintenance of an effective PSM organizational structure, planning, implementation, and performance measurement at the site, business, and corporate level. This paper will review the current global structure and work processes associated with process safety management in the DuPont Company, including PSM competency team and regional organizations, functional roles and responsibilities, leadership involvement, management of internal PSM standards, technical training, risk communications, leveraging common tools and applications, and PSM performance measurement. Current challenges and future improvement activities associated with global PSM managing systems will also be discussed.

1. Introduction and Background on DuPont

Development and sustainability of a comprehensive global process safety management program is critical within a large, multi-national company such as DuPont in order to manage a diverse set of process risks and positively contribute to business success (1). The program should include specific mandatory technical elements to help ensure consistency, while at the same time, enable flexibility and decision making by knowledgeable personnel to address local or regional technical issues and cultural variations. A global PSM management system also includes an extensive set of work processes that address corporate governance and oversight needs. The following aspects will be discussed in more detail as one means to accomplish these objectives:

Background on DuPont
Global PSM Principles and Philosophy
Global PSM Team Structure and Work Processes
Process Safety Policies and Internal Standards

Process Safety Training, Communications and Stakeholder Input
Performance Measurement and Auditing
Special Focus Areas and Management Systems
Logistics with Global Implementation
Issues and Future Focus Areas

About DuPont

DuPont is a global science based product and services company that operationally is organized around five main business platforms, each with several groupings of businesses, which serve over 200,000 customers worldwide with annual revenues of $32 billion. Founded in 1802, the company operates at more than 175 manufacturing sites in 70 countries involving 64,000 employees.

DuPont operates many different manufacturing technologies involving both higher hazard processes and lower hazard operations as illustrated by the examples in Table 1. Manufacturing technologies, products, and associated process safety hazards (fires, toxics, explosions, reactives, and mechanical energies) vary widely across each business.

Business	**Products/Process Hazards**
Chemical Solutions Enterprise	oleum, aniline, sodium, cyanides, specialty chemicals, methylamines, dimethyl sulfate, catalysts, lubricants, phenylenediamine
Engineering polymers	polyester and acetyl resins, glass laminates
Fluoroproducts	refrigerants, electronic gases, fluoropolymer solutions
Agricultural products	herbicides, fungicides, insecticides
Advanced Fiber Systems/Nonwovens	Tyvek®, Sontara, Nomex®, Kevlar®
Performance Coatings	Automotive paints, powder coatings
Titanium Technologies	Titanium dioxide, Titanium Tetrachloride
Surfaces	Corian® and solid surface materials
Pioneer Seed and Solae Soy Protein	Corn, soybeans; soy protein food products

Table 1: Examples of DuPont Products and Manufacturing Technologies for PSM

Operations are also segmented geographically into four worldwide regions including North America, Latin America, Europe, and Asia Pacific.

2. Global PSM Principles and Philosophy

DuPont has adopted a global policy where internal PSM standards and elements are applied uniformly at all manufacturing locations worldwide which involve process equipment and associated hazards through one set of corporate policies and procedures. This includes all locations where DuPont or a subsidiary owns a majority interest (more than 50 percent interest in the entity) and all businesses where DuPont employees operate another entity's assets as a condition of venture formation, lease, or other contractual obligations.

Process safety program implementation at each site adheres to a detailed set of corporate program elements and requirements. Each manufacturing process is classified as a Higher Hazard Process (HHP) or Lower Hazard Operation (LHO) based on the size, complexity, location, and inherent hazards due to accidental release, based on the following definitions

Higher Hazard Processes includes any activity manufacturing, handling, storing, or using hazardous substances that, when released or ignited, can result in death or irreversible human health effects, significant property or environmental damage, or off-site impacts due to acute toxicity, flammability, explosiveness, corrosiveness, thermal instability, latent heat, or compression. Examples of higher hazard processes including handling flammable gases and liquids, combustibles above their flash point, toxics, strong acids and bases, reactives, combustible solids, and steam generation operations, All processes covered under PSM regulations such as US DOL OSHA and SEVESO are considered higher hazard processes by definition; however, many processes not covered under existing regulations are also considered HHP's.

Lower Hazard Operations includes any activity that exclusively manufactures, handles, stores, or uses any substances with low potential for death or irreversible human health effects, significant property or environmental damage, or off-site impacts due to toxicity, physical or mechanical hazards, or asphyxiation. For LHO's, the primary hazard is localized to the immediate areas of the operation and to personnel working in close proximity to the operation. Examples of lower hazard processes include waste water treatment systems, steam distribution systems, cryogenic gas utilities, and mechanical hazards such a high speed rotating equipment (centrifuges, turbines, etc). LHO's are typically not governed by any process safety regulations due to less significant inherent risks.

3. Global PSM Team Structure and Work Processes

DuPont has established a global process safety competency team which is comprised of PSM representatives from each operating region and PSM technology team leaders from key functional roles involving specific PSM elements. This team has overall stewardship and technical responsibility for the global PSM program elements and continuous improvement initiatives and is comprised of the following positions:

Global PSM Competency Leader
Four Regional PSM Competency Leaders (United States and Canada, Europe, Latin America, and Asia Pacific)
Process Hazards Analysis Technology Team Leader
Mechanical Integrity/Quality Assurance Technology Team Leader
Incident Investigation Technology Team Leader
Process Technology Information Technology Team Leader
PSM Auditing Technology Team Leader
Management of Change – Personnel Team Leader
Chemical Reactivity Technology Team Leader
Operational Discipline (OD) Technology Team Leader
Highly Toxic Material (HTM) Technology Team Leader
PSM Resource, DuPont Safety Resources Business
Safety and Health Counsel, Corporate Legal

The Global and Regional Competency Leaders serve as full time resources dedicated to PSM program activities. The role of Technology team leaders typically involve part time, leveraged activities (approximately 15%) using otherwise full time technical resources in functional roles that support PSM in a corporate function or region. The competency team and associated technology teams operate in a decentralized, matrixed structure which minimizes the size of a central corporate organization.

The global PSM team conducts monthly strategy and work session conference calls with at least one three day face-to-face annual meeting. The team maintains an electronic team room for collaboration and issues management. This team also provides ongoing engagement with corporate engineering, sourcing and logistics, human relations, operations, legal, and governmental affairs as needed to collaborate on common PSM issues which requires broad review and support. .

Each PSM Technology team is comprised of subject matter experts and regional contacts for the applicable PSM element in order to maintain active participation from each geographical region. Technology teams have responsibility for the relevant PSM element including:

- Technical content of internal PSM standards
- Training and education on the PSM element
- Evaluation of key learnings and trends associated with incidents involving the PSM element
- Stewardship and updates for applicable sections of the Global PSM website

Regional PSM teams (led by the applicable Regional PSM Leader) have also been established to provide additional focus and technical support for PSM implementation across sites in each region. These teams of typically comprised of a small number of full time PSM resources and a network of site based PSM contacts who have local PSM responsibilities.

Senior Leadership Involvement

Active leadership involvement is a critical component of a global PSM program to articulate expectations, support new initiatives, and monitor performance. Within DuPont's PSM Competency structure, the Executive Vice President of Operations serves as the global team sponsor. Similarly, the Vice-President of Operations within each of the five Business Platforms serves as a sponsor for specific PSM Technology teams

In addition to sponsorship, Operations and Business leadership conducts an annual PSM program review within each of the five business platforms, and most recently, has convened an annual PSM (face to face) Leadership team meeting involving all platforms and regions.

Connectivity to Internal Engineering Groups and Resources

DuPont Engineering maintains an extensive group of experienced technical resources on specific types of equipment and technologies who provide support to sites during projects, normal operations, and troubleshooting. Connecting these subject matter experts to global PSM program initiatives and management systems (particularly for internal standards and training) is a key step that may be overlooked or not adequately structured in a large organization. Central Engineering groups include the following:

Process Safety and Fire Protection (complex PHA's, facility siting, consequence analysis, quantitative risk analysis)
Process Equipment and Technology (Piping, Reactors, Process Engineering)
Energy Engineering (Boilers, HVAC)
Pressure Relief Technology
Rotating Machinery
Non-Destructive Testing and Fitness-For-Service
Pumps and Seals
Control Systems and Instrumentation
Engineering Mechanics (stress analysis, heat transfer, fluid flow, process simulation)

Annual Goals and Objectives

The development and completion of an explicit set of annual goals and objectives within the PSM Competency team is an important element to ensure continuous improvement and strengthening of existing management systems. This may include short term and multi-year work streams involving one or several PSM elements, and are typically complemented with regional specific objectives. Synchronization of global, region and business level goals across the company can create unintended complexities and is important to minimize overlap and ensure consistency in purpose and outcomes.

Electronic Team Rooms and Data Sources (global and regional)

Networking and collaboration tools are critical to the success of the global and regional PSM Competency efforts since these team by nature are decentralized and matrixed across the corporation. Electronic "team rooms" and data sources help share and retain information and knowledge, track year over year initiatives, communicate new information in an efficient manner, enhance collaboration across diverse groups, provide leveraged tools and solutions, and assist in compliance with internal policies on recordkeeping and documentation. These electronic tools typically take the form of one of four applications and functions in DuPont:

Global PSM Intranet web portal - electronic access and downloading of global PSM standards, communications of incident learnings and changes to internal standards, electronic registration to classroom and webinar PSM training sessions, periodic updates on PSM Leading and Lagging Metrics, PSM related tools and solutions, and access to PSM Links – internal and external websites

Lotus Notes Collaborative team rooms for technology teams - meeting activities and follow up action items, collaboration on documents and strategies under development prior to approval and posting on the web portal, leveraging and retaining specific PSM information for sister plant technology groups

Regional Databases - unique PSM guidance for regions and local countries (for example, regulations, documents in local languages, regional issues), regional resources and experts, connections to external regional organizations

Commercial vendor software products - second party PSM audit software, checklists, and global recommendation tracking system; corporate licenses to external resources (NFPA standards, CCPS guidelines on Knovel®, commercially published auditing protocols), electronic PHA software and dispersion modeling tools

4. Process Safety Policies and Internal Standards

The core direction of DuPont is Sustainable Growth—the creation of shareholder and societal value while we reduce our environmental footprint along the value chains in which we operate. This involves a publically stated commitment which includes the following:

Strengthening businesses by making safety, health and environmental issues an integral part of all business activities and by continuously striving to align businesses with public expectations.

Driving a passionate belief that all injuries and occupational illnesses, as well as safety and environmental incidents, are preventable, and our goal for all of them is zero

We will continuously analyze and improve our practices, processes and products to reduce their risk and impact throughout the product life cycle.

Development and adherence to one common set of internal policies and requirements is a key foundational element for all process safety elements. The technical details of audit protocols, performance metrics, training materials, and incident investigation analyses are all based on the specific requirements within internal standards. These internal elements typically exceed most local country regulatory requirements; however, they are not intended to address every local regulation with each of the four global regions. Sites are expected to understand local regulatory requirements and build those aspects into site procedures and local audit mechanisms.

Internal PSM standards are frequently updated to reflect new information which is identified in emerging external codes and standards, technologies, and either internal or external incident reports. These periodic updates serve as the principle means to implement further improvements in risk management and institutionalize new information and practices in a consistent manner. At a minimum, each global PSM standard is reviewed and updated on a five year frequency. Once an internal standard is revised, an internal training and communication package is disseminated to all sites, and the accompanying PSM audit protocols (both first and second party audits) are also amended to reflect new items which can then be evaluated during subsequent audits. Table 2 provides a list of internal DuPont PSM policies and standards and Table 3 provides a supplementary list of examples of internal Engineering Standards and Technical Guidelines for specific types of equipment that support process safety efforts.

Corporate PSM Standard	Description
Process Safety Management	Responsibilities and technical requirements for corporate and site PSM systems (all fourteen PSM elements)
Management of Highly Toxic Materials	Requirements for highly toxic materials (materials of construction, leak detection, procedures, equipment design)
Process Technology	Hazards, process design, equipment design
Process Hazards Analysis	Responsibilities, technical elements and forms/checklists for PHA's
Consequence Analysis Procedures for Identifying Hazardous Events	Technical elements and documentation for qualitative and quantitative Consequence Analyses
Design and Siting of Buildings	Location of buildings in relationship to hazardous operations
Protection of Personnel from Toxic Gas Releases	Technical elements for safe havens and emergency response
2nd Party PSM Auditing	PSM audit requirements involving scheduling, auditing, and reporting
SHE Incident Investigation	Standard data elements and technical requirements for PSM incident investigations
Process Safety Incident Reporting and Classification	Technical elements for PSM incident scoring and internal reporting
Interconnection of Process and Service Piping	Technical requirements for backflow protection for interconnections of processes and utilities

Table 2: Global Process Safety Management Policies and Standards

PSM Related Engineering and Technical/Design Standards	Description
Combustion Safeguard and Burner Management Systems	Internal requirements for selection, installation and operation of combustion safeguard systems
Boiler and Vaporizer Process Indication, Alarms and Interlocks	Consistent requirements to protect against overheating of boiler/heaters caused by water level/flow loss.
Safety Interlock Design and Human- Machine Interface	Internal requirements for design of safety interlocks and the HMI aspects of Safety Instrumented Systems
Requirements for Periodic Testing and Inspection of Safety Interlock Systems	Detailed requirements for testing and inspection of safety interlocks (procedures, training ,documentation)
Combustion and Steam Generator PSM Guideline	Steam Generation systems including boilers, economizers, and fuel systems (gas, off gas, VOC's, LPG, wastes)
Steam, Boiler Feed water and Condensate PSM Guideline	PSM guidance for steam distribution systems, boiler feed water and condensate (all pressures)
PSM for Heat Transfer Fluid Systems	PSM guidance for high temperature heat transfer fluid systems (vaporizers, heaters, condensate, piping)
Flexible Chemical Hoses for Process Safety Management Applications	Life cycle practices to manage reliability of metallic and polymeric flexible hoses in chemical services
Minimum Essential Receiving Inspections for MRO Materials – PSM Components	Receipt inspection procedures for bolts, studs, gaskets, valves, pipe fittings, hoses, and other small components
Corporate Pipe Codes and Standards	Guidance on selection of appropriate piping materials and suppliers for various materials of construction
Engineering Guide to Project Management (PSM Requirements)	Design and construction guidance for major projects with PSM and Front End Loading
Risk Based Inspections (RBI)	Principles, guidelines and inspection strategies for RBI
Grounding and Bonding	Design and installation guidance to manage static electrical hazards with various equipment and facilities
Pressure Relief System Design and Documentation	Internal guidelines and practices involving the design and documentation of pressure relief systems
Mechanical Integrity/Quality Assurance Manual	Internal guidance and practices for key elements of the Mechanical Integrity and Quality Assurance system

Table 3: Examples of Engineering Standards and Guidelines which address PSM topics

5. Process Safety Training, Communications and Stakeholder Input

One of DuPont's key Safety principles is that "employees are our most valuable asset" in order to develop and maintain effective safety, health and environment processes. This certainly holds true for process safety management and for employees at sites, in regions and in the business who manage process risks on a routine basis. DuPont's PSM philosophy is to develop strong capabilities and knowledge within the site line organization to address routine PSM needs as much as possible, and supplement complex technical or training needs with subject matter experts from central roles where necessary. Internal PSM related technical training is provided in several venues in each of the four regions including:

Classroom training

The technical details of key process safety topics require active participation and student interactions as part of adult learning; therefore, many training courses which are considered foundational must be provided in a classroom setting. Table 3 outlines current classroom based PSM training courses which are offered periodically each year in each region. Most courses are targeted to plant personnel in operations, technical, maintenance, and support functions in order to develop and maintain capabilities on site. Each region typically develops an annual training plan and schedule based on year over year objectives and site needs.

Virtual Webinars

DuPont schedules and conducts more that 30 different technical webinars and virtual training sessions on various elements which are open for all sites to participate. Multiple sessions are typically scheduled each year and are led by internal subject matter experts. Some sessions may include a written post-training challenge to demonstrate competency. Table 4 provides examples of selective technical webinars which have been delivered internally. In specific cases, a presentation may also be recorded and provided for repetitive distribution in the form of a video or podcast.

Specialized Training Delivery Systems

DuPont has also developed and implemented specialized training delivery mechanisms on specific topics which are web based to leverage training tools in the most effective and efficient manner. Examples include:

College of Reliability and Maintenance - an on-line library of training materials, self directed computer based training packages, and competency challenges used for maintenance education on common maintaining skills and equipment specific tasks (both initial and refresher instructions)

Learning Management System - a web based library of training materials with focus on site operators to provide on –line education on basic skills (math, reading drawings, computers) and equipment design and functionality.

Training Subject	Duration	Target Audience
Process Safety and Risk Management	2 days	New technical and maintenance engineers , operations line management, PSM resources
Process Safety Management Auditing (initial and refresher)	2 days	1st and 2nd party PSM auditors
Process Technology	1 day	Manufacturing technical resources
Process Hazards Analysis Facilitation (initial and refresher)	5 days/2 days	Site and region PHA resources
Consequence Analysis for Fires and Explosions	5 days	PHA resources
Consequence Analysis for Toxic Releases	5 days	PHA resources
Chemical Reactivity Evaluation and Management	2 days	Manufacturing technical resources
Incident Investigation	1 day	All personnel
Layer of Protection Analysis and PFD Calculations	2.5 days	PHA, E&I, process control and manufacturing technical resources
Inherently Safer Technologies	1 day	R&D , manufacturing technical
Contractor Safety Management	1 day	Site contractor coordinators
Operational Discipline Workshop	1 day	All personnel
Safety Instrumented Systems Overview and PFD Calculations	3 days	PHA, E&I, process control and manufacturing technical resources
Pressure Relief System Design	3 days	Manufacturing technical and design
Combustible Dust Hazards	1 day	Manufacturing technical resources
Furnace and Fuel Safety	1.5 days	Site Operations and technical personnel who operate burner management systems
Fire Protection Systems Management and Fire Technology courses	Variable	Site Operations, Fire Safety , and PSM resources

Table 4: Examples of PSM Related Classroom Training

Webinar Technical Subject	Target Audience
PSM Refresher Audit training	Site 1st and 2nd party PSM auditors
Safely Conducting Pressure Tests	Basic safety principles and background technology for conducting hydrostatic or pneumatic pressure testing of vessels, tanks and piping systems
Refrigeration Maintenance	Site maintenance personnel and contractors
Process Design Review Methodologies for Safety Instrumented Systems	E&I , process control engineers who complete design reviews as part of cyclic PHA's
Microbiologically Influenced Corrosion (MIC)	Site Reliability engineers
Conducting effective PSSR's	Site operations, maintenance and technical personnel
Mothballing Plant Operations Safely	Site operations, maintenance and technical personnel
MI Inspection Planning Toolbox	Site maintenance and planning personnel
Chemical Reactivity Hazards	Site manufacturing technical personnel
Pump Selection Tool and Pump System Modeling	Design and maintenance personnel using AFT Fathom Software for hydraulic modeling
Operability of Centrifuge pumps	Site manufacturing technical/maintenance personnel
Operation of Variable Speed Fluid Drives	Site manufacturing technical/maintenance personnel
Screening and Identification of PSM Critical Rotating Equipment	Site Operations and Maintenance personnel who may identify credible failure modes and use ballistic tools
Minimum Essential Receiving Inspections	Site warehouse and sourcing personnel who inspect incoming spare parts used in process areas
Inspection of Shell and Tube Heat Exchangers	Site maintenance and reliability personnel

Table 5: Examples of Virtual Webinars which Support PSM Implementation

PSM Communications and Stakeholder Input

In addition to formal classroom and web based training sessions, routine PSM related communications to and from sites, as well as mechanisms to solicit feedback from internal stakeholders is also important as part of feedback and continuous improvement.

The global and regional structure includes a formal mechanism to distribute PSM related communication out to all of the 175 operating sites in an effective manner. These internal communications may include distribution of serious incident reports, trends analyses and key learnings, updates and new requirements in global standards, technical alerts, training offerings, and new tools and leveraged solutions that may be posted on the PSM website.

Periodic "Voice of the Customer" surveys (VOC's) are conducted to collect feedback from sites and other internal stakeholders on the current state of PSM. These surveys may be designed and completed using existing commercial electronic tools such as SurveyMonkey™ or Zoomerang ™ which are very cost effective, and when designed properly, can collect input from a large portion of the organization within a short timeframe. Collecting customer feedback is a key to identify improvements, new tools and guidance, clarify internal technical approaches, develop longer terms objectives, or confirm the effectiveness of existing processes (customer satisfaction).

External Participation with Industry Groups

Global participation with external PSM groups is also a key activity to participate in standards development, share risk management approaches, conduct benchmarking, and influence governmental agencies involving effective technical requirements. DuPont has historically played an active role in the American Chemistry Council (ACC) PSM Committee, the AIChE Center for Chemical Process Safety (CCPS) Technical Steering Committee, and other special focus industry organizations such as the Chlorine Institute and standards organizations and committees which are principally focused on applications with the U.S. (NFPA, ANSI, IEC, etc)

Participation in other external organizations outside of the U.S. is also important from a global perspective, both for well developed regions such as Europe, and in emerging regions such as Asia Pacific. The DuPont Regional PSM Leader has the primary responsibility for participation, advocacy, and oversight on these regional groups.

6. Performance Measurement and Auditing

The ability to routinely monitor the status and progress of process safety management across the global organization is a critical need and success factor in order to maintain and improve program results. A combination of leading and lagging indicators may be utilized to accomplish this objective; however, measurement systems should include all global organizations and sites. DuPont has established a broad set of leading and lagging PSM program indicators that are used to measure ongoing performance (2), including:

PSM Quarterly Metrics

DuPont maintains a set of leading indicators and PSM metrics that are reported and analyzed by each manufacturing location on a quarterly basis. These metrics measure each site's status on internal actions and commitments (typically a report on the number of open and overdue action items across multiple PSM elements).

Incidents

DuPont maintains a global incident classification and reporting process to evaluate events based on severity and key learnings. This includes a global process to collect and analyze incidents for trends and performance improvement opportunities.

PSM Audit Results

Second Party PSM audits are scheduled and conducted within each operating unit at all global manufacturing locations. Audits conducted based on the schedule are reported annually and changes to audit dates must be approved at a business level. A composite score and PSM element scores are generated for each audit which can be evaluated for trends within regions and businesses. All PSM audit recommendations are tracked in a global database are reported quarterly.

External (Third Party) PSM Program Reviews

In addition to internal metrics and audit processes, DuPont also participates with third party audit programs

Audit program Review - A third party independent review of the global second party PSM auditing program is conducted each year to assess internal performance opposite ten criteria (e.g., adequate staffing and qualifications of audit teams, quality of audit reports, pre and post audit activities). The results of this audit program review are provided to DuPont Executive leadership and posted on the DuPont internet for public review.

ACC Responsible Care - As a member of the American Chemistry Council, DuPont sites in the U.S. and the corporate SHE/PSM competency center receives an independent Responsible Care evaluation every three years which involves recertification of PSM management systems and performance.

7. Special Focus Areas and Management Systems

Over the years, DuPont has established and maintained a number of specific focus PSM teams to provide additional emphasis and drive global standardization in key areas of risk management. These include:

Highly Toxic Material Guardian Teams

DuPont has adopted an additional layer of process safety oversight for materials which are classified as Highly Toxic Materials (HTM's) such as chlorine, phosgene, sulfur trioxide, anhydrous HCL, and hydrogen fluoride. This involves a total of twenty distinct HTM Safety Guardian teams which are comprised of all global sites that handle the applicable toxic material. Each team has established a detailed HTM guardian manual which provides additional prescriptive requirements for the substance beyond generic PSM requirements such as materials of construction, design of fixed and portable leak detection systems, equipment design, and mechanical integrity tests and inspections. Guardian teams conduct periodic meetings to review internal and external data, and conduct periodic audits of each facility which handles that applicable toxic material. An annual review of the guardian team results and performance is conducted within each business which sponsors the team.

Sister Plant Technology (SPT) Networks

Over the past three years, DuPont has provided additional emphasis and internal policies on leveraging and applying critical PSM related information across sister plant technologies (i.e., all global sites who operate identical or similar manufacturing technologies and products). This includes the leveraged applications of cyclic PHA findings, consistent design of key layers of protection such as Safety Instrumented Systems, maintenance practices, standard operating conditions and incident learnings. Formal and sustainable processes to share and leverage knowledge and information globally across each SPT site is a critical step for improving risk management performance (3).

Special Hazards and Focus Area Teams

Dust Explosion and Fire Safety Special Emphasis Network – a global team across multiple businesses within DuPont which handle combustible solids in order to provide tools and best practices to analyze and solve issues pertaining to combustible dusts, including hazard prevention and control technologies, testing and evaluation, and provision of technical training.

Chemical Reactivity Management Teams - specific teams have been established to share learnings and approaches with a focus on select substances or processes with significant reactivity hazards. This includes chemical reactivity teams for Polymer Coatings Reactor Safety, Tetrafluoroethylene, Butadiene Monomer Handling, and Nitration Reactions. .

Sourcing and Logistics Supplier Audits

DuPont Sourcing and Logistics conducted periodic site visits to equipment and parts suppliers to ensure materials are properly manufactured, labeled and shipped to sites where they involve PSM critical components. This may include hoses, valves, piping system components, and similar materials. In the current global supply chain, these quality control audits may be conducted in all regions and provide a means to authorize new qualified suppliers in emerging countries. There is also a global process to report issues and manufacturing deficiencies with specific vendor components in order to investigate and correct defects that may be identified in the supply system.

8. Logistics with Global Implementation

There are a number of logistical issues and challenges associated with the global implementation of a process safety management program which should be addressed in order to maximize the effectiveness of any approach. These include:

Local Customs, Language, and Translations

DuPont operates in many countries in Europe, Asia Pacific, and Latin America where translation requirements both from local language to English and from English to local languages can be a significant issue. Translating corporate standards, incident reports with key learnings, and training materials into local language are specific needs in order to deliver information throughout the line organization. Second party PSM audits by policy are always written and issued in English to assist in business leadership review; however, certain audit reports may also required translation into local language. Translation of "U.S.based" terminology and expressions in all PSM documents should be minimized and reviewed carefully to ensure a change in meaning or purpose has not been inadvertently included.

The adage "think globally, act locally" also applies to the nexus of corporate PSM policies and requirements and local site culture and organization. For example, communications, consensus building and change management, employee participation, and the site's willingness to be self-critical in incident reports and audits in many countries may contain subtle and significant differences versus the U.S. which can impact PSM implementation and must be anticipated and managed.

Local regulations and Codes and Standards

Local state, country and regional regulations and external codes and standard must be considered as part of a site PSM program. Internal DuPont PSM policies require each site to clearly identify what local regulations and laws apply in addition to the internal global requirements. Similarly, local, country and regional external codes and standards should also be identified and integrated into local site practices. Where conflicts may exist between internal corporate guidance and practices versus local guidance, these must be reviewed and resolved proactively.

Logistics - Time Zones for Collaboration or Training and Development

A majority of the centralized technical expertise and subject matter experts involving process safety management in DuPont Engineering are based in the United States; however, these resources provide consulting and training expertise to all regions, Technical assistance, collaboration, and virtual training and development sessions for sites outside of the U.S. region must be scheduled to accommodate local times for meeting participants (for example, there are multiple time zones across the Asia Pacific region which need to be considered)

9. Issues and Future Focus Areas

Standardized Work Flow and Forms

As previously stated, DuPont has developed and maintains a detailed set of global internal PSM policies and procedures which are applied at all sites. These internal standards typically define the "what's" but not the "how's" of a site based PSM program. Current internal initiatives include several improvement items to increase global standardization and efficiencies associated with PSM elements, including:

Standardized incident investigation form and data fields
Standardized Process Hazards Analysis template and reporting format
A common global corrective action tracking system
Standardized formats for Operating Procedure and Safe Operating Limits

Internal PSM Intelligence Sharing

Site level experienced operations, technical, maintenance and PSM support resources hold a significant amount of knowledge and insights which can benefit other locations to further improve risk management programs; however, sharing knowledge with others can be time consuming and difficult to maintain. Personnel turnover magnifies the challenges associated with knowledge sharing activities. Development of a sustainable means for employees to share experiences, archive data and reports, and provide search engine capabilities to find key information is a future focus area to improve corporate memory beyond the current practices of sharing incident investigation reports and periodic updates of standards.

Personnel Development and Succession Planning

Managing the turnover/loss of personnel, training and development of new personnel, and long term succession planning are key attributes of a strong PSM personnel process. DuPont has maintained a detailed management of change – personnel process for many years; however, the current system does not focus adequate attention on long term personnel changes, expertise, and succession planning. This is an area of current assessment.

10. Summary

The development and implementation of a global PSM management system is complex and requires active involvement and sustained actions across the entire company and operations group in order to achieve the desired outcome – an unwavering focus and passion to achieve and maintain PSM excellence. This includes an overarching philosophy on risk management that lives and breathes within senior leadership and at all sites. The system should include specific roles and responsibilities, detailed internal policies and procedures, a comprehensive program of metrics and auditing to measure performance on an ongoing basis, extensive training and development to maintain organizational capacity, and special focus areas and teams, where needed, to supplement general PSM applications.

DuPont has historically maintained a clearly articulated corporate mission and philosophy involving safety which provides a foundation for PSM - safety is a core value, not a priority, since priorities can change over time (4). DuPont has utilized a four-step (sequential) approach in order to develop, implement and maintain an effective and sustainable PSM process at the site level. This includes:

Stage 1 - Establishing the "Safety" Culture at all Levels
Stage 2 - Leadership Commitment and Involvement
Stage 3 - Implementing the Fourteen Technical Elements of PSM
Stage 4 - Operational Excellence through Operational Discipline

All stages must be fully in place to achieve the desired outcomes for an effective and reliable risk management process to prevent incidents and ultimately protect employees, the environment, and the communities in which we operate. More recently, a fifth stage has been the internal focus across the corporation and at sites which have well developed and "mature" PSM programs to pursue increased effectiveness and efficiencies throughout the organization: The fifth stage includes:

Stage 5 - Continuous Improvement through Leveraging and Knowledge Management

Beyond the site based stages and focus, there are at least two broad elements necessary within a corporation to globally manage and sustain the PSM program. This includes

A detailed series of PSM technical elements and requirements consistently implemented at all sites with a high degree of operational discipline. This should also include an objective to go well beyond minimum compliance

A comprehensive approach for corporate PSM governance and management oversight processes which results in a constant review of the system and all its parts

Without clear and detailed technical elements applied at each site in a consistent manner, local risk management results will be compromised and likely will not produce the desired results over time. Substantial organizational energy and focus at the site throughout the line organization is required in pursuit of the goal of zero incidents. Within DuPont, process safety is said to be "embedded in the DNA of the organization and each individual". Consider the following quote:

We are what we repeatedly do; excellence, then, is not an act but a habit.
—Aristotle

Without an overarching and sustainable PSM governance process across the entire organization, the long term effectiveness and ability to maintain and improve process safety performance is either lost, partially controlled, or left to chance. Once a program has been developed, it must be monitored and assessed over time as the organization changes (e.g., new processes, mergers and acquisitions, changes in senior leadership) to measure performance and identify new strategies to further improve risk management. Organizations who do not address the need for continuous monitoring through a variety of internal and external assessments and performance indicators are likely to fall into cycles of PSM strengthening and deterioration over time (the "hand on" and "hand off" phenomenon) which is not effective in managing low probability, high impact PSM events. Consider the following quote:

> *Even if you're on the right track, you'll get run over if you just sit there.*
> —Will Rogers

Comments or questions on this paper or DuPont Process Safety Management program elements may be directed to the author or through DuPont Safety Resources at the following web address: http://www2.dupont.com/Consulting_Services/en_US/safetycontact.html

References

1. Guidelines for Implementing Process Safety Management Systems, Center for Chemical Process Safety, American Institute of Chemical Engineers, 1994, ISBN 0-8169-0590-8

2. Cummings, David E*; The Evolution and Current Status of Process Safety Management Metrics within DuPont* 23rd CCPS Conference – New Orleans, LA, April 8, 2008

3. Cummings, David E., *PSM Knowledge Management and Applications across Common or Sister Plant Technologies*, 10th Process Plant Safety Symposium – New Orleans, LA April 8, 2008

4. Burk, Arthur F *Achieving Excellence in Process Safety and Risk Management*, Managing Safety and Health in the Workplace, Chicago, June 13-14, 1994

Accident Investigation Using Process Control Event Diagrams

Delmar "Trey" Morrison, III, John D. Martens,

Russell A. Ogle, and Joel T. Cowells

Exponent, Inc.

ABSTRACT

The automation of chemical processes has both positive and negative features. Computer automation can be a powerful tool for safety management. On the other hand, it provides new opportunities for human error both in design and in operation. The interaction of process equipment, control systems, and operators can be complex. In this paper we investigate an accident with an automatically controlled batch dryer that resulted in an explosion and fire.

The batch dryer contained a heat transfer subsystem. The heat transfer subsystem consisted of combustible oil circulated continuously in a closed loop. The purpose of the heat transfer subsystem was to maintain a plate heat exchanger at a constant temperature. The process dryer was programmed to dry a batch of material in contact with the slab by advancing through several stages of operation. The dryer was controlled by a programmable logic controller (PLC) and used a human-machine interface (HMI) to interact with the process. During one of these stages the dryer encountered an abnormal operating condition. The control system placed the dryer into a standby mode and tripped an alarm at the HMI console. The operator responded to the alarm in a manner consistent with the operating procedures.

The standby mode was a fail-dangerous condition. The control system automatically advanced the dryer into a more hazardous state. The process measurements and alarms transmitted to the HMI indicated an abnormal condition but did not indicate an imminent explosion/fire hazard. Three maintenance workers attempting to diagnose the cause of the abnormal operating condition were injured when the unit exploded.

The interaction of the operator and the control system was analyzed using process control event diagrams (PCEDs). The PCED provides a model for representing the interactions between a chemical process, the control system logic, and process operation. This diagram can serve as the basis for a systematic hazard analysis. PCEDs were used to trace and document the changes in equipment states caused by the process dynamics, control logic, and operator actions. This structured approach to failure analysis demonstrated several inadequate sensor uses, sections of control logic, and final control elements. Through the use of this tool, the fail-dangerous state was shown to be a consequence of several design deficiencies in the dryer system and not the result of operator error.

The lessons learned from this incident can be applied by facilities to evaluate similar safety issues that may arise with computer-controlled process systems.

1. Introduction

We applied PCED-based analysis to this accident investigation. In our original analysis, we used several tools to understand the incident and the events that led to it. Some of the tools that we used included: fault tree analysis, event tree analysis, failure mode and effects analysis, state diagrams, state transition matrices, and causal event diagrams. Though we do not present the results of all of the methods we used, we provide comments on the various methods in a later section. This paper focuses on the application of PCED to the incident. The hypothesis we adopted was that the PCED technique might be able to supplement, or even replace, the numerous tools that we applied in the original investigation.

2. Background

The overpressure hazard caused by the thermal expansion of trapped liquid is a well known hazard in the refrigeration and heat transfer industry. The preferred safeguard to prevent overpressure is to employ pressure relief devices with any heat transfer component or segment of piping that can become isolated from the expansion tank (reservoir for the liquid being used as the heat transfer medium). Pressure relief devices, such as pressure relief valves and rupture disks, are mechanical devices. They only require periodic maintenance to ensure their reliability. Generally speaking, it is not acceptable to replace pressure relief devices solely with computer-controlled functions. Potential exceptions to this rule would include the use of safety instrumented systems with a certified safety integrity level.

In this application, the manufacturer designed and fabricated a skid-mounted cryogenic refrigeration system with no mechanical pressure relief devices located at an electric resistance immersion heater (see Figure 1). Under at least two different accident scenarios, it was possible to trap the heat transfer liquid inside the heater. The only protection from overpressure within the heater assembly was two types of overtemperature protection, a safety function controlled by the system programmable logic controller (PLC) and a high temperature switch.

The refrigeration skid was designed for unattended operation. The primary responsibilities of the operator were to select the recipe, load the process chamber, start the refrigeration operation, unload the process chamber when the cycle was completed, and fill out the paperwork. The operator was given little if any training on how to handle abnormal operating conditions and the manufacturer's operating manual provided little guidance for fault diagnosis and management.

3. Incident Summary

The refrigeration system was a skid-mounted package unit that relied upon liquid nitrogen as the cryogen, a heat transfer oil as the secondary refrigerant, and an electrical resistance heater as a heat source to provide split-range control over the heat transfer oil temperature. The portion of the heating and cooling circuit is shown below in Figure 1.

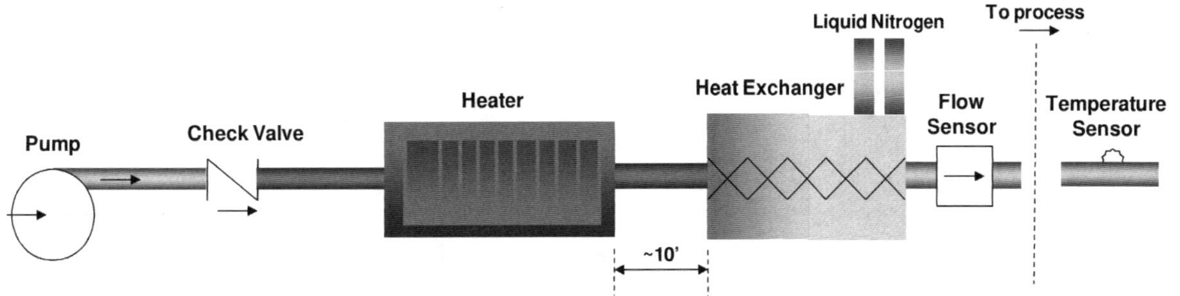

Figure 1. Simplified schematic for process oil heating and cooling.

The cryogenic refrigeration system was designed to work in a batch mode with a typical run taking several days. During the run related to this incident, a temporary failure occurred in one of the control system power supplies. This failure, in turn, caused the temporary failure of several other electrical components. During this process upset, the heat transfer oil stopped circulating and, as a result, the cryogenic coolant caused the heat transfer oil to freeze inside the heat exchanger. The frozen heat exchanger acted like a closed valve and trapped the heat transfer oil in the heater.

Eventually the control system power supply recovered, and its associated electrical equipment resumed operation. The frozen heat exchanger prevented the heat transfer oil from circulating within the refrigerant piping. Following its programmed instructions, the PLC commanded the electric heater to turn on, heating the trapped heat transfer liquid. As the temperature of the trapped heat transfer oil increased, it exerted an ever-increasing pressure against the housing of the heater assembly until the housing violently exploded. The heat transfer oil released by the explosion was ignited, resulting in a fireball.

4. Introduction to PCEDs

Hazard analysis techniques such as hazard and operability studies (HAZOP), fault tree analysis (FTA), event tree analysis (ETA), and failure mode and effects analyses (FMEA) have been used to logically test process systems for safety and as tools to investigate complex accidents. A particular difficulty with many of these methods is incorporating computer logic, complex computer system failures or inappropriate responses from the system (Chung et al. 1999; Yang and Chung 1998). In some instances, experienced individuals who are familiar with the process and the computer-based instrumentation analyze the safety of logic-controlled functions in an informal manner. These individuals try to detect potential causes for hazards by reasoning about possible scenarios. Logic design errors, particularly those that are sequential in nature or that require several independent events, may escape detection without a structured analytical approach (Yang et. al. 2001).

Other methods have been created for this specific application of hazard analysis of logic-based control systems such as a Computer HAZOP (CHAZOP[1]) and the Hazard Analysis for Programmable System (HAZAPS). The CHAZOP procedure analyzes the role of the computer control systems by gradually building a detailed view of how the system will respond to failures. In contrast, the HAZAPS procedure starts with a generic functional model of an incident to which the safety requirements are applied as base tasks. These methods as well as other computer-based HAZOP procedures may not appropriately address controls logic (Yang and Chung 1998). The computer controls industry and related standardization organizations have promulgated standards in this area such as ISA S84, which proposes an integrity-based structure of the sequence control logic for batch processes, and IEC 61508, which includes a model for the software development lifecycle of safety-related software (Kowalewski et al. 1999).

The PCED can serve as the basis for a systematic hazard analysis or accident investigation. A PCED is a multi-level representation of the process, its interactions with the control system, and the hierarchy of the control system. Each functional level in the

[1] The acronym CHAZOP may refer to either Computer Hazard and Operability Analysis or Control Hazard and Operability Analysis.

diagram represents a component level in the control system architecture. The basic structure consists of levels representing the "Operator," "HID & Display," "Communication," "Computer," "Sensor & Actuator," and "Process." Given this structure, the process, the control system, and the operator can be represented in a compact form. Input/output levels to the control system are separated into two types: HID & Display (Human Input Device, e.g. computerized operating terminal) and Sensor & Actuator. The former only interact with the human operator while the latter only interact with the process. Nodes represent objects within the system such as controllers, equipment, states, and input/output devices. Time is generally represented on the horizontal axis starting on the left. Labeled arrows connecting nodes within the diagram represent events. The direction of the arrow represents the direction of information flow, and the events are ordered in the horizontal direction. The following section will develop the basic PCED for the system studied here.

5. Basic Process Control Event Diagram

The process dryer involved in this accident case study used a combustible oil, operating between approximately -70°C and +70°C. The process and instrumentation diagram (P&ID[2]) illustrating the oil heat transfer utility is provided in Figure 2. The oil was circulated in a closed-loop system by means of a centrifugal pump (P01). P01 was controlled at a constant speed and included a "pump is running" interlock with the PLC control system. The oil temperature was maintained at the process set-point (Tsp) through split-range control. For split-range control, the system only heated or cooled, but was not designed to take both actions simultaneously. For cooling, the PLC control system automatically modulated the cooling control valve position (0 to 100% open) while keeping the heater off. The cooling medium was a cryogenic liquid used to provide fast and efficient cooling response. For heating, the PLC automatically modulated the power to the resistance heating element in the oil heater with the cooling control valve closed.

[2] P&ID is also referred to as piping and instrumentation diagram.

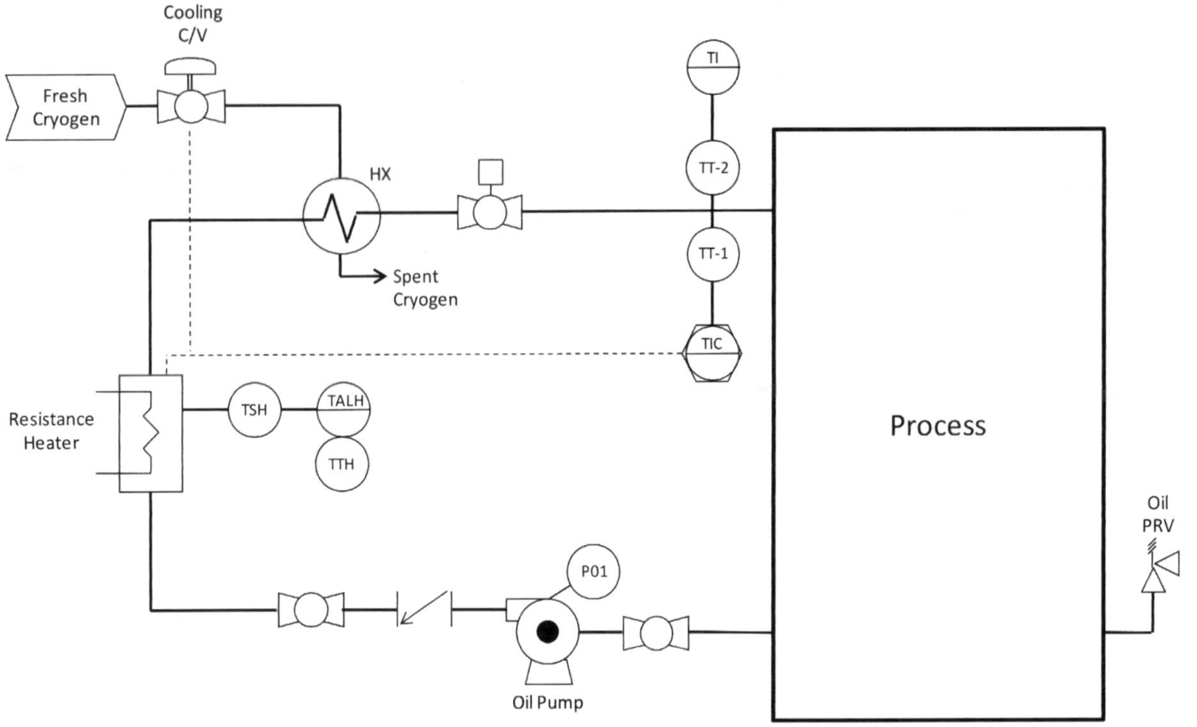

Figure 2. P&ID for process oil heating and cooling.

The oil temperature was monitored by a pair of thermocouple sensors, each with its own transmitter (TT1 and TT2) at the entrance to the process dryer unit (Process). The two temperature transmitters provided redundant readings of the oil temperature; however, two important features of the design should be noted. The system was designed so that the standard operation required one transmitter/sensor pair to be used for automatic control and the second transmitter/sensor pair was used to provide temperature readings to the operator interface (HID). The oil heater was also equipped with a high-temperature shut-off switch (TSH).

Figure 3, below, shows a simple PCED representing the call for cooling of the heat transfer oil. The analysis of this PCED starts from the lower left corner of the diagram where the temperature sensor element senses a current oil temperature that exceeds the oil temperature setpoint. Moving up on the leftmost path of the diagram shows that the interlock conditions for opening the cooling valve are met. Moving to the right on the diagram represents a step forward in time. The middle vertical path on the diagram shows the PLC control system receiving the temperature process variable and determining that the oil temperature exceeds the setpoint. The PLC also communicates the setpoint and process variable to the HID for display to the operator. The final vertical path on the right of the diagram shows that the cooling valve is commanded open by the PLC and the process responds with a lower oil temperature.

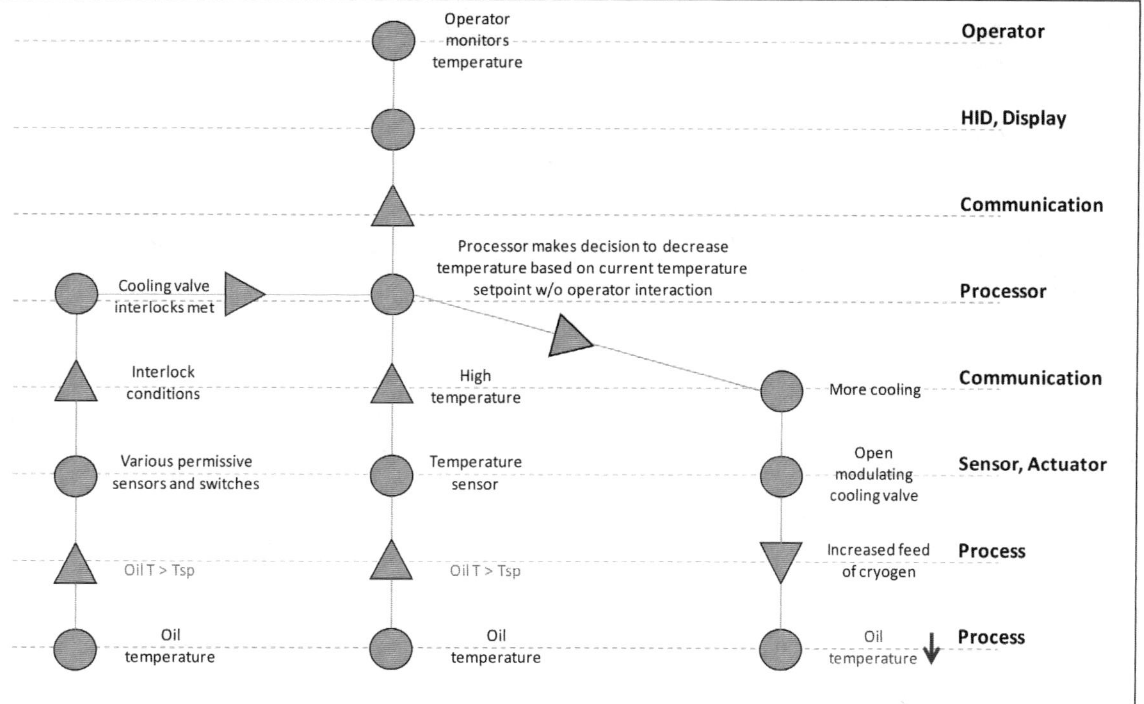

Figure 3. PCED for basic oil cooling with one oil temperature sensor.

6. Incident Investigation using PCEDs

After beginning investigation into the accident and the process design, it was determined that the simple PCED in Figure 3 did not accurately represent the physical process unit and controls. It was only through detailed investigation of the PLC programming and control wiring that the significant distinction of having two separate temperature sensors for control was noted. From the physical investigation and a review of the process data and control logic, it was concluded that the blockage of oil flow was caused by a frozen heat exchanger. Some of the several hypotheses that were evaluated are listed below.

- Hypothesis 1: temperature sensor failure called for constant cooling
- Hypothesis 2: cooling valve stuck open
- Hypothesis 3: failure of PLC rack power supply combined with operator interaction

After significant analysis of the controls, including detailed analysis of the PLC program and actual in-field wiring, the investigation team concluded that Hypothesis 3 was the most likely. The sequence of events are depicted in Figure 4 (at the end of the paper) and will be discussed below.

As shown in Figure 4, the following process steps were identified leading to the explosion and fire:

1. Oil stops flowing

   As a result of overheating, a PLC rack power supply stopped supplying power to the I/O cards in the rack. As a result, a 0 mA signal was sent to the variable frequency drive of the heat transfer oil, stopping the flow of oil.

2. No heating available

   Another result of the PLC rack power supply interruption was that a 0 mA signal was sent to the heater SCR, meaning no heat was available to the oil regardless of what the PLC control system called for. The setpoint was above ambient and, due to the loss of the heater, the oil temperature began to drift down towards ambient.

3. Operator response

   The operator observed the abnormal process data during the loss of the PLC rack power supply and determined, per the operating instructions, that action was required. Two temperature transmitters were available to the PLC for control of the heat transfer oil temperature. The operator changed the temperature transmitter being read by the control system from the one coming into the PLC rack with the bad power supply (reading at minimum of range) to the other temperature transmitter being input to the PLC in a different rack. Switching the temperature transmitters allowed the PLC to read the correct temperature. The operator also decided to advance the phase, which lowered the setpoint to a value that was lower than the actual oil temperature.

4. Control response

   Upon receiving a new setpoint and being able to read the correct temperature, the PLC determined that the heat transfer oil required cooling. The PLC called for opening of the control valve for the cryogenic liquid.

5. Heat exchanger freezes

   With the control valve for the cryogenic liquid open and traveling into the heat exchanger but the heat transfer oil not flowing through the other passages of the heat exchanger, the heat transfer oil in the heat exchanger freezes.

6. Recovery of control

   After cooling, the PLC rack power supply regained power. The heater and pumps were again controllable by the PLC control system.

7. Blocked flow

   With normal power regained, the PLC once again sent a non-zero command to the VFD for the heat transfer oil circulation pump. However, the heat exchanger was frozen and the flow of oil was dead-headed at the inlet to the heat exchanger.

8. Oil temperature drifts

   With the heat exchanger frozen and the flow of oil dead-headed at the inlet to the heat exchanger, the oil temperature drifted back toward ambient.

9. Heating of trapped oil

   At this point in the process, the oil temperature was seen as too low by the PLC due to the freeze-up of the heat exchanger. The PLC sent a non-zero command to the heater SCR in an attempt to heat the oil back to the setpoint. The PLC also commanded the cryogenic liquid valve to close.

10. Operator not warned of abnormal conditions

    During this time, the HMI did not warn the operator with alarms or any other indication that danger was imminent.

11. Oil overheats

    Because the PLC was calling for heat and the flow of oil was blocked, the stagnant oil in the heater continued to heat up. A high-temperature shut-off switch (TSH) safety device was wired in series with the heater contactor from the PLC. This device was mounted to the shell of the heater and consisted of a capillary tube sensing element. It was designed to open the circuit, shutting off the heater, if the heater temperature exceeded its internal setpoint. This safety device was not properly installed and did not perform as intended in this incident.

12. Heater Explodes

    The oil in the heater continued to heat up and eventually the heater shell could no longer contain the built-up oil pressure. The heater explodes.

7. The Role of the PCED method

Four necessary sequential events (i.e., semi-dependent failures) led to the accident:

1. Intermittent PLC power supply failure.
2. Operator action to change TIC from TT1 to TT2.
3. Return of PLC power supply function.
4. Actual oil temperature less than the set-point temperature for the oil.

The PCED tool aided the investigation team in communicating these findings. As described below, in order to effectively analyze the incident, the PCED tool required input from other analysis tools. The PCED was useful at graphically presenting the complex interactions between processes, sensors, controllers, actuators, displays, and operators.

8. Comparison of PCED to other analysis tools

Several tools exist for the analyzing failures. While each tool has its benefits and drawbacks, a topic that is outside the scope of this paper, we have found that using a combination of techniques is often the most helpful. In general, the tools serve two purposes.

The first purpose is to help analyze the available accident data in a structured format that allows for the drawing of logical conclusions. The process of compiling the data in the format needed for the various tools in itself is an important step in the analysis. The structured approaches (bottom-up, top-down, etc.) allow us to approach the incident from several different ways, each of which provides insight into the conditions that can lead to an accident.

The second purpose is to communicate complex findings in simple ways. We have discovered that while various tools may be used to analyze the available data, the tools that are used for analysis are not always the tools that are best for communicating the findings to others. Most tools have recognizable visual features: Fault Tree Analysis' tree structure, Event Tree Analysis' tree structure, Failure Mode and Effects Analysis' matrix structure, state diagrams' arrows and circles structure, and transition tables' matrix structure. Each tool may communicate different aspects and levels of detail and often more than one tool is needed to describe complex accidents.

In our application of PCED to this accident investigation, we found the process to be helpful in organizing the available data. However, given the complexity of the control system, we found that several tools were required in order to accurately understand the accident scenario. In order to put together an effective PCED, other tools were required in order to obtain a sufficient understanding of the control system. For example, appropriate state diagrams and state transition tables were required as input to the PCED method as well as input to our FTA approach.

9. Conclusions

The PCED method was one of the tools we used to analyze the incident and present our findings. We found that several tools were required to effectively model the incident and communicate our findings. The PCED diagram encompasses the process, the sensors, the controls, the displays, and the operator in a single tool. As such, we found that it can be useful to display the interactions of these components as they relate to an accident involving complex systems.

10. References

P.W.H. Chung, S. H. Yang, and D. W. Edwards, "Hazard Identification in Batch and Continuous Computer-Controller Plants," *Ind. Eng. Chem. Res.* 1999, 38, 4359-4371.

S.H. Yang and P.W.H. Chung, "Life Cycle Hazard Analysis for Computer Controlled Processes," *Computers Chem. Eng.* Vol. 22, Suppl., pp. S483-S490, 1998.

ANSI/ISA S.84.01 Application of Safety Instrumented Systems for the Process Industries, the Instrumentation, Systems and Automation Society (1996).

IEC 61508 Functional Safety of Electrical/Electronic/Programmable Electronic Safety-Related Systems, International Electrotechnical Commission (1998).

S. Kowalewski et al. "Verification of logic controllers for continuous plants using timed condition/event-system models," *Automatica*, Vol. 35, pp. 505-518, 1999.

S.H. Yang and P.W.H. Chung, "Hazard analysis and support tool for computer controlled processes, J. Loss Prevention Proc. Ind., Vol. 11, pp. 333-345,1998.

Additional references not cited:

Leveson, N.G., Safeware: System Safety and Computers, Addison Wesley (1995).

HSE, Out of Control: Why Control Systems Go Wrong and How to Prevent Failure, Health & Safety Executive (2003).

ISA 88.01, *Batch Control Part 1: Models and Terminology*, Research Triangle Park, NC (2006).

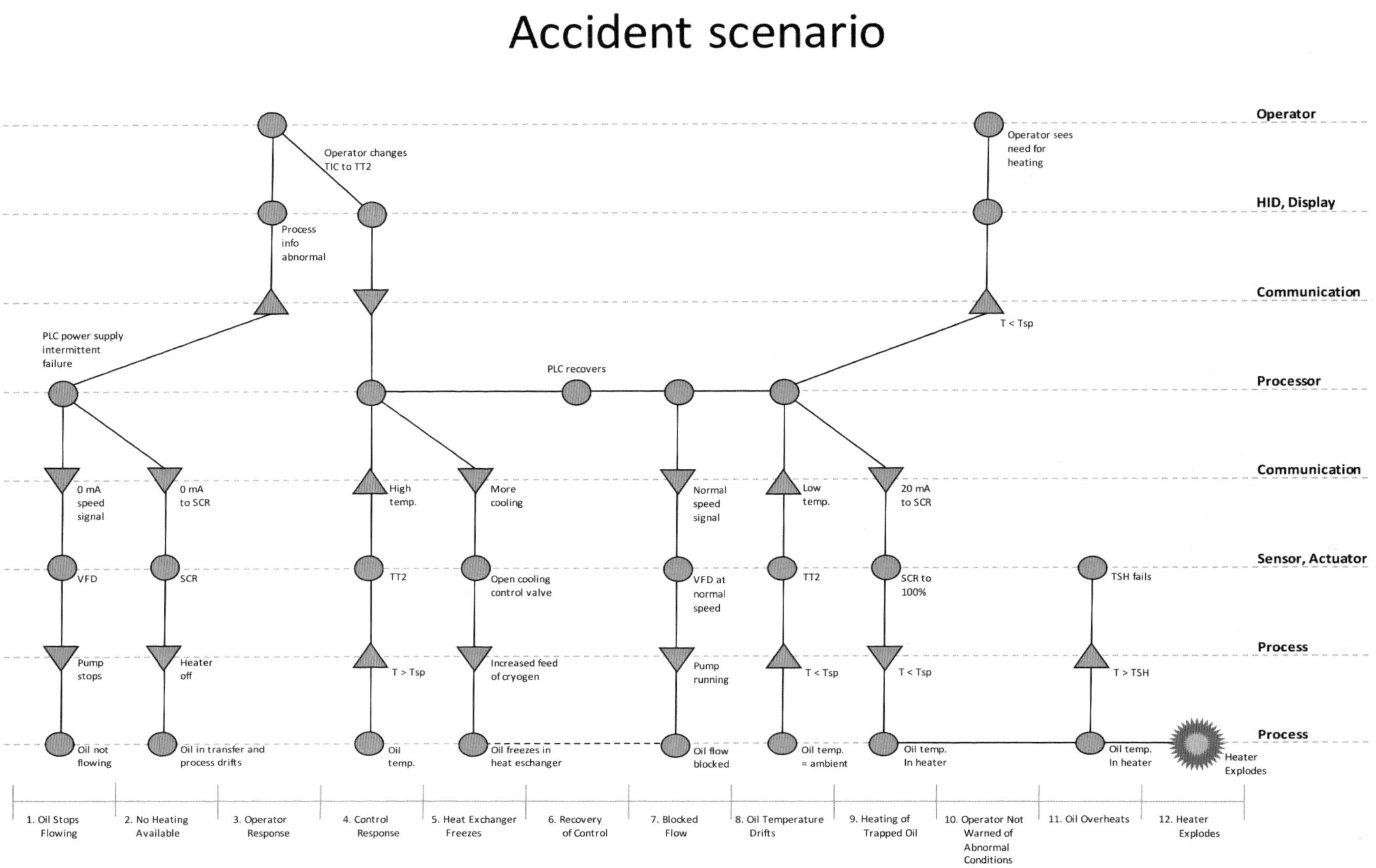

Figure 4. PCED depicting the sequence of events (process, controls, operator) comprising the incident.

The Retro-Synthesis of Risk Assessment Tools for Hazard Evaluation and Screening Analysis

Fred Henselwood, NOVA Chemicals, Calgary, Alberta, Canada
henselfw@novachem.com
Robert Wasileski, NOVA Chemicals, Moon Township, Pennsylvania, USA
wasiler@novachem.com

ABSTRACT

Risk screening tools can be very useful for evaluating hazards and establishing priorities for risk reduction. Further, these tools can provide educational value to field staff when they are distributed to individuals outside of traditional process safety roles. This paper describes a retro-synthesis technique whereby risk screening tools can be readily generated for these types of hazard evaluation and priority setting activities. Further, by utilizing a retro-synthesis approach, data gathered at the field level can often be drawn upon at a later point if a more detailed risk assessment is required. The retro-synthesis approach involves working backwards from traditional risk functions (relationships) so as to define the predominant elements for incorporation within the desired risk screening tool.

For example, risk can be defined as a function of "likelihood" and "severity", where "severity" may be a function of the "energy released" and the "exposure group". Similarly, the "energy released" is often a function of the "properties of the material involved" and the "amount of material released". Through identifying these relationships, "risk" in this example can be characterized as a function of "likelihood", "exposure group", "properties of the material involved", and "amount of material released". Further, by highlighting the various influential parameters, the relationships between each of these parameters, as related to risk, can also be understood. Once the various terms and functions are defined, screening questions can then be developed to quantify each of the parameters so as to allow for an approximation of the relative risk involved for the given scenarios.

This approach benefits from being simple to develop, maintains alignment with the mathematics required for more detailed quantitative risk assessments, and generates screening questions which can be given to field staff. Through careful construction of the screening questions field staff can then rapidly, and in a reproducible manner, screen a large number of scenarios so as to identify those situations which represent priority hazards.

INTRODUCTION

In an ideal world all risks would be studied in detail and all potential threats to various stakeholders would be identified, investigated, and managed. However, resource constraints do not allow for this level of study to be broadly applied. As such, risk screening becomes an important tool whereby scenarios of potential concern can be quickly evaluated so as to best identify those scenarios which need to be further investigated and managed. Consequently, the objective of the risk screening process becomes one of identification and, where warranted, prioritization of those scenarios which are of potential concern. Once identified, choices to act on the threat, not act on the threat, or to study the threat further then need to be made and justified. As such there are two primary goals behind risk screening. First, is to ensure that the risk screening process is efficient, thereby allowing as many threats as possible to be reviewed. Second, is to ensure that the risk screening process is structured in such a way so as to support the required decision making processes (with respect to the need to address the reviewed scenarios). Secondary to the primary goals but also of importance is that the risk screening process be structured such that the information that is generated can be documented to allow for detailed risk assessment activities when called upon in the future.

As is illustrated in the CCPS Layers of Protection Analysis (LOPA)[1] book risk screening is typically conducted in a qualitative manner followed by more detailed studies which tend to be semi-quantitative in nature (such as LOPA) and, in some cases, subsequently followed by detailed quantitative studies[2]. The typical qualitative approach applied at a risk screening level reflects the primary need for this study step to be efficient, thereby minimizing the resource needs of the business. These qualitative approaches excel at meeting the first primary requirement for risk screening whereby the number of scenarios which may be assessed can be maximized. Despite this benefit, the relative ambiguity of the results generated by these types of approaches does not necessarily ensure the success of this technique when judged against the second primary goal. Specifically, qualitative approaches often generate results which push towards further study at the expense of taking or not taking action relative to the assessed threat. With respect to the secondary goal of generating information which may assist in further detailed studies this type of qualitative assessment often provides little insight which may be of use at the semi-quantitative or quantitative level. For example, a qualitative assessment might simply state the scenario is "expected" and that the consequences would be considered to be "adverse". By comparison a well documented LOPA study can be of great benefit if a detailed risk assessment is found to be necessary as the LOPA study will provide a significant starting point for the subsequent detailed risk assessment study.

Again recognizing these primary and secondary goals, a case can be made to move away from qualitative risk screening approaches to semi-quantitative approaches for situations where the commonality of the scenarios being assessed supports this transition (i.e. there needs to be a substantial number of scenarios to assess such that the preparation time required to develop a semi-quantitative risk tool will yield clear benefits, relative to simply qualitatively assessing these scenarios in the minimum amount of time utilizing existing risk screening methodologies). These desired benefits typically come from improvements in the decision making step of the risk screening process and by improvements in the nature of the information gathered, which could then be passed along to more detailed assessment methods if called upon. These improvements relate to the fact that through using a semi-quantitative approach the numerical output generated typically allows for the prioritization and/or ranking of the scenarios assessed. To be beneficial a semi-quantitative risk screening tool needs to prioritize the scenarios assessed and be conducted in the minimum amount of time. Further, the data collected as part of the semi-quantitative risk screening tool should be structured such that it is readily transferrable and applicable if further in depth risk studies are to be conducted. An additional requirement of the data collecting stage is that the data can be readily obtained and clearly and consistently assessed. This paper will focus on a retro-synthesis approach whereby semi-quantitative risk screening tools can be readily generated while providing the benefits discussed above.

RETRO-SYNTHESIS APPROACH

The approach we have termed "Retro-Synthesis" involves the identification of numerical relationships between various parameters which influence risk. Through identifying these parameters in combination with the underlying relationships between these parameters, it is then possible to estimate semi-quantitatively the risk associated with each scenario. The retro-synthesis concept as described in this paper is that one must work backwards from the desired risk measure so as to identify various predominant parameters which combine to yield risk rather than working from a collection of possible parameters and then trying to select those parameters which may contribute to risk. In many ways this approach is analogous to retro-synthesis concepts as applied to organic chemistry synthesis[3] in that risk represents the target molecule while various parameters represent the required functional groups that are needed to assemble the target molecule.

For example, risk can be defined as a function of "likelihood" and "severity", where "severity" may be defined as a function of the "energy released" and the "exposure group". Similarly, the "energy released" may be defined as a function of the "properties of the material involved" and the "amount of material released". As such risk can be approximated using a function involving measures of "likelihood", "properties of the material involved", "amount of material released", and "exposure group". The benefit of the retro-synthesis approach is that these relationships are often intuitively understood by risk practitioners and as such, these relationships can be quickly constructed. Further, the identification step only needs to proceed until the point where sufficient information can be generated which will allow for the differentiation of the various scenarios that are to be reviewed. Figure 1 illustrates this technique as applied against the need to prioritize a collection of product transportation risks.

Figure 1. Parameters Relating to "Risk to Company"

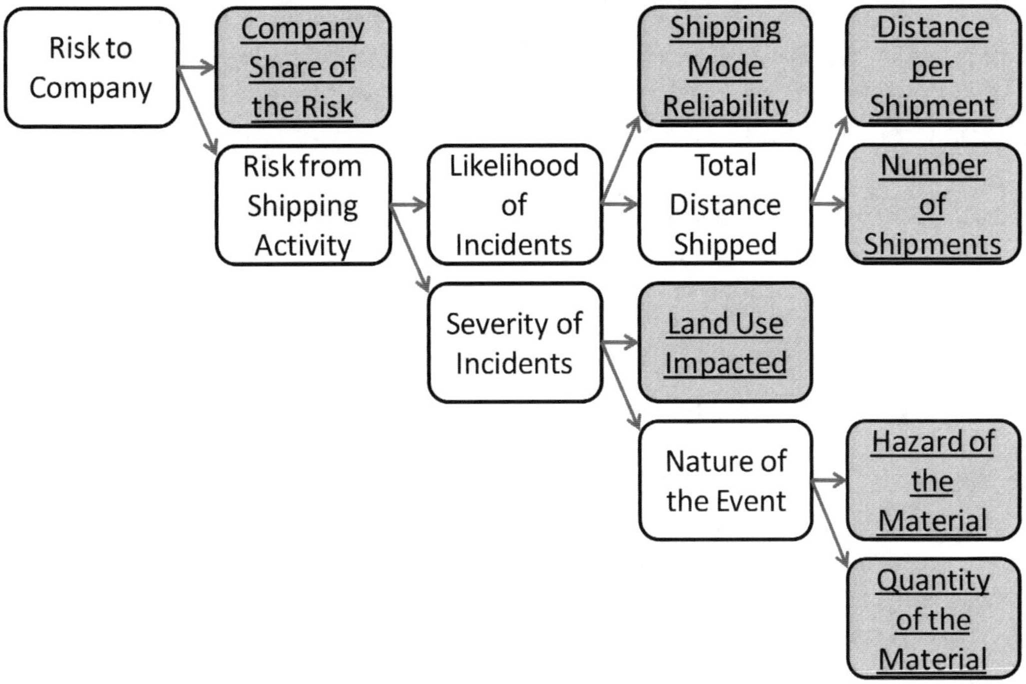

Figure 1 illustrates how by starting at "Risk to Company" one can identify the various parameters which contribute to the "Risk to Company". Then, by assigning values for each of the end parameters (those in shaded boxes featuring underlined text) one can then assess semi-quantitatively "Risk to Company".

Once having identified the predominant parameters through the retro-synthesis analysis of the underlying relationships, the next step is to then design a methodology for scoring each parameter. Parameter scoring should reflect quantified values as related to the overall risk function. For example, the "Distance per Shipment" scales directly with "Risk to the Company" in that if you double the "Distance per Shipment" the overall "Risk to the Company" should also double, therefore in this example the actual "Distance per Shipment" could be used as the assessed value. Alternatively "Distance per Shipment" could be evaluated using an order-of-magnitude approach as illustrated in Table 1. (It is important to note that the intent is to approximate semi-quantitatively the "Risk to Company" and as such exact numerical assignments are not required.)

Table 1. Distance per Shipment Scoring (Linear)

Distance	Score
Up to 300 km	100
300 km to 3,000 km	1,000
More than 3,000 km	10,000

Table 1 illustrates a potential scoring model for the "Distance per Shipment" parameter used in the assessment of "Risk to the Company". Alternatively a logarithmic scale could be applied as shown in Table 2.

Table 2. Distance per Shipment Scoring (Logarithmic)

Distance	Score
Up to 300 km	1
300 km to 3,000 km	2
More than 3,000 km	3

Table 2 illustrates a second potential scoring model for the "Distance per Shipment" parameter used in the assessment of "Risk to the Company". In this case a base 10 logarithmic function has been applied to the assigned scores. Generally, to keep the overall math simple if any one parameter is to be defined using a certain relationship then all other parameters should also be assessed using the same relationship. In cases where logarithmic functions have been utilized, the final risk parameter is calculated using the sum of all of the contributing parameters. Alternatively, where an order-of-magnitude approach is utilized, the final risk parameter is calculated using the product of all of the contributing parameters.

In developing approaches to quantify each of the identified predominant parameters three goals should be addressed. First, an emphasis should be placed on ensuring that the category scoring is reproducible in that different people scoring the same situation should generate the same result. This reproducibility is critical if risks, as scored by different individuals, are to be ranked against each other. Second, it should be recognized that this type of risk screening approach is only of limited accuracy, so the assignment of parameter scores does not need to be perfect. Recognizing that the goal is to approximate risk some distortion of the underlying mathematics may occur. Further, the categorization of inputs may also introduce some inaccuracy which can act to limit the precision of this type of risk screening tool. Third, the scoring within each category should be based on information which is readily available, yet allows for the differentiation of different scenarios. In some cases where data is not attainable one option is to reduce the number of parameters identified by the retro-synthesis analysis. Alternatively, data concerns can be addressed by the assignment of a default value for a given parameter; typically these defaults should represent a pessimistic value but not necessarily a "worst case" value. (i.e. just because the data isn't available doesn't mean the situation represents a "worst case" scenario but at the same time one should err on the side of caution.)

After the development of screening questions for each parameter the required questions (often in multiple-choice form) can be combined onto a single form and data collection can be initiated for each of the common scenarios which are to be evaluated. Data, once collected, can then be compiled within a simple Excel spreadsheet or other tool allowing for the scoring of individual scenarios and for the relative ranking of scenarios against one another. This ranking of scenarios then allows for priorities to be set. Upon careful examination of the results generated it may be possible to identify those specific scenarios which obviously require mitigating actions directly at this level of resolution, rather than requiring additional detailed study. Further, through returning to the underlying risk models it may be possible to readily identify those mitigating actions which will most effectively reduce the identified risks. In other cases it may be necessary to re-evaluate some scenarios using more detailed risk assessment techniques (LOPA, quantitative risk assessment) so as to assess if changes are required. This is particularly applicable to those "marginal scenarios", where the relative risk ranking falls near the corporate risk criteria. A benefit to developing a prioritized list of scenarios of potential concern is that it may be possible to spot check a sampling of the scenarios of concern using a more detailed risk assessment technique instead of investigating all scenarios of concern in further detail. In cases where parallels can be drawn between common scenarios this type of approach has the potential to return significant value to the organization through time savings, consistent decision making, and potentially common solutions.

HYPOTHETICAL STUDY: CARBON DIOXIDE ASPHYXIATION

The methodology presented in this paper is illustrative of one of the approaches utilized by NOVA Chemicals for risk screening. Ultimately, the purpose of this paper is the discussion of tools for hazard evaluation rather than the actual assessment of a specific hazard. NOVA Chemicals is presenting this model so as to further the development of tools and procedures for assessing risk rather than giving guidance on the acceptability of specific process risks.

It is generally understood in the processing industries that inert gas asphyxiation represents a significant hazard which needs to be both assessed and managed. The U.S. Chemical Safety Board has investigated other asphyxiation hazards[4], and reports that these incidents occur regularly within the process industry and can occur in a number of situations. However, when presented with a large number of potential threats spread over multiple manufacturing locations, an organization may struggle with approaching these challenges. As part of evaluating and managing these types of hazards a screening tool could be developed which utilizes the retro-synthesis concepts.

Risk Model:

Starting from "Risk Due to Carbon Dioxide Exposure", one can readily define this term as scaling with "Severity of Exposures" and "Frequency of Exposures". Similarly, "Severity of Exposures" can be related back to the outcome of concern, which in this case is a combination of "Fatalities" and the "Number of People Impacted". When looking at "Frequency of Exposures" one can relate this term to the "Frequency of Hazard Formation" and the "Possibility of Causing Harm". The "Frequency of Hazard Formation" can be assessed as relating to the "Frequency of Leaks" and the "Expected Size of Leaks". The "Possibility of Causing Harm" is then assessed as relating to the "Degree of Exposure to People" and the "Confined Nature of the Area". The "Degree of Exposure to People" can be estimated as reflecting the "Possibility of Detection", the "Possibility of Escape", and the "Contact with the Hazard".

At this point it is likely that the retro-synthesis analysis has identified a sufficient number of predominant parameters so as to allow for a risk screening model to be developed. Further, it is felt that these parameters would allow for the differentiation of the various scenarios being assessed, so as to establish clear priorities for further assessment and mitigation. Figure 2 illustrates the relationships identified in a retro-synthesis analysis of "Risk Due to Carbon Dioxide Exposure". Clearly, the presented approach represents only one option out of many whereby the "Risk Due to Carbon Dioxide Exposure" could be evaluated.

Having identified the appropriate parameters the next step is to establish a quantification scheme for each parameter. In this particular example a decision was made to work with a logarithmic scale and as such "Risk Due to Carbon Dioxide Exposure" would be estimated through the summation of the scores assessed for each identified parameter. As a logarithmic scale was to be utilized, each increase in the assessed value by one point, for a given parameter, equates to a situation in which the overall "Risk Due to Carbon Dioxide Exposure" would be viewed to have increased by an order of magnitude (i.e., in assessing each parameter a value of 1 equates to a minor concern, a value of 2 equates to something 10 times as bad as a minor concern, and a value of 3 equates to something 100 times as bad as a minor concern).

Figure 2. Parameters Relating to "Risk to Company"

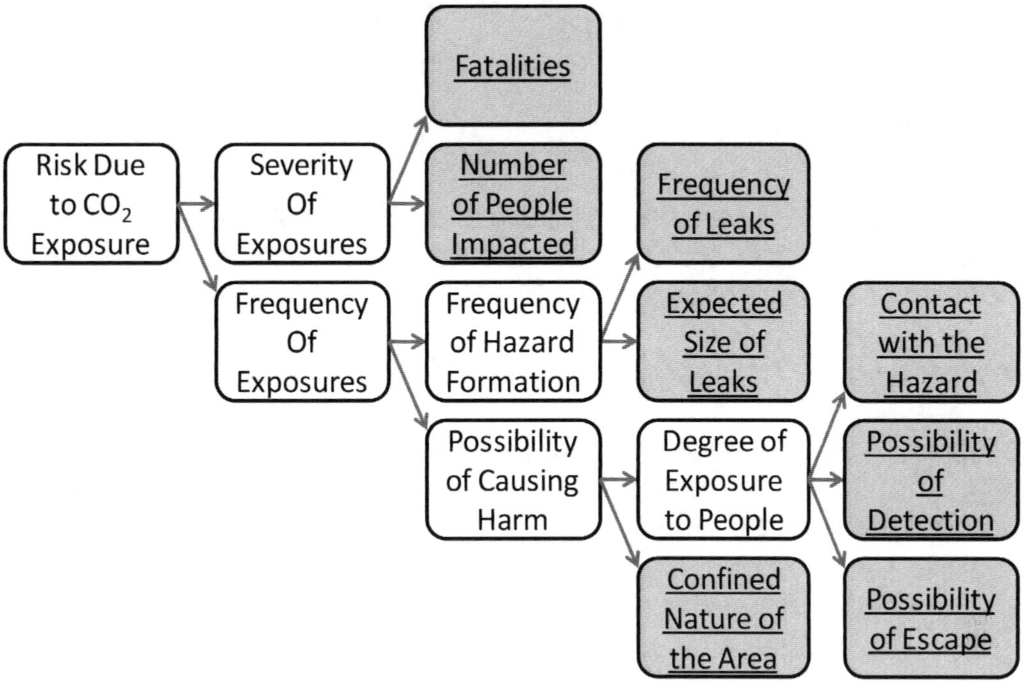

Figure 2 illustrates how by starting at "Risk Due to Carbon Dioxide Exposure" one can identify the various risk related parameters which predominately contribute to this hazard.

Fatalities:

In this sample model "Fatality" was the only outcome considered and as such this term drops out of the overall risk equation as it does not provide for differentiation between scenarios (i.e., no scenario was envisioned where an adverse outcome to an individual would be 10 times worse or 10 times better than a fatality). The assumption being that impacted individuals either survive or they don't survive.

Number of People Impacted:

The "Number of People Impacted" was identified as relating to the severity branch from the retro-synthesis assessment of "Risk Due to Carbon Dioxide Exposure". In assessing the severity related aspect a non-linear relationship is often used for multiple fatality events. The relationship proposed for this model utilizes the square of the number of people impacted (i.e., 2 fatalities in a single event was viewed as being 4 times worse than one fatality in a single event). Table 3 was therefore developed using the square of the number of fatalities.

Table 3. Number of People Impacted Scoring

Number of People Impacted	Score
1-2	1
3-6	2
7 or more	3

In order to assess the "Number of People Impacted" the following definition could be applied "*the number of persons that can reasonably be expected to be present in the area of concern at the time of exposure*".

Frequency of Leaks:

The "Frequency of Leaks" assesses the frequency that releases could occur and result in the formation of a hazardous atmosphere within a given area. For this category scoring was as per that proposed in Table 4.

Table 4. Frequency of Leaks Scoring

Number/Nature of Leak	Score
High	3
Medium	2
Low	1

High would indicate that there were a significant number of leak points present or leak points were present that were associated with a high likelihood of release. Low would indicate that there were few leak points present and those leak points which were present were not associated with a high likelihood of release. Medium would then be used to score situations which were not clearly either High or Low. Alternatively this section could have been scored based on historic failure rate data for the area (i.e., 3 – regular/monthly event, 2 – annual event, 1 – remote chance of occurrence…). A scenario assessed as High would be considered to be 100 times more hazardous than a scenario assessed as Low.

Size of Leaks:

The "Size of Leaks" assesses the possibility that a hazardous atmosphere could develop following a release of carbon dioxide to an open area. (Assessment of the size of the leak was considered based on an open area as elements related to confinement will be captured in a subsequent category.) For this category scoring was constructed as per Table 5.

Table 5. Size of Leaks Scoring

Size of Leak	Score
Large	3
Medium	2
Small	1

Large would indicate that a carbon dioxide release would likely result in the formation of a large hazardous area. Small would indicate that a carbon dioxide release would likely not form a significant hazardous area. Medium was then used to score situations which were not clearly either Large or Small. (Although not done within this model the assessment of Large, Medium, and Small could have been based on various release rates, i.e., Small releases could represents releases of less than X kg/second.)

Confined Nature of the Area:

The "Confined Nature of the Area" captures how readily the area involved in a carbon dioxide release could contribute to the formation of a potentially hazardous atmosphere. As an example, enclosed areas may prevent the dispersion of the inert gases making hazard formation more likely. For this category scoring was as per that proposed in Table 6.

Table 6. Confined Nature of the Area Scoring

Confined Nature of the Area	Score
Inside/Confined	3
Inside/Unconfined	2
Outside	1

Examples of Inside/Confined include: small rooms, trenches, manholes, inside vessels or railcars, inside or next to confined spaces. Examples of Inside/Unconfined include: inside very large buildings or rooms.

Contact with the Hazard:

The "Contact with the Hazard" assesses the possibility that personnel could be exposed to the hazardous atmosphere, given its presence. For example, personnel working on systems utilizing carbon dioxide were considered to be in closer contact with the hazard than personnel working in areas containing carbon dioxide piping. For this category scoring was as per that proposed in Table 7.

Table 7. Contact with the Hazard Scoring

Contact with the Hazard	Score
High	3
Medium	2
Low	1

High would indicate that individuals would likely be involved in situations where there could be direct contact between a potential release and the individuals, such as situations where individuals would enter into the path of the escaping carbon dioxide gas. Low would indicate that individuals would likely not be involved with situations where there could be direct contact between the release and the individuals, such as situations where factors like wind direction or failure of ventilation systems would be required before individuals would come into contact with the potentially hazardous atmosphere. Medium was then used to score situations which were not clearly either High or Low.

Possibility of Detection:

The "Possibility of Detection" assesses the likelihood that personnel could be warned of the possibility of entering into a hazardous atmosphere so as to minimize the chances of the individuals experiencing harm. For this category scoring was as per Table 8.

Table 8. Possibility of Detection Scoring

Possibility of Detection	Score
High	1
Medium	2
Low	3

High indicated that individuals would clearly be warned of situations where there would be a potential for the existence of a hazardous atmosphere. For example, warning lights on an operator shelter indicating the presence of a hazardous atmosphere within the structure. Low indicated that individuals would not receive any warning, either through warning systems or audibly through hearing the release, that they may be entering a hazardous atmosphere. Medium was then used to score situations which were not clearly either High or Low.

Possibility of Escape:

The "Possibility of Escape" assesses the possibility that personnel could escape from, or be successfully rescued, in the event of an exposure to a hazardous atmosphere. For this category scoring was as per that presented in Table 9.

Table 9. Possibility of Escape Scoring

Possibility of Escape	Score
High	1
Low	3

High would indicate that individuals would be able to escape the hazard and in the event that they were unable to escape a clear, functional, and timely rescue plan would be in place (ability to isolate the leak, safety watch with alternate breathing air, ability to physically remove the individual from the situation...) so as to significantly minimize the likelihood of the individuals experiencing harm. Low was then used for all other situations. For this category it was anticipated that most situations would be assessed as being High.

Risk Due To Carbon Dioxide Exposure:

Within this risk screening example "Risk Due to Carbon Dioxide Exposure" for each scenario could be quantified by calculating the summation of the individual scores obtained for each of the categorized parameters. As a result, high risk scenarios could readily be identified relative to low risk scenarios allowing for prioritization. As an example, multiple fatality events associated with indoor locations typically would be expected to score much higher than single fatality events occurring at outdoor locations. In addition to completing the above questions field staff could also be asked to identify existing protective measures (using a provided list) relative to the potential carbon dioxide hazards being assessed during the screening process.

The high risk (top ranked) scenarios could then be further assessed using a LOPA methodology with much of the data already collected during the risk screening steps. Then through the application of the LOPA methodology to specific scenarios it would then be readily possible to determine which scenarios need to be further addressed and which do not. As the LOPA studies are being conducted patterns may emerge in the risk screening data which could potentially allow for the LOPA studies to be completed much more quickly, deriving further benefits from using this structured approach. In addition the quality of the prioritization list may allow for detailed studies only needing to be applied to a few scenarios as it may be possible to draw conclusions on a relative basis through comparisons using the ranking model against the handful of scenarios which have been assessed further. Another benefit may also include that the screening work can be done by site operations staff rather than process safety personnel, while still maintaining the quality and consistency within the study allowing for a much better determination as to which scenarios are a priority.

CONCLUSIONS

Risk screening tools can be readily developed through the application of the described retro-synthesis approach. Typically these risk screening tools are structured such that they capture readily available or well understood information with respect to the predominate parameters. Further, these parameters can be evaluated in such a way that the integrity of the underlying equations defining risk can be maintained. In addition to being applicable to the examples shown within this paper this approach has been used for screening risks associated with scenarios as diverse as control room fires, corporate security threats, and information technology projects. We have found the methodology to be most suitable to situations were a large number of common threats are to be assessed at one time.

Although there is a need to invest time upfront to develop this type of semi-quantitative risk tool for each collection of scenarios, we have found a significant return on this investment with respect to the quality of the risk screening work which is completed, the quality of the data which is collected, and the ability to reduce the number of scenarios which require further analysis using techniques such as LOPA and quantitative risk assessment. Further we have found that field staff readily identify with the models and typically understand what the recommendations are likely to be resulting from the study process. As the field staff, and site leadership, understand and identify with the process they have been found to be much more accepting of the resulting recommendations generating further benefits for this approach.

REFERENCES

1. Center for Chemical Process Safety, *Layer Of Protection Analysis: Simplified Process Risk Assessment,* ISBN 0-8169-0811-7, American Institute of Chemical Engineers, New York, NY, 2001.

2. Center for Chemical Process Safety, *Guidelines for Chemical Process Quantitative Risk Analysis, 2nd Edition,* ISBN 0-8169-0720-x, American Institute of Chemical Engineers, New York, NY, 2000.

3. Warren, S. and Wyatt, P., *Organic Synthesis: The Disconnection Approach, 2nd Edition*, ISBN 978-0-470-71237-5, John Wiley and Sons, United Kingdom, 2008.

4. U.S. Chemical Safety and Hazard Investigation Board, "*Hazards of Nitrogen Asphyxiation*", Safety Bulletin No. 2003-10-B, June 2003.

A Practical Approach to Hazard Identification
For Operations and Maintenance Workers

Christy Franklyn, RRS/Schirmer, League City, Texas, USA
Dennis Attwood, RRS/Schirmer, League City, Texas, USA
Tim Humbke, Shell Canada Ltd., Fort Saskatchewan, Alberta, Canada
Robert Wasileski, NOVA Chemicals Inc., Moon Township, Pennsylvania, USA
Brian Kelly, Bririsk Consulting Ltd., Calgary, Alberta, Canada

Prepared for Presentation at
American Institute of Chemical Engineers
2009 Spring National Meeting
5th Global Congress on Process Safety
24th Annual CCPS International Conference
Tampa, Florida
April 26–30, 2009

UNPUBLISHED

ABSTRACT

Hazard identification has long been recognized as a mechanism for improving safety and reducing incidents. The Center for Chemical Process Safety (CCPS) of the American Institute of Chemical Engineers (AIChE) is developing a new Concept Book that illustrates and extends existing hazard recognition practices and programs to identify process hazards. This new Concept Book will focus on providing operations and maintenance personnel with tools to enhance the hazard recognition process as part of their everyday observations.

To develop the material, a series of workshops will be conducted to review, analyze and recommend existing instruments that are used in industry for hazard identification. Photographs and examples will be used to provide operations and maintenance personnel with guidance on how to apply this knowledge at the field level and achieve results.

This paper and presentation will focus on sharing innovative solutions to providing operations and maintenance personnel with practical tools to identify process hazards and physical hazards.

INTRODUCTION

The motivation for identifying and managing hazards is simple:

- Everyone wants themselves, their friends, and coworkers to go home safely every day;
- Everyone wants to keep their job and process incidents can cause a company to go out of business;
- Process incidents can have consequences for communities located around them. Worker's families and friends often live in these communities;
- We all share the environment and events can cause long-term damage to the environment and the living things that coexist in the same spaces.

An effective hazard identification and control program should instill a culture at the facility where personnel are motivated to accept personal responsibility and manage their own safety. Personnel should be empowered to immediately stop any situations or behaviors that could put people and equipment in danger. To achieve this objective, employees must be empowered to recognize hazards throughout the facility and providing solutions to correct them.

HAZARD IDENTIFICATION IS NOT OPTIONAL.
IT'S AN ESSENTIAL PART OF DAY TO DAY ACTIVITIES.

The goal is to establish a workplace built on a sense of achievement, recognition, responsibility, group decision making, and job enrichment. By building motivation into the work environment, people will become more involved and take ownership in the hazard management objectives.

The new Concept Book provides guidance for identifying and controlling hazards in the workplace to help:

- Improve workers' ability to detect hazards;
- Prevent injuries and accidents;
- Raise hazard recognition awareness; and,
- Empower workers to take action and follow-up.

This book is primarily intended to provide operations and maintenance personnel in all industries working around process hazards (including chemical, petroleum, steel, mining, and manufacturing) with practical methods for identifying and addressing physical and process hazards.

This book should also provide benefit for those persons who:

- Are planning on participating in a formal process hazards analysis or safety review;
- Occasionally enter a process facility and have not received formal training;
- Implement new designs in an existing operating facility;
- Are new employees; or,
- Are responsible for providing resources for hazard control and elimination.

To ensure maximum benefit for the target audience, it will be written in plain language, explain concepts in a step by step straightforward fashion, and utilize pictures extensively.

BACKGROUND

Failure to identify hazards can lead to unforeseen consequences. Risk reduction begins with hazard identification. However, this is just the first step - multiple tools (methods and techniques) must be used to broaden and deepen hazard awareness and control.

There are many books already published on identifying physical hazards. This book extends hazard recognition practices and techniques to include identification and mitigation of *process hazards*. Process hazards are different from occupational hazards because they have the potential to impact multiple persons and the community. Process hazards are often the result of equipment or systems operating outside their safe operating limits.

The process for managing risk through hazard identification and elimination/mitigation include the following steps:

- Understanding basic concepts;
- Teaching employees how to recognize hazards;
- Using tools to identify hazards;
- Understanding different types of hazards and their severity;
- Evaluating the risk of physical and process hazards;
- Providing resources and solutions to eliminate the hazard or mitigate the risk;
- Ensuring we learn from past mistakes; and,
- Implementing a hazard management program and making it part of a facility's culture;

This paper will highlight some of the approaches to eliminate or mitigate the hazard or risk.

HAZARD RECOGNITION

Identifying hazards often depends on our ability to recognize the hazard. Hazards are occasionally recognized using our basic senses:

- Sight
- Sound
- Smell
- Touch

The Concept Book will provide guidance on how our senses are used to detect hazards in the workplace – as well as the limitations of our senses.

HAZARD IDENTIFICATION AND ASSESSMENT

There are a number of hazard identification methods that have been developed in the process industry over the last 20 years. They can be categorized into three groups:

1. <u>Field Surveys</u>: Tools that are designed to collect data in the field either through walk-through surveys using custom-designed checklists, or through the observation of personnel as they perform operations or maintenance tasks.
2. <u>Pre-job Assessments</u>: Tools that are designed to evaluate the job site to identify hazards in the area of interest.
3. <u>Plant Assessments</u>: Tools that are designed to identify hazards either before the plant is designed or in an existing plant with the thought of improving the design of facilities or equipment.

The Concept Book will contain:

- A description of each of the tools;
- Literature references for published tools; and,
- An overview of how each of the tools can be applied.

Field Surveys

Identifying hazards can be a difficult task when there is less information available. Typically, workplace hazards are identified from incident records and near miss reports. Plants with good safety performance typically have fewer incidents, so there is less to learn from unless they learn from others.

Field surveys and worker observations can fill this void since they can be conducted as many times as necessary to identify hazardous situations. This section reviews field survey methods that are published and have been produced by process industries.

Behavior Observation

Behavior is defined as what workers do or say - i.e., their actions - not what they think or feel. Thus, behavior is an objective observable concept. This section, reviews the use of behavior observation as a method of identifying workplace hazards.

Observing the behavior of operators can provide insight on:

- Errors that are committed during the performance of the task;
- Time required to perform each activity according to procedures (this will make it more evident when shortcuts are taken in the future);
- Difficulty or ease with which the task is performed;
- Improved ways of performing the task or alternative tools that could be used to perform more safely or efficiently; and,
- Quality of procedures.

Observations are typically performed in three different ways:

1. Direct observation: Where an observer deliberately observes another person (or persons) performing a task. This is arguably the most frequent method used and is key to behavior-based safety programs.
2. Indirect observation: Where those being observed do not know they are being observed. Traffic and pedestrian surveys fall into this category.
3. Participatory observers: Where the observer is the one performing the task.

For each type of observation scheme, preparation is the key. Some items to consider during preparation include:

What activities will be observed?
What is the instrument that is used for observation?
Who is being observed?
Who is making the observations?
What data are being collected?

Facility Walkthrough Checklists

Facility walkthroughs are a common method of looking for hazards and are typically conducted by operations and supervisor/management:

- Operations plant rounds - Plant rounds are conducted by operators several times per shift to monitor equipment operation and observe that the equipment is operating satisfactorily, e.g., no leaks, no vibration. The rounds may be guided by a data collection sheet on which values of process variables are recorded.
- Plant management walkthroughs - Plant walkthroughs can also be conducted by plant management to look for safety hazards such as broken insulation, steam leaks, tripping hazards, water on grade, hot water dripping on walkways, poor housekeeping, broken equipment, broken lighting, etc.

Checklists have a history of being poorly designed and poorly used. Often the words on a checklist are ambiguous or misleading. For example, "Valve actuator is potentially hazardous" leaves a great deal to interpretation. Is the actuator positioned too high or too low? Is it difficult to turn? Is it difficult to access? Many times, the checklist user is not familiar with checklist terms. And, often, the checklist is not asking the right questions, e.g., "List the gauge pressure _________ bar" (when in fact the gauge is reading psi).

Figure 1. Example Plant Walkthrough Hazard Identification Checklist

Checklist Question	Illustration	Present (√)	Responsible Group			Potential Mitigation
			Maint.	Oper.	Design	
Potential adverse outcome from inadvertent leaning against control panel, switches, etc.?						
Labels on critical switches, valves, piping and vessels inadequate?						

Checklist Question	Illustration	Present (√)	Responsible Group			Potential Mitigation
			Maint.	Oper.	Design	
Valve access inadequate?						
Electrical connections and enclosures loose or open?						
Insulation inadequate?						
Sample point inaccessible?						
Eye-wash station is obstructed and difficult to access?						

Checklist Question	Illustration	Present (√)	Responsible Group			Potential Mitigation
			Maint.	Oper.	Design	
Corrosion leading to failure?	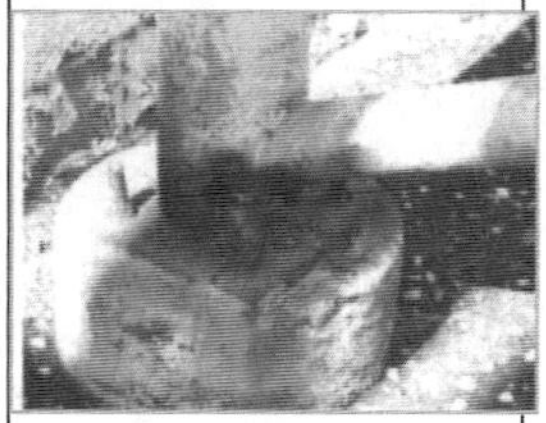					

Pre-Job Assessments

Job Hazard Analysis

A Job Hazard Analysis (JHA) is a technique that is used to identify and assess the hazards in a job before they occur. The JHA focuses on a specific job and examines the:

- Steps required to perform the job; and,
- The relationship between the job and the worker(s), the tools used by the worker(s) and the work environment;

It evaluates the risk of injury involved in the task and specifies interventions to reduce the risk. Other terms used to describe this technique are Job Safety Analysis (JSA) and Job Hazard Breakdown (JHB).

Pre-Job Planning and Permitting

In most plants, hazard identification, recognition and analysis includes many different programs that are conducted to meet varying needs. The Job Hazard Analysis that is described above, for example, is conducted to identify hazards associated with process tasks. The JHA is often used to review jobs before they are performed to familiarize plant personnel with the generic hazards. This is often part of a pre-job planning session.

Pre-job planning can be performed for every task that is performed in the plant. Often, however, it is conducted for those jobs that are non-routine or require a work permit.

If the job already has a JHA/JSA written for it, pre-job planning merely involves the work team and supervisor reviewing the JHA/JSA to ensure that the precautions noted in the analysis are implemented. These could include:

- PPE in addition to routine equipment (e.g., fall protection);
- Hot or cold work permits;
- Special equipment, such as cranes or barricades; or,
- Special precautions (e.g., fire truck onsite).

If the job does not have a JHA/JSA, then one must be prepared by the work team to ensure that all hazards are identified and preparations and precautions are implemented.

Risk Assessment At The Work Location

Once personnel arrive at the job site, they will often find that the work conditions that were assumed when the JHA/JSA was developed have changed. The site could be rain or snow covered, construction could be ongoing in the area, and minor changes could have been made to the equipment. Consequently, the JHA/JSA might not identify all the hazards on the job site.

The type of risk assessment is carried out by the individuals who are performing the onsite work. The assessment is conducted at the job site just before performing work. The objectives of the process are to:

1. Improve operators' hazard recognition and awareness abilities;
2. Retain awareness of the local hazards throughout the task;
3. Reduce the number and severity of incidents and illnesses in all onsite activities; and,
4. Over the long term, to change operators' mindset such that the site risk assessments are done automatically without the need for a paper-based system.

The site risk assessment works best if:

1. Operators (including contractors):
   - Take responsibility for their safety and the safety of those they are working with; and,
   - Consistently follow the LMHRA process for all onsite permitted jobs (especially for non-routine, high risk operations).

2. Supervisors support the operators by:
   - Training all operators and contractors in the technique;
   - Encouraging operators to complete worksite risk assessments to convince them that the technique is beneficial in reducing accidents and to promote a "want to do" approach rather than a "have to do";
   - Encouraging the completion of worksite risk assessment to a consistently high standard;

- Committing to lead by example; and,
- Ensuring compliance by review and follow-up;

3. Plant management supports the program by:
   - Providing funding to enable the work site risk assessment to be implemented;
   - Stewarding the progress of the program; and,
   - Ensuring that compliance is consistent across shifts.

The following are some suggestions to ensure that work site risk assessments are effectively conducted:

- Ensure that all work site hazards are identified;
- Provide enough time for the workers to complete the assessments; and,
- Ensure that the workers understand the hazards.

Plant Assessments

Critical Task Identification and Analysis

Critical Task Identification Analysis (CTIA) is a systematic method of identifying critical tasks within a process plant, prioritizing their importance, analyzing those tasks that are considered most critical and identifying appropriate interventions to mitigate the risk. Critical in this context means high risk. The tasks that may have the potential for personal injury, equipment or facility loss, environmental release, community exposure, or business interruption are defined as "critical". These tasks represent a hazard and the goal is to identify the hazards and correct them before they turn into incidents.

Experience has shown that the analysis is best conducted by a small group of process personnel (site team) who, together, have the knowledge necessary to complete the task. The suggested site team consists of:

- Process Technician
- Maintenance Technician
- Plant Engineer
- Health and Safety Specialist
- Human Factors Specialist

Step 1 – Create Process Flow Diagram. The analysis starts by specifying the plant processes and equipment and creating a process flow diagram.

Step 2 - Identify Process Significant Tasks. The Critical Task Identification process is a 'brainstorming' technique that is designed to encourage the site team to identify the 'critical' tasks that they could be required to perform in each area of the process plant.

A critical task could occur during a process upset or other abnormal situation or it could be a task that is difficult to perform or performed frequently. At each block of the PFS, the team is asked to identify the 'critical' tasks that he or she performs by placing them into one of seven different categories including:

1. Safety-critical tasks (e.g., responding to the failure of a pump flange)
2. Quality-critical tasks (e.g., responding to an out-of-limits lab analysis)
3. Production-critical tasks (e.g., responding to the loss of an exchanger which reduces output)
4. Most frequently performed tasks (e.g., filter changes)
5. Difficult to perform or complex tasks
6. Time critical tasks (e.g. responding to tasks that, in a short period of time, will cause a larger upset. An example is a compressor shutdown)
7. Environmentally-critical tasks (e.g., responding to an out-of-limit environmental deviation. For example, a high SO_2 reading on the stack of a fired heater)

Step 3 - Identify the Risk of a Loss Event. No company can afford to spend money and commit resources on issues that have little benefit to the operation. Thus, it is essential to identify the tasks which, when improved, will reduce risk, reduce cost or improve environmental protection. The approach used at this stage is to estimate the risk associated with each task identified in the above exercise. Most companies have developed their own risk matrix. The next step in this exercise is to prioritize each 'critical' task in terms of its risk to the plant.

Step 4 - Identify Highest Risk Tasks. The tasks that most require analysis are those with the highest risk. A risk ranking matrix is typically used to categorize risk.

Step 5 - The final step in the process is to mitigate risk.

TYPES OF PROTECTION

Processing facilities require a variety of strategies and approaches to keep process risk at tolerable levels. By using diverse risk reduction measures, the chance of common safeguarding weaknesses can be minimized and preparedness for loss events can be enhanced.

The following strategies provide measures for hazard control (in order of decreasing reliability and robustness):

1. Inherently Safer Designs
2. Engineering Controls
3. Administrative Controls
4. Personal Protective Equipment

Hazard Elimination or Moderation (Inherently Safer Designs)

When possible, elimination of the hazard is the best strategy for risk management. For example, many facilities who traditionally used one-ton cylinders of chlorine gas for water treatment at cooling towers have switched to liquid sodium hypochlorite. Chlorine is immediately dangerous to life and health (IDLH) at 10 ppm. The Level 2 Emergency Response Planning Guideline (ERPG-2) is 3 ppm. A release of one ton of chlorine will produce a vapor cloud that may travel 3 miles and deliver concentrations that are hazardous to the community and require evacuation and/or shelter-in-place. Alternatively, exposure to sodium hypochlorite can cause topical burns to exposed workers, but switching from chlorine to sodium hypochlorite provides substantial risk reduction.

Engineering Controls

To manage and control risks, industry has incorporated many engineered safety systems into our processing facilities. When hazard elimination is not possible, the next best options are engineered solutions. These solutions can be either passive or active.

A passive approach uses safety solutions that do not require an action to be taken. Examples of passive solutions are curbs around equipment, fire or blast walls, or robust design of pressure vessels to handle the expected overpressure. Another example is an enclosure around a piece of noisy equipment to reduce the noise exposure to personnel.

Active approaches require that a physical action occur in response to a process event. An example is a Burner Management System (BMS) that does not allow lighting of a furnace until the firebox has been purged. Another example is a Safety Instrumented System (SIS) that brings a process to a safe state when a process deviation exceeds the safe operating conditions. Active solutions are generally less reliable than passive solutions.

Administrative Controls

Administrative controls provide another layer of protection, but should not be solely relied on if there are engineered options that can also be put in place. Administrative approaches that require human action can increase the likelihood of human error. Examples of administrative controls are procedural checks, operator actions in response to an alarm, and emergency response following a loss event.

Many companies use administrative controls for operations that require personnel to follow specific steps. This could be for either operational or maintenance procedures.

Another example of administrative control is lockout/tagout, e.g., a car seal on the inlet of a pressure relief device that is car sealed open.

An administrative procedure can be used to control the amount of material in storage so that in the event of a release is not a sufficient quantity of material to cause an offsite issue with the community.

CONCLUSION

Hazard identification is an essential aspect of risk management programs in an array of industries, from chemical and petroleum to the steel and mining industries. The specific types of hazards founds in these industries may vary across the spectrum. However, they all share one thing in common, in that operations and maintenance workers play a key role in the hazard identification process. As front-line personnel, they are often in a position to be the first to recognize the presence of a hazard. As a result, these employees first need to be educated on the types of hazards present in their field of employment, and understand how to recognize these hazards. Furthermore, they must be provided with practical tools and techniques, such as those described in this document, so as to facilitate their commitment to risk management. Through the cooperative efforts of industry organizations, such as the Center for Chemical Process Safety (CCPS), valuable tools for hazard identification can be broadly expanded for use in wide range of industries.

REFERENCES

None.

Cheddar or Swiss?
How Strong are your Barriers?
(One Company's Experience with Process Safety Metrics)

Michael P. Broadribb
BP International
Sunbury-on-Thames, UK

Bill Boyle
BP International (retired)
Sunbury-on-Thames, UK

Steven J. Tanzi
PricewaterhouseCoopers LLP
Chicago, IL, USA

1. BACKGROUND

The process operations of oil and gas production, petroleum refining, petrochemical manufacture and marketing contain many Health, Safety and Environment (HSE) hazards, not least of which is the highly flammable and explosive nature of the substances handled. Over many years the industry has continued to develop and strengthen control barriers to prevent potential catastrophic consequences to people or the environment resulting from accidental release of hydrocarbons which may cause fires or explosions.

In March 2005 BP experienced a major incident at its Texas City Refinery involving a release of hydrocarbons that ignited resulting in an explosion, which took the lives of 15 people and injured many more. An internal investigation (ref. 1, 2) identified that, although the site had numerous measures for tracking operational and safety performance, these measures did not focus on leading indicators that would provide early warning of potential major incidents. Similar recommendations were made by the Independent Review Panel (often referred to as the 'Baker Panel') in their report (ref. 3) on the safety culture in BP's five US Refineries. These documents recognized that weaknesses in established process safety barriers may collectively unlock the door to incidents with catastrophic consequences. While industry history tells us that such high consequence events are infrequent, this significant process safety risk is always present in oil and gas operations, and thus needs to be continuously assessed and reduced through systematic safety management. Within BP, this recognition has resulted in a substantial strengthening of process safety risk management through a more comprehensive system of controls embedded within its new Operating Management System (OMS), which is being implemented at the site level across all of its operations globally (ref. 4).

The BP OMS and its process safety elements provide systematic processes so that risk assessment outcomes feed into plans and controls to mitigate the risks within operations. Based on the old adage "you can only manage what you can measure", it is clear that the new OMS needs to be supported by good indicators to measure, evaluate and correct performance. The Independent Panel, in particular, urged BP to develop and implement Process Safety Performance Indicators (PSPI), and then subsequently work within the industry to encourage their further development and adoption (ref. 3).

A useful model to use when considering the role of the management system in relation to process safety and other HSE risks is the 'Swiss-Cheese' model, which was first articulated though the work of James Reason (ref. 5). Figure 1 below illustrates how failure to contain a hazard could result in a serious consequence if the process safety barriers and controls have weaknesses – the holes in the slices of 'Swiss-Cheese'. For example, in order to prevent the risk of a 'top event', such as hydrocarbon release, a number of barriers (i.e. risk reduction measures in the management system involving plant, process and people) are established. Further 'escalation controls' are also in place to respond to the 'top event', such as fire and gas detection, shutdown systems, and emergency response to control and/or mitigate the consequences of the top event. If consequences do result, then this demonstrates that all the barriers have holes (weaknesses).

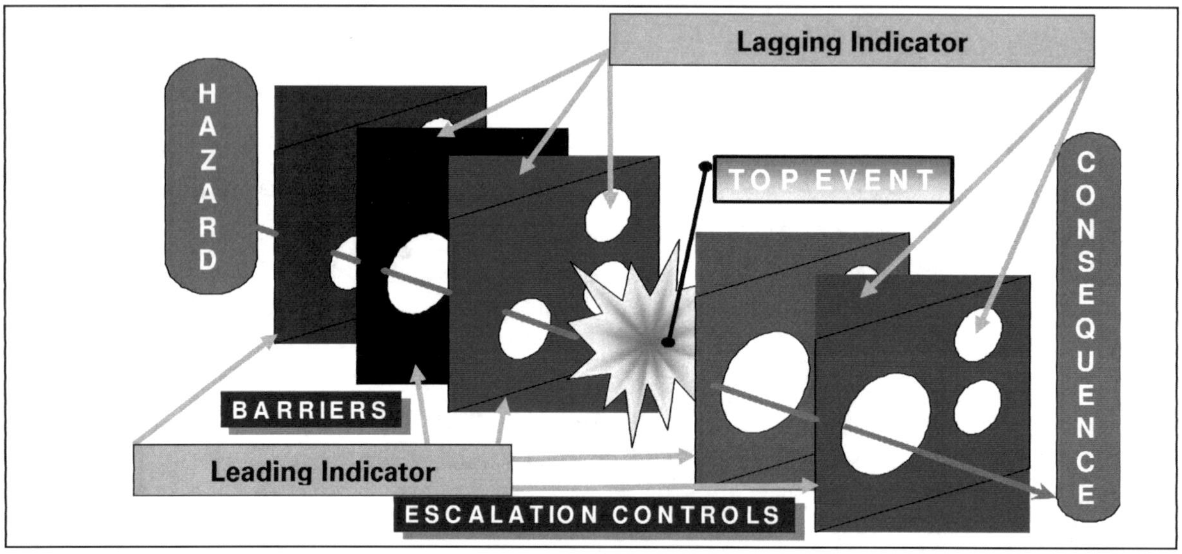

Fig. 1: The 'Swiss-Cheese' Model (after J. Reason)

Traditional measures of performance have focused on failures of the overall system and count the number of 'top events', e.g. spills or releases, or the number and severity of consequences, such as fatalities, injuries or environmental impacts. However, to systematically reduce incidents, especially low probability high consequence process safety events, it is necessary to focus on prevention, and specifically the strength of the barriers in the 'Swiss-Cheese' model. The approach described in this paper therefore aims to create a combination of PSPI that lead to stronger more robust barriers, improving the cheese slices from 'Swiss' to 'Cheddar'.

Another factor related to the PSPI challenge is that large multi-national corporations are complex organizations with many different activities and stakeholders including employees, business partners, local communities and regulators. These stakeholders have the aspiration, and indeed expectation, that the same level of risk management and performance is achieved across the entire range of the company's operations. The potential social and environmental impacts of the industry also drive expectations of transparency and comparability. Thus metrics need to be well defined to serve multi-user communities, internally at site and corporate levels, and externally for public disclosure.

In order to facilitate both internal and external reporting, as well as performance benchmarking, BP prefers, wherever possible, to evolve and implement metrics which are based on established or emerging industry good practice, ideally with international acceptance. On process safety metrics, BP has been working with many national and international organizations including the UK Health & Safety Executive (UK H&SE), Center for Chemical Process Safety (CCPS), the American Petroleum Institute (API) and the European Process Safety Centre (EPSC).

This paper explores the ongoing methodology and approach being used by BP to select a limited but crucial set of PSPI to enable effective measurement and management of process safety performance.

2. BP METHODOLOGY AND APPROACH

It is challenging to select, develop and establish good performance indicators. Any engineer will tell you that there is a no lack of things to measure on a process plant, but there are potential pitfalls in gathering too much data on too many metrics. The burden on the company's workforce could become disproportionate to their other day-to-day tasks to ensure safe and reliable operations. The resultant plethora of data may also be complex to analyze and the key trends from the most insightful metrics may be obscured. Thus care must be taken to regularly review the effectiveness of even well-established industry metrics to ensure that they remain relevant and insightful, otherwise their use may detract attention away from the currently most material and important risks. Consequently, the BP approach follows another recommendation of the

Independent Panel to ensure that a process was established to not only develop metrics but also subject them to regular analysis and review.

Therefore the challenge for BP is to carefully select a limited number of process safety and other OMS metrics that together provide clear management information on how well the plant design, management processes and people are addressing the most important operational risks to the business. Such application of management system metrics is analogous to regular health checks with your doctor. These key indicators can then be reviewed by management to support their decisions on more detailed assessments and planned improvements to further mitigate risk.

BP has had, for many years, a set of established safety metrics to understand and compare performance at site and corporate level. These are generally outcome based measures. The most commonly accepted metrics previously employed by BP to judge past performance use industry accepted aggregate metrics based on recording and investigation of HSE incidents. These metrics are fundamentally measures of actual or potential harm, such as pollution or injuries, resulting from unintended events, such as oil spills or fires. Although these are after-the-event measures, they are essential metrics because they provide a transparent estimate of the effectiveness of the management system, and provide key opportunities to learn from past events and avoid recurrence of loss or damage. The aim is to eliminate such unintended events, whether the event resulted in harm or was a near-miss. Improving performance for these reactive metrics should generally be indicated by a downward trend towards zero. These indicators are commonly referred to within BP as *lagging*, although as noted below, it is recognized that low severity or near-miss events are to some extent *leading* when compared to incidents of higher severity.

In the mid-90s, BP decided to implement the US Occupational Safety and Health Administration (OSHA) injury reporting requirements (ref. 6) as a global standard to provide comparable measures of lagging safety performance. These metrics are included in the classic safety incident triangle of fatalities, days away from work cases, recordable injuries/illnesses, and minor injuries/near-misses (fig. 2). Because less consequential incidents are statistically more probable than higher severity ones, the lower part of this triangle is considered to some extent *leading,* and is argued to provide some predictive correlation to the lower probability events, particularly fatalities. The majority of the events that result in injuries are the result of failings in systems designed to ensure personal safety. Often referred to as 'slips, trips and falls'; these are invariably related more to the behavior of people in the workplace rather than the operation of the plant. While less severe injuries can be considered *leading* information, within BP it is recognized that any harm to people is effectively a *lagging* outcome, and it will become more appropriate to monitor near-misses and other input measures to be more predictive of personal safety incident rates. As personal safety in the workplace relates to generally different hazards than process safety in oil and gas facilities, the PSPI also need to be somewhat different.

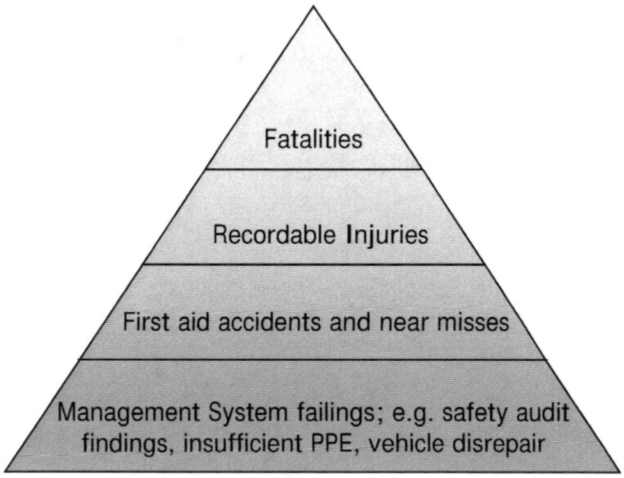

Fig. 2: Personal Safety Incident Triangle

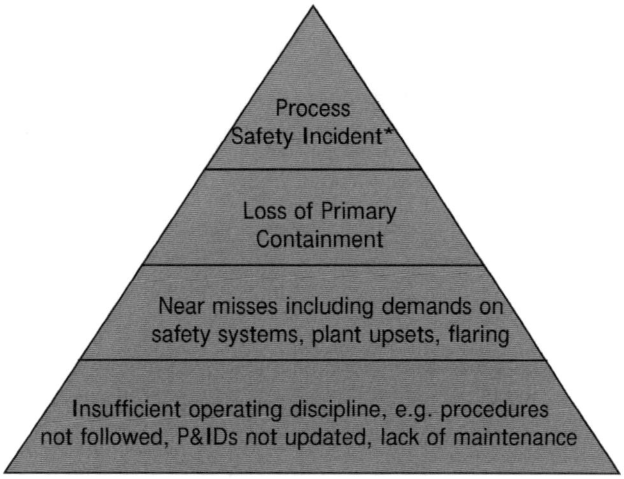

* Note: Process Safety Incidents are defined as loss of containment of hazardous material or energy, which has resulted in severe consequences including fatalities/injuries, environmental or property damage

Fig. 3: Process Safety Triangle

Following the Texas City accident, it was recognized by BP and by the Independent Panel, that process safety incidents need to be specifically identified to facilitate early warning, learning and intervention before a major incident occurs (ref. 1, 2, 3). Within BP this has been interpreted as all incidents, near-misses, events or other measures that relate to actual or potential losses of containment of either hazardous materials or energy. This also recognizes that the long-established personal safety triangle (fig. 2) does not contain sufficiently specific data to correlate with the less frequent but potentially more severe process safety incidents, such as fires or explosions resulting from hydrocarbon releases. For this reason, there has been recognition by CCPS (ref. 7), API (ref. 8) and others that process safety incidents need to be better defined and that process safety incidents of differing levels of severity can form a similar, but different, triangle (fig. 3).

During 2007 BP participated with CCPS, API and others in work programs that generated new industry guidance on PSPI. Both the CCPS (ref. 7) and API (ref. 8) publications provided an aligned industry definition for a high severity 'Process Safety Incident' (top of the triangle). CCPS also provided a preliminary approach to assessing and scoring the severity of incidents so that an index could be created which would weight the most severe incidents. The Independent Panel had also suggested such a weighted index to ensure appropriate emphasis was placed on the incidents that most informed company management on where to focus attention and corrective action (ref. 3). BP has adopted and developed this approach to produce a set of lagging indicators that are process safety specific, including a PSI Index (Table 1).

Table 1: Existing Lagging Process Safety Metrics

Quarterly Metric	Brief Description
Process Safety MIA (2 Levels)	Number of Major Incident Announcements that were process safety related
Process Safety Incident Index (4 Levels)	Process safety incidents scored in a weighted index by severity of harmful consequences to people, environment, equipment or LOPC release
Fires & Explosions (4 Levels)	Number of all fires and explosions, ranked by severity
Loss of Primary Containment (4 Levels)	Unplanned / unintended releases of gas or liquid hydrocarbons or other hazardous materials, ranked by severity
Flammable Gas Releases (4 Levels)	Unplanned / unintended releases of gaseous hydrocarbons or other hazardous materials, ranked by severity
Oil Spills (4 Levels)	Number and volume of unplanned / unintended releases of liquid hydrocarbons or other hazardous materials, ranked by severity

Note: Lagging indicators are assessed and categorized against a number of levels corresponding to severity within the BP HSE risk matrix

Following the Texas City explosion, BP recognized the need for better leading indicators that track process safety performance inputs and the long term health of the OMS. The Independent Panel also placed considerable emphasis on the lack of such indicators not only for BP, but more widely across industry. Apart from the use of low severity incidents (HiPo) to provide indications of potential for higher severity events as mentioned above, the most leading indicators are input-based and track levels of activity that aim to prevent incidents.

Table 2: Existing Leading Process Safety Metrics

Quarterly Metric	Brief Description
Process HiPo's (2 Levels)	Number of High Potential near-misses that could have resulted in an MIA and were process safety related
MIA & HiPo Lessons Learned Reports Issued	Number of MIA and HiPo investigations completed with lessons learned shared, including number completed within target period of 30 days
Overdue Plant Inspections & Tests	Number & percentage of scheduled safety critical inspections and tests completed / overdue
MAR Assessments Completed	Number & percentage of planned Major Accident Risk Assessments completed and authorized
MAR Action Closures	Percentage progress on closure of actions from Major Accident Risk Assessments
ISO 14001 certification of major sites / operations	Number & percentage of major operations (e.g. refineries) with ISO 14001 certification complete and up-to-date
Safety & Operations Audit Actions	Numbers of actions open, overdue and delinquent from independent corporate audits of management systems including process safety
Number of Approved Audit Action Changes	Number of approved changes to action deadlines from the independent corporate audits of management systems including process safety
Incident Investigation - Action Closure	Tracks change in percentage of actions resulting from MIA or HiPo incident investigations closed on time within the past 4 quarters
Elimination of HTALHC Blow-down stacks	Tracks commitment to take all heavier than air, light hydrocarbon blow-down stacks out of service (i.e. type involved in the Texas City explosion)
Elimination of Occupied Portable Buildings in Red Zones	Tracks commitment to prevent use of portable buildings to accommodate workers within high risk areas within or close to process unit
Implementation of BP standards	Tracks aggregated progress of sites to implement strengthened mandatory BP operating practices, including Integrity Management, Control of Work and Regulatory Compliance
Leadership competence	Ensures that assessment of BP operational management has been assessed at a general level in terms of experience and knowledge

Leading PSPI are typically focused on barriers derived from management system processes, such as inherently safer design, inspection/testing and maintenance programs or process safety training. Assessments of behavioral inputs, such as safety culture, also provide leading management information. BP regards these types of leading metrics as increasingly important because they provide a means to assess the strength of the risk reduction barriers in the new OMS. In contrast to lagging metrics (including the PSI Index), improving performance of leading management system metrics is usually indicated by an upward trend towards a desired level of completeness (100%) or quality, rather than an objective of zero undesirable events. BP has developed a set of leading indicators that are process safety specific (Table 2).

2.1 Selection and Development of New Metrics

In developing its approach, BP recognized that PSPI can be applied at different levels of the organization. At corporate level, a relatively small number of metrics focuses on key HSE risks to the BP Group. This compares to site level where metrics need to incorporate corporate and local regulatory requirements, but additionally involve metrics tailored to local risks specific to the operations including plant design, operating procedures and staff capability.

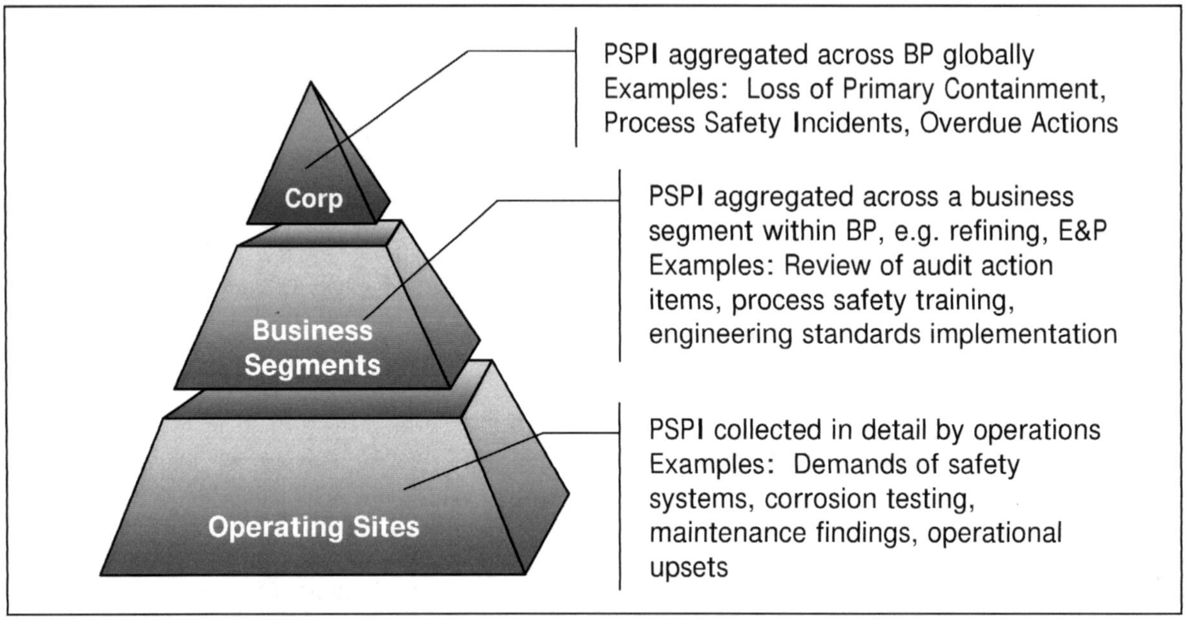

Fig. 4: Process Safety Metrics at Different Organizational Levels

Between the extremes of local and global application, there is also value in recognizing that particular sites within business segments, such as refining or exploration and production, have similar equipment and processes that result in common risks, which can be managed systematically through aggregated information or metrics. This approach is aligned with the BP OMS that is being implemented at the local level, but has common operating practices within each business segment and an overarching framework supported by standards to manage key risks at the corporate level. Examples of PSPI related to these organizational levels are illustrated in Figure 4.

At each organizational level, in order to assess whether the OMS is successfully addressing process safety risks, three different approaches can be used to select and develop PSPI. These approaches are referred to as 'pro-active', 'reactive' and 'external learning'. Figure 5 below illustrates how these approaches link to barriers and indicators in the 'Swiss-Cheese' model.

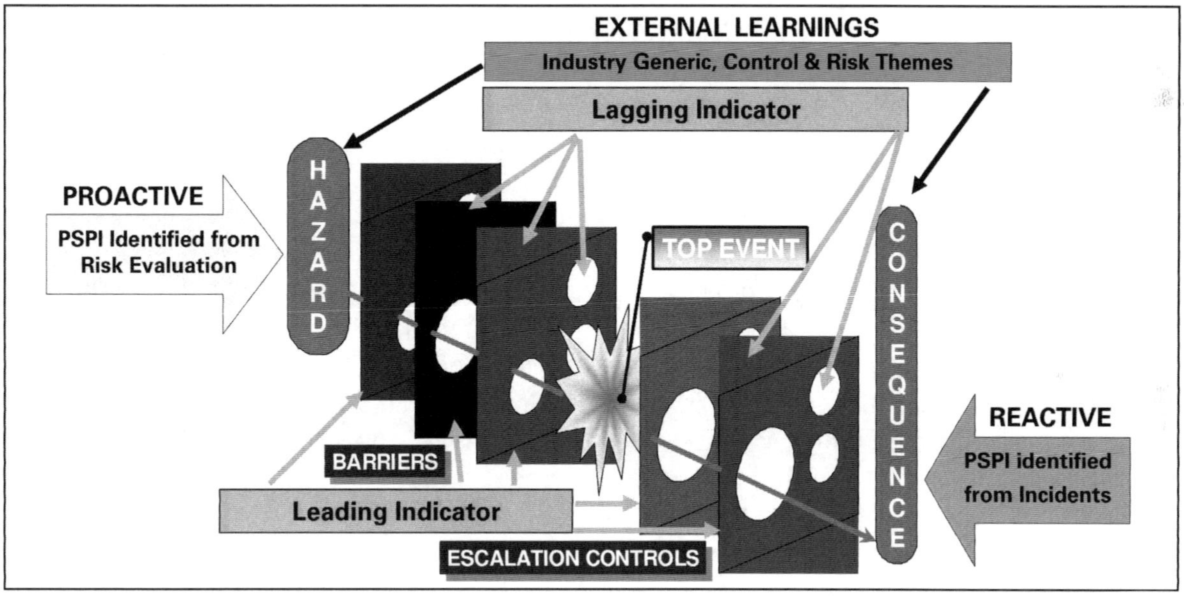

Fig. 5: Selecting Barriers for PSPI Development

These three approaches are simply described, as follows:

Pro-active input relies upon identification of hazards and risks, and closely follows UK H&SE guidance to select metrics based on barriers that address the identified risks. This is particularly appropriate for operating site level indicators, and an example of this approach is provided in Section 3 below based on work in the UK.

External learning takes account of experience in the oil and gas or other industries in terms of past incident or events, best practice risk reduction systems, and experience of applying metrics. External learning therefore provides an important approach to provide a broader platform of experience against which barriers and metrics can be identified. Texas City Refinery made substantial use of this approach and provides a site level example in Section 4 below, while the adoption of the new lagging Process Safety Incident metric proposed by CCPS and API provides a good example of this approach at business segment and corporate level.

Reactive input is based upon learning from process safety incidents and near-misses, and may be verified by comparison with audit findings. BP investigates all major incidents to determine root causes. By correlation of the data from incidents with OMS elements, barriers can be identified on a reactive basis to focus on prevention of similar incidents. This approach is statistically most relevant at corporate level and is illustrated by an example in Section 5.

From any or all of these three approaches, the next step is to identify the most important barriers which require measurement. A useful starting point is the guidance published by UK H&SE on developing process safety indicators (ref. 9). This 2006 publication starts with the 'Swiss-Cheese' model and then provides a step-by-step guide based primarily on identification of a plant's most important 'risk control systems', i.e. the process safety barriers. It then recommends, for each key barrier, selecting one lagging and one leading indicator to assess the performance of the barrier. Here lagging refers to measuring undesirable outcomes or weaknesses of a barrier and leading monitors key elements of activity that maintain a strong and complete barrier. Because these lagging and leading measures together test barrier strength, BP has generally regarded these in combination as providing "leading" information on the management system when compared to traditional consequence based lagging indicators, such as the CCPS Process Safety Incident metric.

As noted previously, care must be taken to ensure that only a limited number of metrics result from the process, and this is carried out in BP at corporate and business segment level through an annual workshop review of the three approaches by subject matter experts (SME) with deep experience on process safety, risk analysis, performance review and metrics. This includes functional experts in health and environment as well as safety and operations. The workshop aims to determine whether the most important barriers are being adequately monitored through metrics or other mechanisms (such as audits) and to recommend changes to senior management on which metrics need to be developed, which can be phased out and which require improvement. For example, BP's most recent reviews identified the need to develop metrics related to competency of operations personnel as a barrier. As a result, work is underway to develop company-wide metrics that will indicate if the requirements of the OMS related to competency are being adequately fulfilled.

As with all metrics, but especially leading indicators, care must be taken to pilot new metrics to ensure that the intended barrier outcome is achieved and unintended consequences do not result. For example, a competency based metric should not prevent investment in people through development moves or recruitment, but it should encourage building of capability for teams to provide improved competency on safety critical tasks.

The next 3 sections of the paper explore examples of the pro-active, external learning and reactive approaches that have been used in BP for metric development.

3. PROACTIVE ANALYSIS

As part of an initiative launched by the UK H&SE, a BP site developed a number of PSPI in accordance with the UK H&SE guidance (ref. 9). A workshop was conducted with operational and technical representatives together with local H&SE inspectors and the author of their guidance, who led the review process.

The plant COMAH reports were used as a basis to identify the major hazard scenarios. While various process units produce different products, the key operational stages are the same, i.e. feed system, reactor section, initial separation and recycling, and final distillation train. The barriers (known as 'risk control systems' in the UK H&SE methodology) were therefore fundamentally similar, but each process unit was reviewed individually. A total of eight different barriers were considered.

For each barrier, two key questions were considered:

1. What are the most important parts of the barrier that must be present all the time?

2. What does successful implementation of the barrier look like?

The answers to the first question provide leading indicators, while answers to the second provide lagging indicators.

For the eight barriers chosen, many lagging and leading indicators were identified, which were then subject to a 'beauty contest' to determine a single leading and lagging indicator for each of five key barriers, as presented in Table 3 below.

Table 3: Agreed Barriers and PSPI

Barrier	Lagging Indicator	Leading Indicator
Equipment Overpressure	Number of high pressure alarms	Number of times the condition of all pressure control loops on the unit falls below set standard
Equipment Overfill	Number of high level alarms	Number of times the condition of all level control loops on the unit falls below set standard
Accidental Leakage	Number of high & medium priority leaks	% completion of planned Process Technician routines
Corrosion	Number of times corrosion rate exceeds predicted rate	Number of UT thickness checks not completed on time
Management of Change (MoC)	Number of incidents with MoC as a root or system cause	Number of risk review comments added at MoC Approval stage

3.1 Example of PSPI Generation - *Equipment Overpressure and Overfill*

For equipment overpressure and overfill, the lagging indicators identified were the number of high pressure or high level alarms that occurred on the process unit and were captured directly from the Distributed Control System (DCS). These alarms are precursors to loss of containment scenarios which could lead to flash fire, vapor cloud explosion or toxic release, and they identify events where actual operation has moved beyond the normal operating envelope. Rather than collecting alarm data for just the top level events, such as reactor overpressure, it was decided that the data would be collected for every high pressure and high level alarm configured on the unit.

The leading indicators for these barriers were chosen as the number of pressure and level control loops that operated outside their normal configuration. This is essentially when the control loop is in (i) manual rather than auto, or (ii) auto local rather than cascade, and also (iii) when the output of the controller is saturated either at 0% or 100% output. In these situations the DCS control is less effective and will not quickly return the process to within its operating envelope.

These indicators have shown a significant improvement in performance since measurement started. One side benefit was an indication that more alarms were occurring than expected in spite of previous experience tracking the EEMUA alarm metric[1] at low levels. This was surprising given that the 'control loop with poor performance' count (leading indicator) was also low, and may indicate that alarm settings are set rather tight.

[1] Number of alarms per operator per 10mins.

3.2 Conclusions on the Pro-Active Process

Overall the pro-active approach has identified a number of areas where further action (alarm reviews) may be beneficial to improve clarity and integrity of operations. There are also a number of metrics that have not worked as well as expected (overdue inspections, execution of operating routines), and will require further work to identify better measures. The methodology defined in the UK H&SE guidance recognizes that this is an ongoing process, and indicators that show little change should be reviewed and others tracked instead.

Due to the nature of the process materials, corrosion is a key mechanism for loss of containment and so a barrier addressing corrosion was selected. The leading indicator identified was the completion of all inspection activities to plan and, since this metric started, there have been no overdue inspections. The suitability of an indicator that is consistently zero is clearly questionable and should be evaluated for potentially better indicators.

The other metric of concern was the execution of operating routines. The leading indicator for accidental release/leakage was difficult to determine but it was agreed that operating technicians should be out checking the plant proactively using their senses (sight, hearing and smell) to ensure any accidental releases are quickly identified and any substandard conditions are addressed before they result in a release. The level of completion of operating routines initially saw a gradual improvement to 100% ("what gets measured gets done"), but now that the metric consistently records 100% its value is also questionable.

One of the requirements of these PSPI is to be able to generate a hierarchy of indicators that can be rolled up from process unit level, through a site, and up to the organizational level. Of the metrics identified in this review, most of them would be valid for other process units at the site and could be captured at a site level. However, rolling these metrics up to an organizational level in a major global company with diverse operations like BP would prove difficult and not all of them would allow a reporting hierarchy to be developed. That said, the process of reviewing the individual plant assets and thinking through what the barriers are and the key elements of their delivery is in itself a beneficial exercise and allows the spotlight to be placed on the key process safety issues.

4. EXTERNAL LEARNING

The External Learning approach identifies PSPI by using information provided from external incident investigations, industry groups, and intra-industry and cross-industry best practice. This approach uses outside expertise and knowledge sharing to find unique ways to measure the controls that mitigate the business' risks. BP's Texas City Refinery and BP corporate & business segments both applied this approach with the support of external consultants to help strengthen their PSPI, using external learning in combination with reactive inputs. This example focuses on how Texas City used external learning to identify specific barriers and PSPI to enhance their existing metrics. The corporate approach is illustrated in Section 5.

Texas City Refinery reached out to a number of external sources to test the robustness of their existing PSPI and to help look at alternatives used in the refining industry as well as other industries to bring a desired external perspective. The external consultants helped to research professional, private, governmental, non-profit, and university organizations for process safety best practices. Each of the external groups selected were heavily involved in process safety. These included the CCPS, API, UK HSE, American Chemistry Council (ACC) and others.

Typically, when looking to attain the most current best practice, it is good to start with many of the industry leaders. Each of the organizations selected is recognized by BP's peers in the industry as leaders for their novel processes or key SME. BP has developed relationships with many of these organizations and industry leaders to gain knowledge of their best practices and to help actively shape the direction of process safety in industry. Working with many of the governmental, non-profit and university organizations has been invaluable because BP has obtained access to SME who have in-depth knowledge in various subject areas, which has helped lead to new insights and enhanced processes within BP.

In this example, Texas City Refinery with the help of their external consultants gathered external incident information from the external groups selected, looked for lessons learned from both current and historical incidents, and reviewed metrics in place or planned. The research into these historical incidents is valuable because there are clear examples where barriers have failed. Experience has shown that more recent events have more relevant technology and findings that can be directly applicable to BP. However, not all the risks in one process may be directly translatable to another similar process. Therefore it is important to transfer the relevant learning and identify PSPI that will help to measure the same barriers that failed in the historical incidents. These PSPI should be leading indicators to ensure the same failures do not manifest themselves.

Table 4: Examples of External Indicators

Metric	Definition	Type	Industry
Number of overloads reported	If a controller is overloaded in terms of the number of movements he/she is handling, they can report an overload	Lagging	Air Traffic Control Industry
Percentage of duty periods in excess of the standard week	The number of duty periods that are exceeding the standard week compared to the total number of duty periods.	Leading	Air Traffic Control Industry
Number of infringements	Number of infringements to controlled airspace, i.e. planes that come within the edges of controlled airspace without permission.	Lagging	Air Traffic Control Industry
Number of infringements followed up	Number of infringements that have been investigated and actions implemented to avoid in the future.	Leading	Air Traffic Control Industry
Process Safety Incidents	An incident that occurs within a Responsible Care company or Responsible Care Partner company in which a chemical process was involved resulting in a significant fire, explosion, chemical release, or injury. A process safety incident is considered reportable if it results in (a) a fire or explosion causing more than $25,000 in property damage, (b) a release of a chemical greater than the CERCLA reportable a quantity for extremely hazardous substances or a release of 5,000 lbs of a flammable or combustible material or (c) a serious injury arising from a fire, explosion, chemical release or a release of energy or material from a process.	Lagging	American Chemistry Council
Major Fires	A major fire is defined as a fire where the following occurs (1) Fatality, or 3 or more hospitalizations (OSHA definition), or Off-site (affecting community) shelter in place or evacuation, or (2) $1million or property damage, or (3) Negatively affects consumer supply, or (4) Of national interest with widespread media reporting	Lagging	American Petroleum Institute
% of safety systems functioning properly	The number of safety systems that are functioning within acceptance criteria compared to total number of safety systems.	Lagging	Nuclear Industry
% of scheduled corrective maintenance programs performed on time	The measure of schedule adherence for maintenance programs that have definitive dates for completion.	Leading	Nuclear Industry
Routine testing, qualifications review and evaluation	A metric that looks to ensure all personnel are tested and evaluated on their qualifications to perform their assigned tasks.	Leading	Nuclear Industry

In the historical examples the common risks and controls from the external incidents were analyzed. In some of the external incidents the information that was available was very general in nature, so the analysis only resulted in general risk and barrier themes that were applicable. The goal of the analysis was to identify as many specifics and commonalities from these incidents within BP. The risks and barriers discovered from this approach were compared to those identified using other approaches. Where gaps were discovered, these were opportunities to identify PSPI to adequately measure the risks and controls. It is recommended that this review of industry incidents should be conducted at least annually.

In addition to historical incidents common PSPI were identified that are utilized by other industries. Examples of some of the external industry indicators collected are contained in Table 4. Some of these metrics can be directly used by BP as PSPI, e.g. fires and explosions, while others point to similar systems and processes that should also be measured, such as number of overloads reported, number of infringements, etc.

Table 5: Potentially Relevant PSPI Identified by Analysis of External Learning

PSPI from 'Process Safety / Loss of Containment' Category	PSPI from Other Areas
Total Near Misses*	Communication Effectiveness
Operating Envelope Exceedences*	Leadership Team Collaboration Ratio
Primary Loss of Containment*	MIA, HiPo Near-Misses
Safety / Environmental PSM Action Items Overdue Rolling %* • Action Item Closure Quality • Average Action Item Closure Time	Follow-up Rate to Procedural Non-conformances
Incident Investigation Action Items Overdue Rolling %*	Executive Stability
• Action Item Closure Quality • Average Action Item Closure Time	HSE Investment
Overdue Policies and Procedures Update and Review* • Action Item Closure Quality • Average Action Item Closure Time	Training effectiveness • Training Completion Rate • Quality of Training • Training Impact
Total Number of Fires	Turnover Rate
LDAR Missed Inspections*	Fatigue
Integrity Management Inspections and Tests Overdue*	PM Compliance
HSE Audit Action Items Overdue Rolling %* • Action Item Closure Quality • Average Action Item Closure Time	Control of Work Action Items • Action Item Closure Quality • Average Action Item Closure Time
PS Minimum Expectation Action Items	Contractor Conformance with Contracts Audit Action Items

*Potential Leading Indicator

At Texas City Refinery, these externally identified indicators were matched with barriers identified from the external incidents. Selected PSPI were incorporated by Texas City into a site dashboard for management to review performance and improve mitigation of their identified risks. Where possible Texas City incorporated the use of common industry indicators as these can be used for benchmarking across companies and industries. Table 5 provides a list of example PSPI from applying the external learning analysis from which Texas City were able to select some process unit or site level metrics.

The last external technique utilized is benchmarking in order to establish a baseline of where the company is in relation to others in the industry. In this example similar metrics are used to compare against other companies in order to continuously improve. BP's practice is to conduct benchmarking against oil and gas, and other industries on a regular basis with at least annual review. After completion of the benchmarking exercise, a company should make appropriate adjustments to its management system to reach the ideal state. The ideal state occurs when all the barriers are functioning optimally, periodic reviews reveal the systems are up to date, and all applied pro-active barriers are fully effective in identifying failures in the barriers before they occur.

In summary, to ensure complete effectiveness, the external learning approach to PSPI development should contain these basic elements:

- A standardized monitoring process and tool for data collection from external sources
- A pool of trained individuals to monitor the system
- A list of industry, private, governmental, non-profit and university contacts
- A well defined benchmarking process

BP has been able to use external learning from intra-industry and cross-industry research on historic incidents, current process safety management topics, metrics used by other industries and current best practices at the corporate level and local level, as illustrated by Texas City, to identify barriers and measurement techniques that can be leveraged for PSPI.

5. REACTIVE ANALYSIS

Reactive analysis uses evidence of barrier weaknesses identified through investigation of process safety incidents and near-misses. BP has used reactive analysis over the past two years in combination with the pro-active and external learning approaches. Information for all three of these approaches is already available through the annual cycle of BP's HSE planning and performance reviews at the corporate level. Reactive analysis is more effective at corporate than site level because there is sufficient data from process safety incidents and near-misses to provide statistically meaningful inputs for barrier identification. It may be possible to perform reactive analysis at a large site providing the number of incidents and near-misses is statistically significant.

BP requires that all HSE incidents or near-misses are reported, and has a single system in place to capture data on a global basis. Within the system, incidents are classified based on the severity of the event's consequences, including harm to people or damage to the environment. The classification corresponds to consequence levels used in the process for assessing HSE risks within the BP OMS. An advantage of this classification is that the reactive analysis inputs to barrier identification align with the inputs from pro-active analysis of risk-based planning at the corporate level.

An event with the highest severity classification, e.g. resulting in one or more fatalities or long term environmental impacts, is notified to BP senior executives with 8 hours of occurrence and is termed a Major Incident Announcement (MIA). A near-miss event which had the potential under slightly different circumstances to result in the same level of harm or damage is notified within 24 hours and termed a High Potential Incident (HiPo). MIA and HiPo events are subject to a thorough investigation procedure involving a team external to the site where the incident took place. Each event is also classified by the nature of the incident, including whether it was process safety related, e.g. a loss of containment of hazardous materials / hydrocarbons. The investigation procedure results in root cause and related information, which can then be used to identify the barriers that failed.

Analysis of the investigation data for more than one hundred process safety related MIA and HiPo events during the last 3-5 years was analyzed and categorized into themes based on risk and the barriers (i.e. system requirements) that would have prevented those risks from being realized. Effectively this allowed identification of barriers within the OMS that have failed, and the frequency of failure provides an indication of which may need to be monitored more closely in the future using leading indicators. One outcome of the work was to draw greater attention to the need for good quality investigation and learning reports from incidents. This has led to improved tracking of incident reporting to ensure that lessons are properly documented and issued in a timely manner.

Figure 6 provides an example of the output obtained from reactive analysis based on MIA and HiPo events related to process safety. Note that one event can result in multiple appearances as a number of barriers are likely to have failed. For example, investigation of the 2005 Texas City accident revealed that more than 10 barriers failed resulting in the loss of containment and explosion (ref. 1, 2).

These barriers were then mapped against HSE and Operating risks which are the result of a bottom-up and top-down risk assessment process that is performed annually. This step helps verify the barriers resulting from the reactive analysis but also ensures that any further key barriers are considered on a pro-active analysis basis. This resulted in a candidate list of barriers for review.

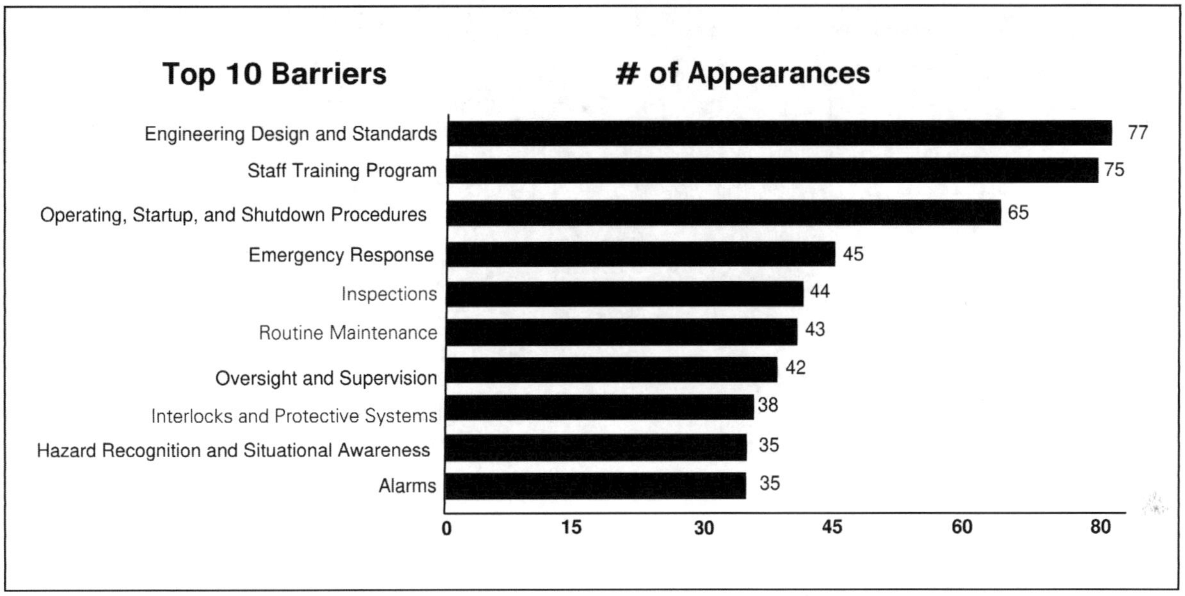

Fig. 6: Barriers Identified by Reactive Analysis

A workshop attended by process safety and other SME reviewed the data to consider candidate indicators. From this review it was determined that engineering design/standards, staff competency, operating procedures and maintenance practices were prevalent barriers that should be considered for performance measurement, and, following UK H&SE guidance, leading and lagging indicators should be considered for each of these barriers (fig. 7).

Consultants also conducted an external learning review of existing documented metrics and knowledge based on:

- Process safety metrics from UK and Texas City experience (as per Sections 3 and 4 above)
- External publications including those from CCPS, API and UK H&SE (ref. 7, 8, 9)
- Research of scientific journals, and industry and professional organization websites
- Information obtained through conversations with representatives of the chemical industry, air traffic control, and Sellafield Ltd (formerly British Nuclear Group).

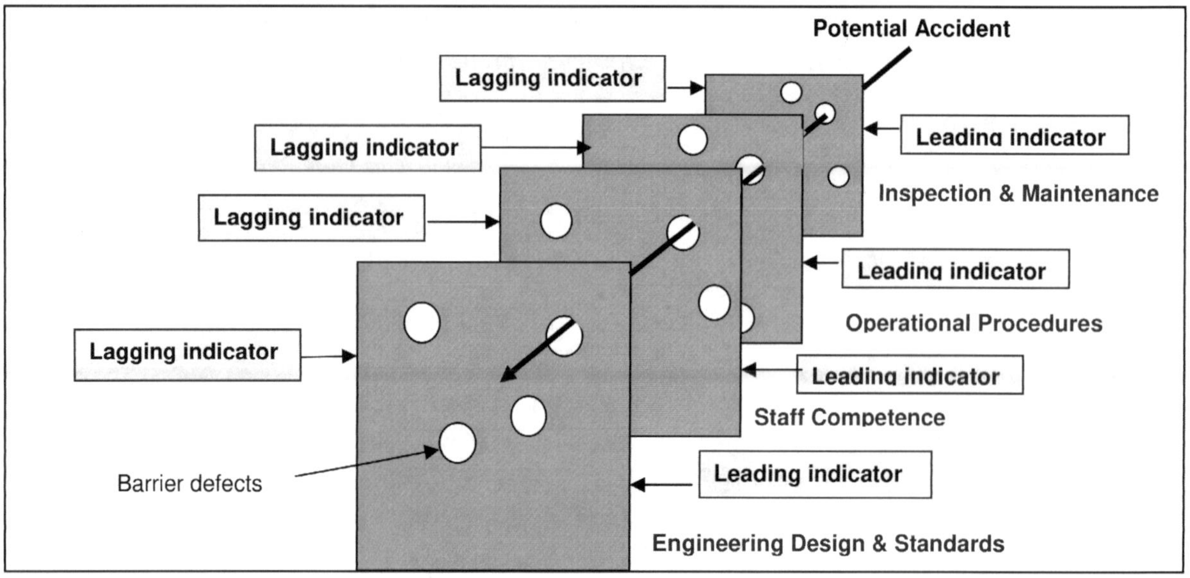

Fig. 7: Reactive Analysis

This provided an initial list of metrics which were then matched to the identified barriers in order to develop a refined list of candidate metrics. Table 6 below summarizes the outcome for the four main barriers.

Table 6: PSPI Identified at Corporate Level

Barrier	Candidate Leading Indicator	Candidate Lagging Indicator
Engineering Design and Standards	Number of site technical practices implemented in conformance with priority BP Engineering Technical Practices	*Proportion of equipment in compliance with site technical practices*
Staff Competency	Number of safety critical roles and proportion of task-based competency assessments completed versus planned level	Proportion of incidents & near-misses where competency or training was a factor
Operating, Start-up, and Shutdown Procedures	Number of safety critical procedures documented, up-to-date and meeting quality criteria	Number of incidents where inadequate procedures is a contributing factor or cause
Inspections, Equipment Tests, Preventative Maintenance	Inspections and tests of safety critical equipment carried out (complete vs. overdue)	*Number of inspections and rests requiring remedial actions*

The metrics have subsequently been reviewed by BP's executives and all metrics are either implemented or under development with the exception of those in italics which are still under consideration, primarily because existing group-wide systems are not yet available to readily capture information of this type.

In summary, reactive analysis, in combination with pro-active analysis and external learning provides valuable input to identify key barriers for which combinations of management system metrics can be developed. Selecting good PSPI remains a challenge, but a systematic process with a range of inputs provides a basis for a robust choice. For BP it is early in the process, since implementation of corporate-wide systems to collect robust, consistent data takes time. Beyond this it is necessary to correlate the trends from these metrics with the process safety incident rate, which is also being established based on the lagging metrics outlined in the introduction to this paper.

6. REPORTING

Organizations need a systematic way of collating and reporting performance indicators and presenting the information to management. There are various approaches to these systems. At an operating site level the use of performance "dashboards" is gaining popularity, although paper reports, posters and PowerPoint presentations are still widely used. Dashboards have the advantage that everyone accessing them will have the latest data, which can often be compiled automatically from the plant DCS. Figure 8 illustrates a typical example of one screen from a dashboard at an operating site tracking a number of process safety metrics, including HAZOPs, audits, action items, and relief valve testing.

Corporate management invariably requires written reports of performance trends. Within BP a quarterly HSE and Operations Integrity Report (*known as the 'Orange Book' because of the color of the cover*) is produced. This report comprises comprehensive management information for monitoring leading and lagging indicators of performance and audit findings associated with safety and operations. As described above, the selection of these PSPI has been influenced through involvement in the CCPS Process Safety Metrics project (ref. 7, 10), and by the work at BP's UK and Texas City sites. These data include a number of leading and lagging metrics, which are monitored at corporate level, such as process safety incidents, overdue inspections, and, new for 2009, a pilot on competency of persons in safety critical roles.

The main purpose of any PSPI is to enable management intervention, if performance declines or does not meet the required standard, before a serious process safety incident occurs. BP has established a Group Operations Risk Committee (GORC) to provide assurance that operational risks are identified and managed in accordance with approved Company policy (ref. 4). It is chaired by the CEO, and membership includes the business segment chief executives and senior functional expertise, such as engineering and HSE. It meets regularly to oversee and build a foundation for consistent, safe, reliable operations, and regularly communicates direction to the Group. The Orange Book is an important tool to assist the GORC in their deliberations. These data are routinely reviewed and allow GORC and the business segment executive teams to view activity at an operating unit level, track progress to close-out, and receive feedback on barriers to successful completion. The company now has better management information to support the delivery of its strategic plans.

Bad Values	Target	PTA 3	PTA 4	PTA 5	UT&WT
Emergency		0	0	0	0
High		6	0	0	0
At Least 3 Days Old	< 1	6	0	0	0
At Least 7 Days Old	< 1	6	0	0	0
Low and Journal		11	0	5	10

No. of Disabled Triconex Points	
PTA 3	0
PTA 4	0
PTA 5	17

Leak Repair Devices	
No. 3	6
No. 4	9
No. 5	10
WT	0
Utilities	9
Red>15, Yellow>8	

	Target Date Aging						Target Date Aging		
12/17/2008 15:07	**Overdue**	**<30**	**<60**	**<90**	**3-12 mo**	**>1 year**	**<1 yr**	**1-2y**	**>2 yr**
PTA Area Summary									
Action Items	0	0	0	0	0	0	0	0	0
HAZOPs	0	7	8	2	40	26	76	12	10
Major Audits	0	1	0	0	2	11			
DecSite (Inc. WT/UT)									
Action Items	0	0	0	0	0	0	0	0	0
HAZOPs	0	0	0	0	0	0	0	0	0
Major Audits	0	5	1	2	20	4			

12/18/2008 7:15	No. 3 Ox	No. 3 PTA	No. 4 Ox	No. 4 PTA	No. 5 Ox	No. 5 PTA	WT	Utilities	Building & Grounds
Next TAR	1-Mar-09		1-Apr-10		1-Apr-11				
Relief Valve Test Summary									
Overdue	0	0	0	0	0	0	0	0	0
0~3 Month	0	0	0	0	3	1	0	0	0
3~6 Months	0	2	0	0	1	0	0	1	6
6~12 Months	68	27	0	0	0	0	0	8	1
Due Before Next TAR	0	0	2	3	4	5			
Safety Shutdown Device Summary									
Overdue	0		0		0		0	0	

Fig. 8: Typical Dashboard Screen

Another purpose for metrics is to provide an objective basis for external communication on process safety performance to various stakeholders, such as government agencies and local communities. BP is working with a number of industry associations who are debating the most appropriate manner of reporting performance externally. Discussions to date appear to favor an approach based upon the CCPS guidance (ref. 7) around process safety incidents.

7. FUTURE

Industry understanding of PSPI is growing rapidly and the forthcoming publication of the CCPS guideline book (ref. 10) will help advance this knowledge.

Future developments in PSPI are likely to involve greater application and use of leading metrics, and measurement of smaller events and near-misses at the bottom of the process safety pyramid (fig. 3). For example, industry may focus on process upsets and small deviations from the intended operating envelope as precursors to more serious incidents. Some assets already collect data on incidents, such as relief valve actuation (essentially a lagging metric), and this is likely to become more common within the industry.

Other developments are likely to include greater application of dashboards for reporting PSPI at site, and perhaps corporate, level. These dashboards are likely to be increasingly automated in a move towards capturing real-time information from the plant DCS and other sources.

8. CONCLUSION

All organizations within the major hazard industries need a systematic approach to selecting, defining, measuring, collecting, reporting and managing PSPI in order to assure safe, reliable operations. Senior management at both site and corporate level need to review these data regularly and make interventions when performance declines or fails to meet required standards.

The use of PSPI is not as mature as metrics that have been used for monitoring personal safety. However organizations need to start the journey and learn, regularly review which PSPI are working or not working, make adjustments, and continuously improve. Different PSPI will be applicable at site and corporate levels within an organization, i.e. some site metrics will not be applicable across all sites within a company.

At an industry level, major manufacturers with international operations require alignment so that one set of corporate PSPI with common definitions are collected throughout the world. This will facilitate understanding, measurement, reporting and benchmarking in the most cost-effective manner. However achieving this alignment will require vested interests in different regions and countries to compromise. The early work by CCPS (ref. 7) should be a foundation for this alignment.

To systematically reduce incidents, especially low probability / high severity process safety events, industry must focus on prevention, and more specifically focus on the strength of barriers, as described by the 'Swiss-Cheese' model. The use of carefully selected and managed PSPI can lead to stronger more robust barriers, in essence changing the cheese slices from 'Swiss' to 'Cheddar'.

9. REFERENCES

1. BP America, Inc. (2005), "Fatal Accident Investigation Report, Isomerization Unit Explosion, Texas City, 23rd March, 2005". (Available for download from www.bpresponse.org)

2. Broadribb, Michael P. (2006), "Lessons from Texas City - A Case History", in the AIChE 40th Annual Loss Prevention Symposium and CCPS 21st Annual International Conference, Orlando, Florida.

3. Baker, James III (2007), "The Report of the BP US Refineries Independent Safety Review Panel".

4. Broadribb, Michael P. (2008), "3 years on from Texas City", in the AIChE 42nd Annual Loss Prevention Symposium and CCPS 23rd Annual International Conference, New Orleans, Louisiana.

5. Reason, James (1990), "Human Error", Cambridge University Press.

6. US Department of Labor, Occupational Safety and Health Administration, "Recording and Reporting Occupational Injuries and Illnesses", 29 CFR 1904.

7. Center for Chemical Process Safety (2007), "Process Safety Leading and Lagging Metrics".

8. American Petroleum Institute (2007), "API Guide to Report Process Safety Incidents (Report Year 2008)", Version 1.1.

9. UK Health and Safety Executive (2006), "Step-by-Step Guide to Developing Process Safety Indicators", HSG 254.

10. Center for Chemical Process Safety (future publication), "Guidelines for Process Safety Metrics".

Process Safety Leadership – A Systematic Approach

Cory D. Shelton
BP plc
Houston, TX USA
281-467-2252

Grainne A. Matthews, PhD
Quality Safety Edge
Houston, TX USA

ABSTRACT

How can leaders strengthen and continually improve their process safety leadership? What leadership behaviors are essential for establishing and sustaining a safe operating culture?

This paper will focus on a systematic approach being developed for use in BP to process safety-related leadership that includes establishing expectations, regular assessment, and personal development. A hard copy of BP's behavioral assessment tool, "Assessing Your Safety Leadership" will be provided to conference attendees.

INTRODUCTION

A focus on improving Process Safety also encompasses the role and impact of leaders on how people 'live' or 'act out' process safety systems.

To be effective, leaders must clearly identify the practices that they rely on to maintain safe operating practices and a safety culture that under-pins and sustains them. However, identifying practices alone does not necessarily ensure that all leaders will adopt or practice them as a matter of course.

Therefore it is suggested that an organisation's approach to the improvement of leadership practices needs to be as systematic as its approach to improving operating processes. The approach should draw on valid research about specific leadership practices that reliably predict safe operations and it should use an organized and logical approach to helping leaders adopt them.

Each organization is responsible for its own approach to managing and improving process safety and leadership. This paper describes an approach developed for use in BP, which will be subject to continuous improvement as learnings emerge from its application in BP. It is offered in case something similar may be considered appropriate for use in other organizations.

LEADERSHIP ASSESSMENT AND CONTINUOUS IMPROVEMENT

Systematic improvement of practices that reliably predict safe operations requires systematic mapping and evaluation of relevant dimensions of process safety leadership.

"Assessing Your Safety Leadership" comprises a number of dimensions based on a review of internal and external management systems, leadership assessments and factors that have a statistical correlation with process safety. Its dimensions distinguish successful leadership, integrating aspects of both the management system and the critical human element that brings it to life.

The approach applies a Behaviorally Anchored Rating Scale (BARS) to assessing leadership. The assessment describes increasingly effective levels of performance and is particularly relevant to a cycle of performance improvement in process safety leadership. The performance improvement cycle (e.g., Plan, Perform, Measure, Improve) is highlighted in Figure 1.

Figure 1. Simplified PPMI Cycle

Performance improvement follows the traditional route of activities include incorporating results in a leader's personal development plan, identifying objectives, establishing specific measures and feedback processes, selecting development options, and seeking frequent feedback on predetermined and specific behaviors that need to be demonstrated.

However, the key approach of the assessment tool is to provide information to drive development and dialogue on often neglected behaviors vital to safety leadership.

The primary application of "Assessing Your Safety Leadership" is intended to be self-assessment. Certainly a supervisor or coach can also provide valuable feedback on the results of a leader's self-assessment to provide additional clarity and objectivity. Once the improvement cycle is completed, then it may be beneficial to expand this assessment by conducting a 360 degree feedback version which could test progress and provide additional insights.

Application of this information to date within BP includes assisting with safety culture assessment results and in training courses for leadership in projects and operations. Future implementation includes applying tailored versions for supervisors and managers in a 360 degree feedback format.

Significant insights and improvements have been reported such as "adding specific contractor engagement focus", "increasing my positive feedback on critical areas of performance", etc. The tool is intended to provide strategic insights into behaviors critical to safety leadership.

PROCESS SAFETY LEADERSHIP

Before we look at the assessment tool, in more detail, it is worth revisiting the context of process safety, for which the following are offered:

"Process Safety is a disciplined framework for managing the integrity of hazardous operating systems and processes by applying good design principles, engineering, and operating practices.

It deals with the prevention and control of incidents that have the potential to release hazardous materials or energy.

Such incidents can cause toxic effects, fire or explosion and could ultimately result in serious injuries, property damage, lost production and environmental impact." (BP process safety definition)

To understand leadership behaviors in the context of process safety, we must first distinguish the difference with personal safety. How does process safety relate or differ from personal safety?

Traditionally, safety has been measured primarily by injury rates (e.g., recordable and lost time injury rates). These rates are primarily personal safety related. Examples of personal injuries such as slips, falls, contact with electricity, heat, and cuts. The precursors or warning signs of these personal injuries might include unsafe conditions and behaviours, and near misses.

Major accidents related to process safety, such as fire, explosions, occur on a less frequent basis and generally are not effectively linked to injury rates. Precursors or warning signs of process safety-related incidents include small leaks, operating envelope excursions (pressure, temp, level, corrosion, etc.), non-compliance with procedures, overdue maintenance and inspection, alarm and trip malfunction. Many of these process safety precursors are less visible than those for personal safety. Systematic process safety leadership relies on an understanding of this distinction in order to maintain appropriate focus.

This distinction would allow a leader to begin to focus on both domains in a manner that allows appropriate focus on both.

"Assessing Your Safety Leadership" has been designed to cover both personal and process safety. For purposes of this paper, we will explore its application to process safety leadership.

WHO ARE YOU AS A LEADER?

Are you a project manager involved in design or construction phases? Are you an operations leader in a new start-up or existing operation? Are you a key support person (e.g., HSE, engineering, construction supervisor)?

Regardless of your role, if you influence people and their delivery of process safety performance, you can apply an approach such as BP's, as described in this paper.

We'll explore categories of behaviors that can, in principle, be applied to process safety leadership regardless of an individual's role. Specific circumstances will need to be considered and used to tailor the results and actions to fit an individual's specific situation.

PROCESS SAFETY CULTURE, MANAGEMENT SYSTEMS AND CONTINUOUS IMPROVEMENT

As a leader, it is vital that you are able to articulate a vision of a positive process safety culture and what you are doing to achieve it. If culture can be defined as "the way we do things around here" (Reason 1997), then certainly we must consider the people, their relationships to key processes (or management systems) and the facilities (or plant) they operate. Leaders create cultures by what they systematically pay attention to and the behaviors they themselves exhibit and support.

High performance organizations apply a continuous improvement approach to understanding and enhancing their performance.

We will be exploring one example of how you can apply continuous improvement to your process safety leadership behaviors.

LAYERS OF PROTECTION

If you are a process safety leader, your success in leading process safety depends in part on the degree to which hazards and necessary controls have been addressed in conceptual design, detailed design, procurement, construction, commissioning, and handover to operations.

Critical decisions made in the early concept design phase establish a baseline for risk. Once established, the residual risk is dependent on variables that relate to such factors as maintenance, management of change, and operating practices. The more inherently safe a facility is in design, the less likely a major process safety incident will occur. Thus the role of process safety leadership in Projects can not be overstated.

Likewise, the role of an operations leader in process safety leadership is to continuously sustain the management of risk during operations through operations practices, people, processes, plant integrity, and sufficiently managing the level of risk during the inevitable changes that occur over the life of the operations.

"ASSESSING YOUR (PROCESS) SAFETY LEADERSHIP"

To further define systematic process safety leadership, we will now explore the eight categories covered in "Assessing Your Safety Leadership".

- Role Model
- Integrate Systems and Provide Resources
- Set Standards
- Motivate
- Communicate
- Support

- Measure
- Improve

The assessment lists one to four expectations for each of the eight categories. Each expectation can be assessed on a behaviorally-anchored rating scale ranging from Level 1 (undesirable performance) to Levels 2-5 (basic to advanced performance). Figure 2 is adapted from the assessment.

Figure 2. Adapted from "Assessing Your Safety Leadership"

rank	1	2	3	4	5	score
			Model positive safety behaviour			
Personal compliance and involvement	Violates policies and procedures, and avoids participating in safety activities.	Follows policies and procedures	Visibly and actively participates in local safety activities.	Visibly and actively participates in contractors' safety activities.	Follows policies and procedures and visibly participates in and leads safety activities.	

The behaviors described are intended to address behaviors relevant from supervisor to senior management. Thus some of the behaviors would not be expected of supervisory-level staff and likewise some of the behaviors would not be appropriate for key support staff (e.g., engineer, safety advisor, etc.). Likewise, if the behavior is not expected for the specific role or level in the organization, then it can be disregarded.

Let's review each of the eight categories by considering some of the key questions a leader could pose against their specific role as it applies to process safety.

Role Model

To deliver systematic process safety leadership, it is essential to serve as a role model. For example, not following procedures or not participating in key process safety related activities will not inspire nor motivate the workforce and it certainly creates a perception of the value placed (or not placed) on process safety.

A. Personal conformance and involvement

What activities related to process safety could a leader be involved in to set a good example (e.g., engineering reviews, HAZOPS, investigations, action tracking, testing quality of implementation, expressing appreciation, etc.)? What process safety-related policies and procedures involve leadership?

B. Bias for action

Under what sort of process safety-related circumstances should a leader consider shutting down their operations or project? How does the leader react when his/her operation or project is shut down for process safety-related concerns?

C. Personal competency

What ongoing technical and leadership skills are required for a leader to carry out their area of responsibility? How active is a leader in pursuing development that would improve their process safety leadership? How transparent is this development activity?

Integrate Systems and Provide Resources

Some commentators suggest that a common underlying contributing factor in major accidents is something described as "group think". Group think is where people tend to seek unanimous agreement in spite of contrary facts pointing to another conclusion. Generally, deference to expertise is an effective antidote for group think.

A. Develop and implement an effective SMS and integrate into the business

How effective are your management systems that relate to process safety? What are the key gaps and how are you driving continuous improvement?

How is process safety integrated into the operational reviews? How vulnerable is the operation to a power failure, instrument malfunction, procedural error, or mechanical failure? What controls or contingency plans are in place to deal with these scenarios?

Who are key contractors who have an impact in process safety? How are their systems and resource allocation regarding process safety?

B. Allocate and develop competent resources

What has been done to ensure key process safety personnel have necessary competency? Who is involved in process safety-related decisions? How well do you seek out and defer to expertise?

Set Standards

A. Establish effective safety goals, standards, and roles

How well are process safety-related goals and expectations established? How well do process safety critical personnel understand their accountabilities in delivering process safety?

B. Establish effective safety performance measures

Are process safety metrics in place? How well do these metrics go beyond strong signals such as leaks, fires, and other loss of containment to address weak signals in the quality of various defenses or layers of protection (e.g. functional instrumentation, maintenance on time, effective management of change, exceedance of safe operating envelopes, corrosion monitoring, and mitigation, etc.)? How well do these metrics drive continuous improvement? How well do they relate to current plant condition and operations? Detailed guidance on one possible approach can be found in "Developing process safety indicators, A step-by-step guide for chemical and major hazard industries", UK HSE.

Motivate

Effective process safety leadership is highly dependent on the people who design, construct, inspect, start-up, operate, and maintain the plant. For people to be effective, skillful leadership helps people to achieve more than minimum expectations by delivering discretionary performance.

A. Give and receive effective performance feedback

What behaviors are you monitoring that relate to effective process safety leadership? How frequent do you provide feedback on these behaviors? How do you respond when you receive process safety-related feedback for your operations? How often do you ask for feedback on your process safety leadership?

B. Reinforce and reward positive safety performance

How do you express your appreciation to individuals and teams for their delivery of process safety leadership and related metrics?

C. Correct and discipline negative safety behavior

What behaviors would be considered negative related to process safety? How do you systematically conduct observations? How do you react to negative behaviors? What criteria would you use to determine whether discipline is warranted? How fair is your delivery of discipline perceived by others in the organisation?

Communicate

One measure of a strong process safety culture is the degree to which everyone is informed. To be informed, leaders must seek out, welcome, and act appropriately to bad news. Skillful communications is another area that could reduce the risks of the "group think" phenomena.

A. Engage in clear, two-way safety communication

How do you seek out diverse points of view on process safety? How do you react to bad news? How relevant and current is process safety information that you provide to your operation? How are relevant employees and contractors involved in making process safety-related decisions?

Support

Extensive statistical analysis of performance data established a direct correlation of safety performance (both personal and process) with the relationship between the supervisor and their employees. Three aspects of this relationship were chosen as indicators of this relationship. A key aspect of process safety culture is related to degree of respect and fairness. These aspects of culture underpin the ability of the organization to be mindful, informed, and to learn.

A. How do you assist personnel in balancing work and personal life?

B. Do you seek out and listen to individuals who raise process safety concerns?
How do you seek out and act on the process safety concerns of your people?

C. How well do you work with people who have backgrounds different from your own?

Measure

To establish and sustain a process safety culture where everyone is mindful of what might go wrong, it is vital to measure the degree to which your teams are applying systems and practices related to process safety. A common pitfall facing leaders has been described as "normalization of deviance" or "degradation of norms". By actively measuring and monitoring practices and systems, a leader can be alert to and address early signs of this common problem.

A. Assess and monitor safety practices and outcomes
How do you use observations, recording, and tracking to continually measure process safety performance? How do you assess and monitor short duration, high risk activities (e.g. start-ups, abnormal conditions, emergency response)?

B. Assess and monitor safety management systems
How clear are management system accountabilities related to process safety? How do you hold people accountable for delivering their roles? How do you use management reviews to assess and improve management system performance?

C. Assess and monitor contractor safety
How do you assure contractor's performance of process safety-related responsibilities? How do you identify key contractor process safety activities and confirm that sufficient planning and monitoring is delivered?

pImprove

A critical aspect of process safety leadership is to focus your culture on learning. Your behaviors "set the tone" on how concerns are raised and how much you value continuously improving your practices and systems.

A. *Ensure organization applies safety lessons*
   How do people who raise process safety concerns expect to be treated? How clear are the criteria for investigating a process safety incident? How are lessons shared and addressed? How are lessons learned from others shared and applied as appropriate?

B. *Promote continuous improvement of the performance of the safety management system.*
   How are process safety-related actions developed, documented, tracked, and closed out? How do you involve employees in improving management systems?

PUTTING RESULTS INTO ACTION

To understand and improve your process safety leadership, it is vital to take the time for self-assessment. Assessing your perception of strengths and gaps is a first step.

As you consider the results, you should keep the following in mind:

- Point of the assessment is to provide insights that facilitate improvement.
- It is important to pay attention to strengths and areas for improvement.
- Even the lowest scoring area may not be a significant problem, and even the highest scoring area will be open to improvement.

- What 2 or 3 areas could you focus on that would have the biggest impact on process safety in your area of responsibility?

- What is your personal style and strength? How could you use these characteristics to help improve your process safety leadership?

Issues to consider are:

- As you look at each section, are any sections identified where results for most of the individual questions show evidence of strength, or challenge?

- Considering the overall profile, are any of the sections clearly stronger, or weaker, than others?

- How could you use your areas identified as "strengths" as a way to achieve improvements in areas identified as "for improvement"?

Testing your perception with others such as your supervisor, peers, and your team is another vital step to ensure your perception is accurate and that your action plan is relevant.

DEVELOPING QUALITY ACTIONS

It is essential to describe the desired behavior or outcome in clear and measurable terms. Remember to breakdown the action to achieve this result in attainable and realistic steps. Ensure that both the frequency and quality of the behavior change is assessed and recorded by those you select to provide you with objective feedback.

How could you enlist the help of others to help ensure that difficult gaps are closed? Consider establishing an "accountability circle" (e.g., supervisor, trusted colleague, key support staff such as an HSE advisor, HR, or an external coach). These individuals can serve as useful sources of feedback and accountability to support and sustain your efforts to change.

The annual performance cycle generally includes personal development plans which can also be used to document and track development objectives.

360 DEGREE FEEDBACK

After testing your perception and working with your supervisor on improvements, it may be helpful to test others perceptions with a 360 degree feedback survey or other similar process.

Specific versions of "Assessing your Safety Leadership" have been derived for supervisors and management in a 360 degree feedback format.

Depending on your relationship with your team, another approach would be to review your assessment results with them and ask for their feedback directly, either individually or in a group.

CONCLUSION

Systematic process safety leadership requires a vast amount of technical and leadership knowledge and skill. Use of the "Assessing Your Safety Leadership" tool can help identify both strengths and gaps in process safety leadership and can be helpful in achieving continuous improvement.

REFERENCES

1. Reason, James (1997), "Managing the Risks of Organizational Accidents"

Implementation of Process Safety Management into Diverse Corporate Cultures

Charles A. Soczek, DuPont Safety Resources, 4417 Lancaster Pike, Barley Mill Plaza 13/1108, Wilmington, Delaware 19805

FORWARD

The successful implementation of a DuPont-based Process Safety Management system in non-US facilities and companies faces cultural challenges. These challenges are expressed in the implementation of the elements of PSM and in the management of PSM at the corporate, business, and site levels. DuPont has developed a transformational model, built upon CCPS PSM Implementation guidance, which addresses these issues and drives sustainable improvement through process, organization, technology, and people. The speaker will present examples that illustrate some of the challenges that have been encountered across the globe, and how they been addressed.

INTRODUCTION

By the 1960s, DuPont had been operating hazardous processes for over 160 years, but an explosion at the Louisville, Kentucky, operations in 1965 told us we needed to do a better job. Our safety engineers took a hard look at our manufacturing operations and began to assemble the management principles and elements of Process Safety Management (PSM). During the 1970s, following several accidents at DuPont operations in New Jersey, PSM began to take the form we see today.

In 1979, DuPont issued the first corporate standard for Process Hazards Management (PHM). This was later changed to Process Safety Management. After the Bhopol accident of 1984, OSHA and several states began developing regulations for control of major accidents involving hazardous chemicals. The DuPont PHM process was used as a starting point for the OSHA regulations. Beginning in 1986, DuPont began working with other companies to develop a proposal for PHM that in 1991 became the US OSHA regulation 1910.119. At that time the overwhelming majority of DuPont manufacturing operations were in the United States, and thus the PSM process reflected a North American corporate culture.

As DuPont began to globalize operations, the company faced the challenges of incorporating PSM into hazardous operations across the globe. Today, DuPont operates in 44 countries, and at each location with hazardous processes, PSM has been implemented consistent with corporate requirements. These requirements describe ***what*** needs to be done. ***How*** PSM is implemented is left up to the various sites, and there are significant differences in the way PSM is put into operation. Some of these differences are a result of cultural adaptations.

In recent years, DuPont has been working with non-DuPont companies to implement PSM at their sites, and this effort has faced additional challenges. This paper will first review the PSM implementation process that we use. It is based on guidance from the Center for Chemical Process Safety (CCPS) and is modified as necessary to meet client needs. This will be followed by examples of obstacles that have arisen.

THE DUPONT IMPLEMENTATION MODEL

Successful PSM implementation requires either adaptations to the local culture or change in the culture, but as Geert Hofstede, the Dutch educator and author who conducted a comprehensive study of how cultures influence values in the workplace, has argued, cultures change slowly if they change at all[1]. Therefore, cultural differences must first be recognized, followed by development of a PSM implementation strategy that takes these differences into account. Inability to culturally adapt the effort can result in an organization that does not fully realize the benefits of Process Safety Management.

As shown in Figure 1, the model we use in DuPont for PSM implementation includes four work streams that address organization, process, people and technology. The objective is to meld the existing management process with U.S.-based management process and technology. After a business case for change is developed, a transformation to sustainable improvement is achieved with a Demming-type process.

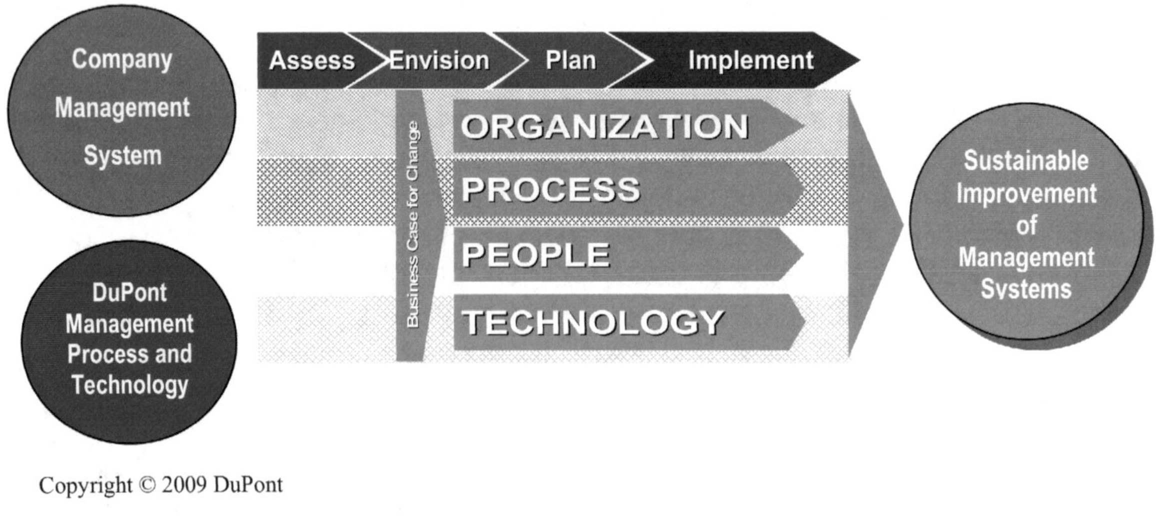

Figure 1. The DuPont PSM Implementation Model

It is important to note that successful implementation of PSM alone cannot guarantee success in eliminating major incidents and achieving a culture of safety. Actions of people are critically important. Employee safety must also be addressed along with any PSM effort.

ADDRESSING CULTURAL ISSUES

In 1994, the CCPS published the book ***Guidelines for Implementing Process Safety Management Systems***. Based on this information, below are key components that are used by DuPont in the implementation process:

- preparation, planning, and communications,
- an implementation framework,
- a specific attributes workshop,
- a site PSM leadership team,
- PSM element implementation teams,
- a PSM implementation plan,
- putting the system in place, and
- a full element implementation plan.

If the general cultural environment is evaluated early in the project, an implementation plan can be developed that will enhance success and avoid false starts. The specific attributes workshop can be used to address some of the cultural issues that arise.

Hofstede studied business cultures around the globe and developed a model based on five dimensions to describe them. These dimensions are:

1. Power Distance Index
2. Individualism
3. Masculinity
4. Uncertainty Avoidance Index
5. Long Term Orientation

This model was used to understand some of the observations that have been made concerning implementation of PSM in various geographic locations. A description of these dimensions can be found at the Geert Hofstede web site, http://www.geert-hofstede.com/ and are presented in Attachment A.

Table 1 shows ratings for the Arab World, Brazil, China, India and Mexico for three of the five Hofstede dimensions. The United States is added for comparison. High means above average and low is below average, compared to world average scores.

Table 1. Ratings Based on Hofstede Model						
Dimension	**Arab World**	**Brazil**	**China**	**India**	**Mexico**	**U.S.**
Power Distance	High	High	High	High	High	Low
Individualism	Low	Low	Low	Low	Low	High
Uncertainty Avoidance	High	High	Low	Low	High	Low

IMPACT OF CULTURE ON PSM IMPLEMENTATION

DuPont has helped companies around the world with various components of PSM. While the implementation model remains constant, there are always adaptations for local culture as DuPont addresses organization, process, people and technology.

Organization: PSM Networks

One organizational issue that is addressed during the implementation of PSM is employee networks. DuPont makes extensive use of these. They can be formal or informal, temporary or permanent. These networks are responsible for identifying or developing new and emerging technologies related to PSM, keeping abreast of regulatory developments and recognized and generally accepted good engineering practice, and incorporating these into DuPont standards. Theses networks also help in developing training required for the implementation of changes.

Some of the key formal standing networks at DuPont include PSM, PHA, Process Technology (process safety information), and Mechanical Integrity/Quality Assurance. When DuPont in the United States implemented the EPA Risk Management Program, a temporary network was formed to assure implementation of the requirements. PSM is also supported by engineering networks such as the Interlock committee, responsible for setting guidance for interlock design. These networks span the corporation, with the majority of resources in North America, but also with global participation.

From the Hofstede dimensions, the low rating for Power Distance and high for Individualism help explain the success of networks in DuPont. People readily take direction from network leaders, and their leadership allows and values network participation. From the corporation perspective, networks allow leveraging of highly skilled resources, particularly for the technical aspects of PSM. This brings business value by allowing the sites to implement all PSM-related requirements with minimum resources. Sites also maintain networks or committees across the site to further leverage resources.

In contrast, the Arab countries, Brazil, China, India, and Mexico have a high rating for Power Distance and low for Individualism. Individuals in these countries accept strong directed leadership but have difficulty accepting direction from others than their direct supervision. Additionally, managers have difficulty allowing their direct reports to spend time doing work directed by others. We have seen cases where people who attend network meetings receive calls from their supervisors asking when they will be back at their "real" job.

In these countries, labor costs are much lower than they are in the United States, so the economic incentive to leverage resources is not as strong. Companies can add or build the necessary PSM expertise within the workgroups, and do not need to rely on network contacts for specialized information or technology. For example, each group can have the expertise necessary within the group to advance dispersion modeling, and not rely on network contacts for guidance.

Process: PSM Standards and Operating Discipline

One of the process issues important for the implementation is PSM governance. The DuPont process for PSM governance starts with a set of policies and standards that describe the corporate requirements for PSM. These documents include the OSHA requirements and provide detail and additional guidance. From these documents, internal audit protocols are developed. These are used for internal corporate audits of site PSM processes.

The low Hofstede ratings in the United States for Power Distance and Uncertainty Avoidance indicate an expected cultural resistance to development and implementation of these requirements. However, the standards are carefully crafted to describe only what needs to be done, and the sites are given considerable flexibility in determining how PSM is implemented. As a result, there is considerable variability for implementation across the United States. Examples include a wide variety of change management processes, with different review requirements, forms, and types of changes reflecting plant process hazards, organization, and past methods of getting things done. This variability is accepted and expected, but the rigorous audit process assures that each of these processes meets the corporate requirements and standards. The audit results and closure of recommendations are closely monitored by leadership. This is done to overcome any cultural resistance to the development of standards and to operating discipline that drives full implementation.

In contrast, high Uncertainty Avoidance and high Power Distance in Arab, Mexican, and Brazilian cultures indicate a strong acceptance of rules. In these cultures, people readily accept the concept of PSM standards, or strict guidance from the company, and implement these requirements better than in other countries.

India and China also rate high in Power Distance but have a low rating for Uncertainty Avoidance. In this situation, organizational relationships tend to be formal but workflow is not[1]. The organization may develop standards, but less formal workflow can result in poor quality and inconsistent implementation of standards. Operating discipline may be poor, with lack of clarity for expectations. DuPont worked with a client in India where fatalities resulted from a traffic accident on the manufacturing site. DuPont recommended the site establish rules with severe consequences, including job loss, if these rules were violated. Site leadership resisted this recommendation.

In the oil and gas operations in the Middle East there is significant use of overseas workers, including engineers, operators, contractors, and laborers. Many of these workers come from India, whereas most of the leadership is local. The differences in Power Distance and Uncertainty Avoidance ratings can result in cultural clashes around the issue of operating discipline. The local leadership in these countries sets the rules and mandates that they be followed[2]. The consequences of not following the rules are serious, resulting in job loss and being sent back to the home country. This is a strong incentive, but does result in the desired discipline.

Technology: Regulatory Compliance. DuPont classifies the PSM elements in three major groups: facilities, personnel, and technology, as shown in Figure 2.

Figure 2. The elements of PSM

These elements differ in two ways from the US OSHA regulations. First is the addition of a requirement for Management of Personnel Change; second is separate Management of Change Processes for Technology and Facilities (identified above as Subtle Changes). Notice the core of the model is management commitment. The importance of leadership for successful implementation of PSM cannot be over-emphasized. The outer ring of was added to the model to stress the importance of Operating Discipline.

Many of the activities within these elements are technical in nature. Companies in the developing economies such as China and India that are initiating a formal PSM process view technology as the intellectual property they want when seeking help on PSM. Implementation of these elements is normally done through the operating, maintenance, and technical engineers in the company. These engineers are highly trained with skills equal or comparable to their U.S. counterparts. They thoroughly understand the technical aspects of PSM.

Yet while the engineers understand the technical aspects of PSM, there can still be formidable difficulties implementing the elements. In counties with high Power Distance and low Individualism, as in many developing economies, engineers look to line management for specific direction, setting of priorities, and resources. In one company, the process developed requires the site manager to sign off every prestart-up safety review (PSSR) and management of change (MOC). This takes time, and when there is pressure to keep production online, poor decisions about what is a change may occur. This can result in a complete by-pass of the change process.

Low Uncertainty Avoidance can also impact technology. Acceptance of uncertainty, coupled with a belief that licensed processes are overdesigned, can result in the modification of process variables beyond those provided in the technology package. In one case, the pressure in a reactor was increased, resulting in a capacity increase. There was no technical documentation to support the change—no MOC or PSSR. Since this action was successful, it could lead to further increases, producing unknown conditions and hazards

Understanding the technical components alone will not result in successful PSM performance. This must be coupled with strong and continuous leadership commitment and operating discipline. These are the areas where the local culture has great impact.

CONCLUSIONS

When implementing PSM, the challenge is not to change the local culture but to implement PSM within the existing environment. There are many examples of successful PSM implementation in cultures outside the United States. Where networks are difficult to form or operate poorly, it may be necessary staff PSM efforts at a higher level. Operating discipline and audit protocols can be established and performance measured. These issues can be overcome with an implementation model that addresses organization, process, people and technology.

The general observations for a particular culture may not fully apply in all situations. There are many factors that affect this, including the influence of western partnerships and joint ventures and education in the West.

References:

1. Hofstede, Geert. "Motivation, Leadership, and Organization: Do American Theories Apply Abroad." Organizational Dynamics. (Summer 1980) 42-63.

2. Salem, Ahmad. Personal discussions. General Electric, Abu Dhabi

3. http://www.geert-hofstede.com/ Feb. 2009

Appendix[3]
Description of Three of the Hofstede Indices

Power Distance Index is the extent to which the less powerful members of organizations and institutions (like a family) accept and expect that power is distributed unequally. This represents inequality (more versus less) but defined from below, not from above. It suggests that society's level of inequality is endorsed by the followers as much as the leaders. Power and inequality, of course, are extremely fundamental facts of any society, and anybody with some international experience will be aware that "all societies are unequal, but some are more unequal than others."

Individualism on the one side is its opposite, collectivism; that is, the degree to which individuals are integrated into groups. On the individualist side we find societies in which the ties between individuals are loose: everyone is expected to look after him- or herself and him- or her immediate family. On the collectivist side, we find societies in which people from birth onwards are integrated into strong, cohesive in-groups, often extended families (with uncles, aunts, and grandparents) which continue protecting them in exchange for unquestioning loyalty. The word "collectivism" in this sense has no political meaning; it refers to the group, not to the state. Again the issue addressed by this dimension is an extremely fundamental one, regarding all societies in the world.

Uncertainty Avoidance Index deals with a society's tolerance for uncertainty and ambiguity; it ultimately refers to man's search for the Truth. It indicates to what extent a culture programs its members to feel either uncomfortable or comfortable in unstructured situations. Unstructured situations may be novel, unknown, surprising, and different from usual. Uncertainty avoiding cultures try to minimize the possibility of such situations by strict laws and rules, safety and security measures and, on the philosophical and religious level, by a belief in absolute Truth; "there can be only one Truth and we have it." People in uncertainty avoiding countries are also more emotional and are motivated by inner nervous energy. The opposite type (uncertainty accepting cultures) are more tolerant of opinions different from what they are used to; they try to have as few rules as possible. On the philosophical and religious level, they are relativist and allow many currents to flow side by side. People within these cultures are more phlegmatic and contemplative and are not expected by their environment to express emotions.

Safety Culture: From Black Art to Real Improvements

By: Cinzia De Santis, Evelyn Mitchell, Diane Chadwick-Jones, Cory Shelton, Matthew Lawrie, Deborah Sortland, Chris Fitch, Karen Moorcroft, Dianne Parker

INTRODUCTION

BP has devoted a significant amount of effort to improving safety culture. A project started 6 years ago identified, through statistical analysis, a correlation between culture and incidents in the work place.

After using the project to characterise a strong safety culture, BP developed a series of tools for assessment and improvement.

Currently, these tools have been applied to a wide range of businesses and offices across the Group in all the segments (upstream, downstream, and alternative energy).

Safety culture as a key topic is included in the development programs both for First Level Leaders and for Senior Leaders in the organization. In addition, the BP Operations Management System specifically includes operating culture as one of the essential elements in managing risk and achieving operating discipline.

The objective of this paper is to explain, through some case studies, some aspects of BP's journey, so far, to improve its safety culture.

WHAT IS SAFETY CULTURE

Culture, in general, can be defined in a simple way as: *"The way we do things around here"*. This description implies an accepted way of doing things among a group of people in an organization.

A robust and sustainable safety culture needs to be based on the right individual and collective values, beliefs and behaviours, a clear commitment from the leaders of the organization, and mechanisms to drive continuous improvement. These organizational aspects need to be aligned and in place to make changes in behaviours sustainable.

CHARACTERISTICS OF A ROBUST SAFETY CULTURE

To identify what a strong safety culture looks like, BP visited several best-in-class organizations with outstanding performance records, also known as high reliability organizations. Consolidating the in-house statistical analysis, the pilots and academia research, BP developed a model to define an ideal safety culture.

This model includes the following characteristics of a robust culture that delivers outstanding safety performance:

- **Informed Culture**
  - Communications are fluid in both ways: the workforce tells management the truth, even when it hurts.
  - Management knows what is really going on and is prepared to provide feedback, even if it is not popular.
  - Everybody is prepared to report their errors and near-misses; attention is paid to the details and small events.
- **Mindful Culture**
  - Everyone thinks about what might go wrong.
  - Information is not discarded.
  - People can deal with the unexpected; they do not just rely on what worked the last time.
  - People listen to those with knowledge.
  - Decisions are delegated to the level where the expertise resides.
- **Learning Culture**
  - Seeking new ways of understanding and working, procedures are under constant scrutiny.
  - Training and more training helps create competence; cross-training supports effective communication.
- **Fair Culture**
  - Everyone knows what is acceptable, since they helped define the standard and it has been properly communicated.
  - Everyone is aware of the consequences of their actions, both positive and negative, and communications are clear in this aspect.
  - There is not a blanket no-blame approach, but blame is reserved for truly egregious behaviour, involving recklessness or malice.
- **Respectful Culture**
  - Workforce is involved in all aspects related to safety and they are encouraged to participate
  - People's ideas are listened to and taken in consideration
  - It is an acknowledged worker's right to work free from unacceptable risks
  - Workers are treated with respect and sensitivity towards local cultures

SAFETY CULTURE ASSESSMENT

To assess safety culture, BP used the Oil and Gas Producers (OGP) model developed by Dr. Patrick Hudson and Dr. Dianne Parker from Leiden and Manchester University. The model was inspired by the work of Dr. Ron Westrun from Michigan University on how information flows in an organization depending on its level of cultural maturity.

Hudson and Parker defined five levels of Safety Culture as described in Figure 1.

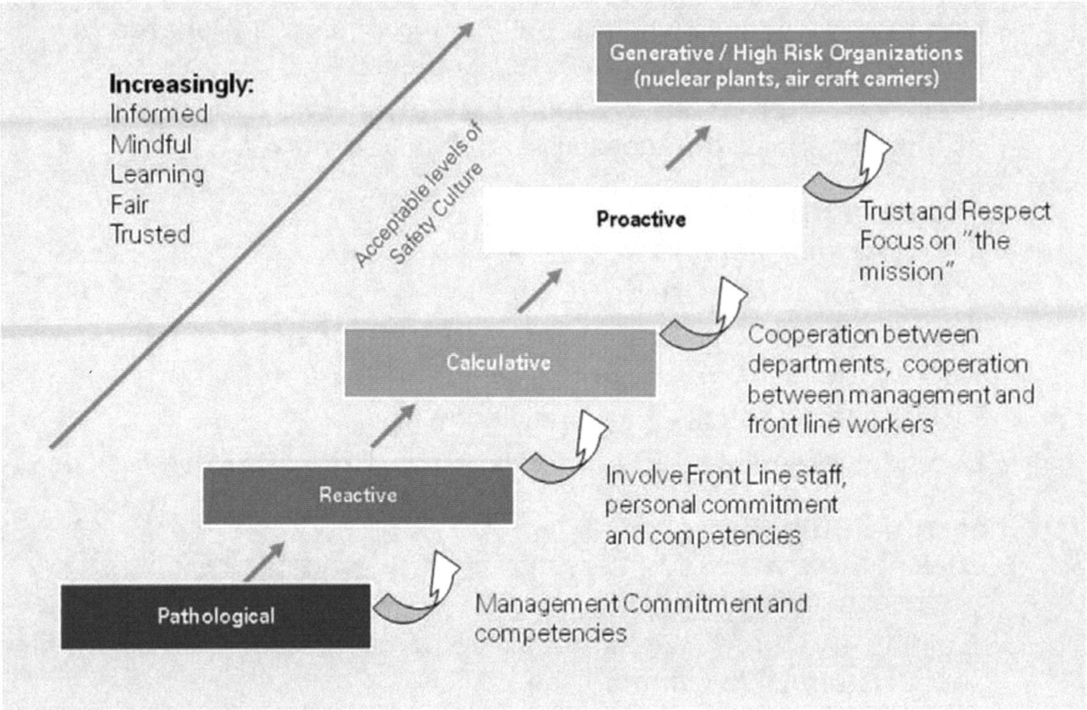

Fig 1: OGP SAFETY CULTURE MODEL

Using this model, two tools were developed to assess safety culture with additional support guides and assessments. Each tool is applied depending on the requirements of the business.

A risk based criterion is generally used when deciding on which tool to use.

1) Surveys

Recommended for higher risk sites and led by the business resources with guidance and steering from the corporate safety team.

2) Workshops

Recommended for medium/ lower risk sites, with less than 80 people (including contractors) and geographically dispersed. This was piloted in the Marine business.

Support guides and assessments include safety culture cards, safety leadership assessment, guide to reward and recognition and behavioural assessments. These are used in support of improvement efforts associated with assessment results and action planning with site leadership.

These materials are supplemented by a comprehensive website and videos.

METHODOLOGY

Whether the survey or workshop approach is used, the methodology to assess Safety Culture is basically the same:

1) Define the scope of the assessment

2) Leaders and workforce, including contractors, answer a series of questions. The data is consolidated in bar charts and is plotted by the percent of answers for each of the level of the safety culture ladder (Figure 1).

3) Results are discussed with the workforce and improvement plans are agreed with the leadership based on feedback from the discussion groups.

A more detailed explanation of how the methodology works is shown in Figure 2.

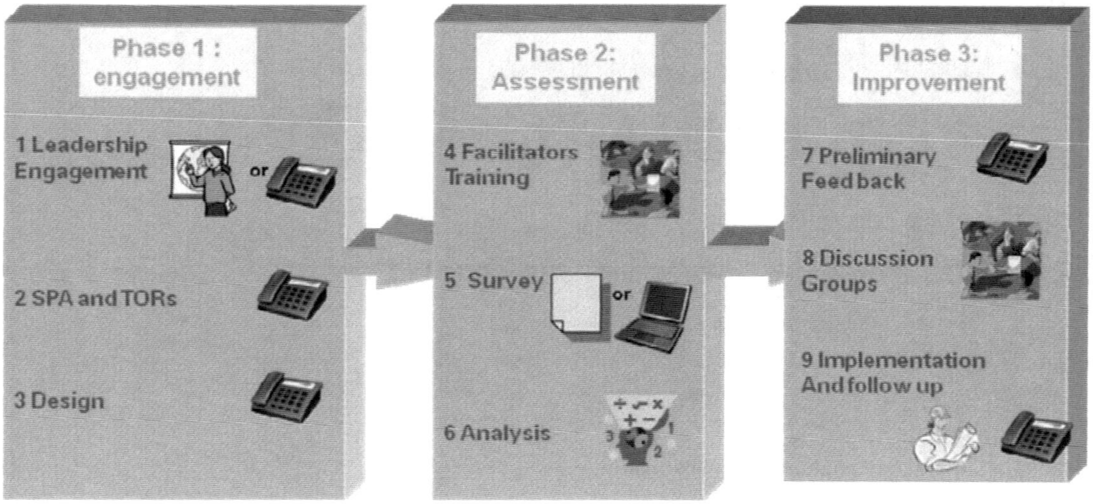

FIG 2: Safety Culture Assessment and Improvement Process

TYPE OF RESULTS

The results help to identify strengths and weaknesses in terms of culture perceptions in a team or organization. A hypothetical example of these results is shown in Figure 3.

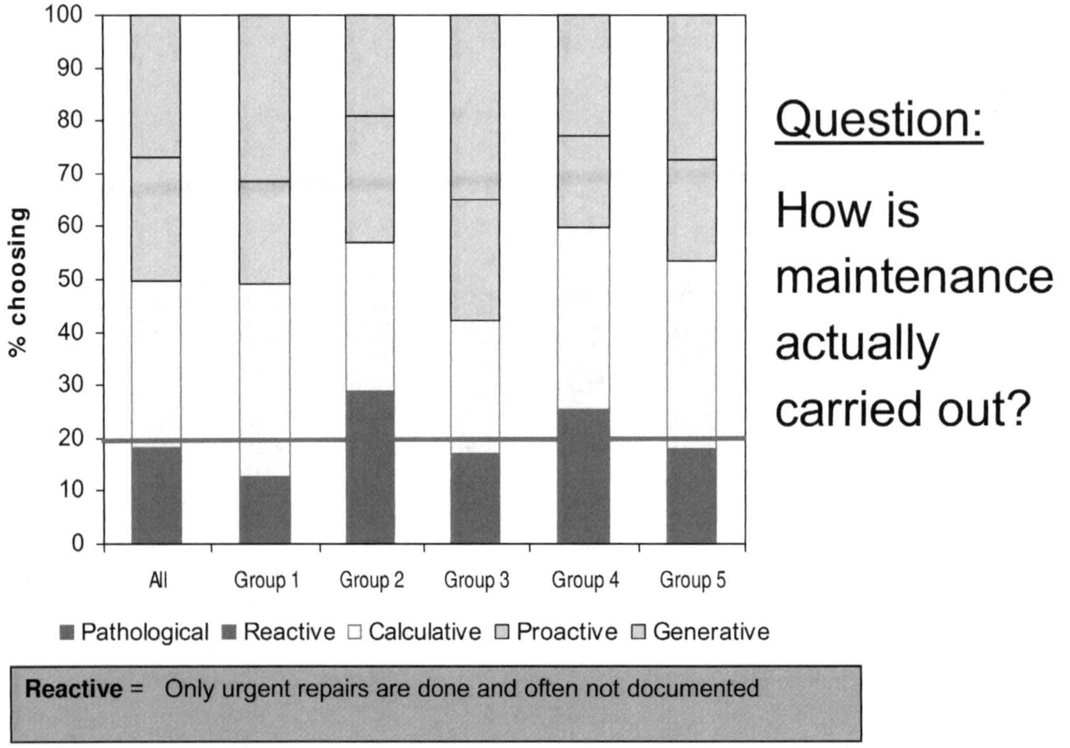

Fig 3: Hypothetical Example of Assessment Results

This type of chart helps identify, for each question, perceptions of specific behaviours in their operation.

Questions are related to topics such as:

- Leadership Commitment
- Relationship with Contractors
- Learning
- Following Procedures
- Trust and Respect
- Life Balance

While all results are carefully considered, if more than 20% of the population responds in the "pathological" or "reactive" categories in a specific question, this is considered an area for significant attention.

DISCUSSION GROUPS

Once the results have been consolidated, they are presented to the workforce to test their validity and to gain input on how things could be improved. This part of the process is vital for the program to achieve the intended positive change.

Once the data from the discussion groups is consolidated, recurrent topics are identified and key areas for improvement developed.

This information is presented to the Leadership Team to prioritise and create an action plan. This process includes use of a prioritisation matrix, which is used to rank key improvement areas according to impact and degree of difficulty.

For the improvement areas selected, an accountable individual, time line, and resources needed to address the issue are identified, and an action plan is created.

SITES THAT HAVE PARTICIPATED IN THE SAFETY CULTURE PROGRAM

As previously mentioned, the safety culture assessment and improvement process has been widely conducted across BP. Fig 4 provides a geographical representation of Business Units that are participating in various activities associated with the program.

Fig 4: Snapshot of Activity

CASE STUDIES

1) Business 1

When the program started, the workforce was quite sceptical about how the results would be used and if there would be any follow-up to the survey. After completing the discussion group work and presenting results to the Leadership Team, an action plan was put in place and the following are some of the actions taken or reinforced within the Business Unit:

- Significant budget provided to address equipment issues (e.g., pulse meter leakage)
- Strengthened contractor vetting process
- HSE audits / IM audits performed at customer level
- New process developed to communicate incidents and lessons learned.
- Enhanced HSSE induction / refresher
- Two-way communication and listening greatly improved: e.g., face-to-face factory presence of leadership team members versus sending e-mails, toolbox problem-solving sessions on specific issues with either leadership team member or team leaders

2) Business 2

In this particular Business Unit, the workforce had unvoiced concerns. Also, while their own responsibilities in new business were understood, those of other colleagues or other departments were not. Thus, for some of the workforce, the complete picture was a blur. Some of the workforce also felt that there was not enough communication of progress of strategy implementation, which led to unnecessary and unvoiced concerns. After the results were explained to the Leadership Team, an action plan was put in place and the following are some of the changes that occurred at the Business Unit

- Clearer roles and responsibilities with Gas Centre contractors
- Less distraction in the organisation through improved communication relating to workforce concerns
- Clearer roles and responsibilities of all staff accounting for new business model
- Improved interface between BP Gas and the Gas Centres through process updates and clear accountability
- Improved HSE performance

3) Business 3

At this site, the issues were related to leadership alignment and understanding of the importance of safety culture. After the assessment, discussion groups and improvement plans were put in place, and the following are some of the improvements reported:

- Improved Leadership Team alignment and cohesion
- Continued embedding of safety culture at the filling plant
- Improved understanding of role of Gas Centre Managers (combined contractors and sales reps)
- New steel pallet system: significant business benefits to be gained through increased supply chain efficiency and improved customer offer

CONCLUSIONS

After the Texas City incident, BP embarked on a journey to improve its safety processes in a detailed and thorough manner. As a result, an integrated Operating Management System has been put in place and is being implemented across BP; a strong audit function has been set up; and, leaders at all levels participate in training programs to deepen their expertise in areas such as process safety and integrity management.

In addition, as part of this journey, a safety culture assessment and improvement process was developed. This process has been very successful in affecting significant change, particularly where it has been fully embraced.

These changes were not only related to improvement of safety but also led to improved operational efficiency. This is aligned with the view that good safety and good business go hand-in-hand and both rely on effective leadership.

The Organisation for Economic Co-operation and Development
Guidance on Developing Safety Performance Indicators related to Chemical Accident Prevention, Preparedness and Response

Kim Jennings, U.S. Environmental Protection Agency (EPA) and OECD Working Group on Chemical Accidents; U.S. EPA, 1200 Pennsylvania, N.W., Mail Code: 5104A, Washington, DC 20460, 202-564-7998, jennings.kim@epa.gov

Francine Schulberg, OECD Group of Experts on Safety Performance Indicators 415-668-2239, fschulberg@aol.com

INTRODUCTION

In October 2008, the Organisation for Economic Co-operation and Development (OECD) – an intergovernmental organization bringing together 30 countries – published *Guidance on Developing Safety Performance Indicators related to Chemical Accident Prevention, Preparedness and Response.*

The *Guidance* is divided into two documents, one directed to Industry and the second directed at government agencies ("Public Authorities") and the Public (in particular, communities located near hazardous installations).

This *Guidance* was designed to allow each of the target groups to analyze their own actions, to determine whether the steps they are taking to support chemical safety are, in fact, achieving their objectives and to help identify where further action is needed. The *Guidance* is a companion to the OECD *Guiding Principles on Chemical Accident Prevention, Preparedness and Response* (2nd ed, 2003).

This *Guidance on Developing Safety Performance Indicators* (SPI) was the culmination of more than ten years' effort, in order to capture existing experience and develop guidance that can be adapted to meet the needs of a variety of industries and companies, as well as government bodies at different levels and with different responsibilities.

The Group of Experts charged with developing the guidance consisted of representatives of countries, industry, labour, non-governmental organizations and international organizations. The Group prepared "interim" *Guidance*, which was published in 2003. In order to get practical experience with its implementation, a pilot program was established. Based on the feedback received and new developments, the Group revised the *Guidance* and prepared the final version published in 2008. (see www.oecd.org/ehs).

The 2008 *Guidance* contains two primary components:

- a step-by-step approach for developing SPI Programs; and
- a menu of possible indicators which addresses the range of issues involved with chemical accident prevention, preparedness and response.

The *Guidance* also includes a number of annexes which provide supporting information such as guidance on developing metrics and a bibliography.

This paper will first provide some background information on the OECD and how the *Guidance* was developed. It will then provide an overview of the *Guidance*, focusing specifically on the steps for creating an SPI Program and the menu of possible elements. This will be illustrated by an example of a small company going through the process of developing an SPI Program. Finally, it includes information on the next steps OECD is taking with respect to the *Guidance* and the opportunities to cooperate with CCPS.

WHY OECD?

What is the OECD: The Organisation for Economic Co-operation and Development (OECD), established in 1961, is an organization of governments that are committed to democracy and a market economy. The OECD currently has 30 members[1] and involves many other countries in its work.

The role of the OECD is to provide a venue for governments to compare experience, seek answers to common problems, identify good practices and, as appropriate, coordinate policies. Work is managed by specialized committees and, under these, working parties and working groups which are composed of experts from member countries (and others, as appropriate).

OECD and Chemical Safety: In 1971, OECD established a program to address chemical safety, focusing in its initial phases on chemical testing and assessment. It has expanded over time to address risk assessment and management, to share the burden of testing certain high-production volume chemicals and to consider the safety of pesticides, biocides and products of biotechnology.

---

[1] OECD member countries are: Australia, Austria, Belgium, Canada, the Czech Republic, Denmark, Finland, France, Germany, Greece, Hungary, Iceland, Ireland, Italy, Japan, Korea, Luxembourg, Mexico, the Netherlands, New Zealand, Norway, Poland, Portugal, the Slovak Republic, Spain, Sweden, Switzerland, Turkey, the United Kingdom and the United States. The European Commission also takes part in the work of the OECD.

Following the Bhopal and Basle accidents, member countries concluded that they should work together to address the issues associated with chemical accident prevention, preparedness and response. This subject was particularly well-suited for a cooperative effort and sharing of experience, as countries and companies were seeking to improve their programs related to accident prevention, preparedness and response in light of these well-publicized accidents. Therefore, a new working group was established in 1987 to share best practices, exchange information about accident case histories, identify opportunities for cooperative activities, analyze issues of mutual concern, develop guidance materials and reach agreement on common policies.

This long-standing Working Group on Chemical Accidents (WGCA) has over twenty years of establishing relationships among member countries and formalizing an international network of industry (through the OECD Business and Industrial Advisory Committee (BIAC)), labor associations (through the OECD Trade Union Advisory Committee (TUAC)) and non-governmental organizations. It works closely with other international organizations involved with issues related to chemical accident prevention, preparedness and response such as the United National Environmental Programme (UNEP), the International Labour Organization (ILO), the World Health Organization (WHO) and United Nations Economic Commission for Europe (UNECE)). With this diverse network of experts from all stakeholder groups, the WGCA has the unique ability to bring together appropriate groups of policy makers and experts from around the world in order to share ideas and lessons learned, and to develop guidance documents through a consensus process. It is within this forum that the OECD *Guidance on Safety Performance Indicators* was developed.

The OECD Chemical Accidents Program has organized more than twenty workshops, each addressing a different issue related to chemical safety and each bringing together experts from government, industry, labour, non-governmental organizations and others.

The OECD *Guiding Principles and Guidance on Safety Performance Indicators*: Using the results of these workshops, the OECD developed the OECD *Guiding Principles for Chemical Accident Prevention, Preparedness and Response*, first published in 1992 and updated in 2003. The *Guiding Principles* provide guidance for all the key stakeholders on the range of activities involved in the safe planning and operation of facilities where hazardous substances are produced, used or handled, as well as guidance for effective land-use planning, emergency preparedness and emergency response.

The *Guiding Principles* have been distributed world-wide, and a number of countries prepared translations so that the guidance can be used at the local level by government authorities, industry and others.[2]

---

[2] Translations can be downloaded from the OECD Web site in Chinese, Czech, French, German, Hungarian, Italian and Korean.

In 1997, the WGCA concluded that they should consider ways to assess whether specific guidance contained in the *Guiding Principles* is successful in achieving its objectives. They decided to create a companion to the *Guiding Principles*, addressed to the same group of stakeholders, to provide the tools for determining the success (or lack thereof) of actions taken to improve accident prevention, preparedness and/or response.

WHAT IS THE VALUE OF AN SPI PROGRAM?

From the perspective of industry, an SPI Program allows a company to:

- assess whether it is implementing appropriate chemical safety programs and policies;
- evaluate whether these programs and policies are achieving their desired objectives; and
- help determine the extent to which such programs and policies are making a difference.

This will allow a company to identify whether there is appropriate emphasis on different aspects of safety management and provide insights needed for setting priorities for future investment of resources.

An SPI Program can also facilitate communication and cooperation with public authorities, other enterprises and the local community.

With respect to public authorities (government agencies), there are also a number of reasons for establishing an SPI Program, perhaps the most important being that it provides a tool to respond to the questions: what is our contribution to improved safety and are our activities leading to overall improvements. Using the output from an SPI Program, agencies can help to identify gaps in regulations and policies.

An SPI Program for government agencies also facilitates cooperation and communication with industry, communities and others, and it may help motivate industry to improve safety. An SPI Program can help authorities to establish priorities for inspections and identify areas that should be considered during inspections and reviews.

HOW WAS THE *GUIDANCE ON SAFETY PERFORMANCE INDICATORS* DEVELOPED?

In 1998, the WGCA established a Group of Experts charged with developing guidance for implementing SPI programs. The Group of Experts consisted of representatives of member countries and the EC, industry, labor, non-governmental organizations and other international organizations.

Their mandate was to prepare flexible guidance that can be adapted for use in different countries, in different industries and in a variety of contexts. It was agreed that the indicators should include both measures of activities/organization of work (leading indicators) and measures of outcome/impact (lagging indicators). The Experts were asked to provide guidance for three target groups: industry, public authorities and communities/public.

> ***Outcome indicators*** help assess whether safety-related actions achieve desired results leading to less likelihood of an accident occurring and/or less adverse impact on human health, the environment and/or property from an accident. They are reactive and measure impact of actions.
>
> ***Activities indicators*** help identify whether enterprises or organizations are taking actions believed necessary to lower risks. They are pro-active and can measure safety performance against tolerance level(s) indicating deviations from safety expectations at specific point in time.

Work was initiated by examining existing experience on performance indicators related to chemical safety. Three experts, one from each of the target groups, took the lead in drafting text for industry, public authorities and the community, respectively.

The Group of Experts met a number of times, reviewing drafts and collecting practical experience from agencies and companies that agreed to test early drafts of the *Guidance*.

In 2003, "interim" *Guidance* was published and made available on the web, to get broad-based feedback on the overall approach as well as on the specific content. This interim *Guidance* was composed mainly of long menus of possible activities indicators and outcome indicators for each of the three target audiences, based on their roles and responsibilities for chemical accident prevention, preparedness and response.

A pilot program was established to test the interim *Guidance*. About twenty companies and agencies agreed to try to establish SPI Programs based on the *Guidance* and provide periodic reports on whether the *Guidance* is useful and how it might be improved. This proved to be an invaluable exercise; the feedback provided the Group of Experts with insights on how to make the *Guidance* more valuable and user-friendly.

In addition, during this period, the UK Health and Safety Executive and the UK Chemical Industries Association were working with companies to develop a generic model for establishing a performance measurement system aimed at senior managers and safety professionals working at hazardous installations. The UK guidance, "Developing Process Safety Indicators: A step-by-step guide for chemical and major hazards industries," was published in 2006.[3] This document, along with the experience of companies working with the UK authorities, provided a basis for the development of a new chapter in the OECD *Guidance* on how to develop an SPI Program.

In 2006, the OECD Group of Experts re-convened to review the feedback from the pilot program and other relevant experience. As a result, the Group concluded that it should:

- add a new chapter describing how to develop an SPI Program (to complement the existing text on what could be the content of such a program);
- include more detailed guidance on developing and using metrics;
- separate the *Guidance* into two publications, with one directed to industry and one to public authorities and communities; and
- provide targeted guidance for emergency responders *(e.g.*, police, fire, hazmat teams, emergency medical personnel).

WHAT IS INCLUDED IN THE OECD *GUIDANCE ON SAFETY PERFORMANCE INDICATORS*?

The revised *Guidance*, taking into account the conclusions identified above, was approved by the WGCA and published in 2008. This new version differs from the 2003 text in a number of ways, specifically:

- The primary focus of the *Guidance* is the new chapter on "How to Develop an SPI Programme," setting out a seven-step process (Chapter 2);
- The previous text containing the menu of possible outcome and activities indicators, along with targets, was updated and revised in light of experience (Chapter 3);
- There is additional material on how to develop and implement metrics, including detailed guidance in a new annex; and
- The *Guidance* was broken into two publications:
  - one for industry defined to include any enterprise that poses a risk of a significant accident involving hazardous substances, irrespective of location, size or nature of ownership; and
  - one for public authorities at all levels that have a role to play in accident prevention, preparedness or response, as well as for communities in the vicinity of hazardous installations. This publication has focused guidance for emergency response personnel.

---

[3] **Health and Safety Executive (UK) and Chemical Industries Association, (2006)** ***Developing Process Safety Indicators: A step-by-step guide for chemical and major hazard industries,*** HGN 254, ISBN 0717661806 (see http://www.hse.gov.uk/pubns/books/hsg254.htm)

WHAT ARE THE STEPS IN THE GUIDANCE FOR CREATING AN SPI PROGRAM?

The *Guidance* sets out a seven-step process for creating an SPI Program, i.e.:

- Step One: Establish the SPI Team
- Step Two: Identify the Key Issues of Concern
- Step Three: Define Outcome Indicator(s) and Related Metrics
- Step Four: Define Activities Indicator(s) and Related Metrics
- Step Five: Collect the Data and Report Indicator Results
- Step Six: Act on Findings from SPIs
- Step Seven: Evaluate and Refine SPIs

This process is illustrated in Figure 1 below, which shows that Steps Three and Four are generally undertaken on an iterative (rather than consecutive) basis, and showing the review cycle incorporating Steps Two – Seven.

Figure 1. Seven Steps to Create and Implement an SPI Program

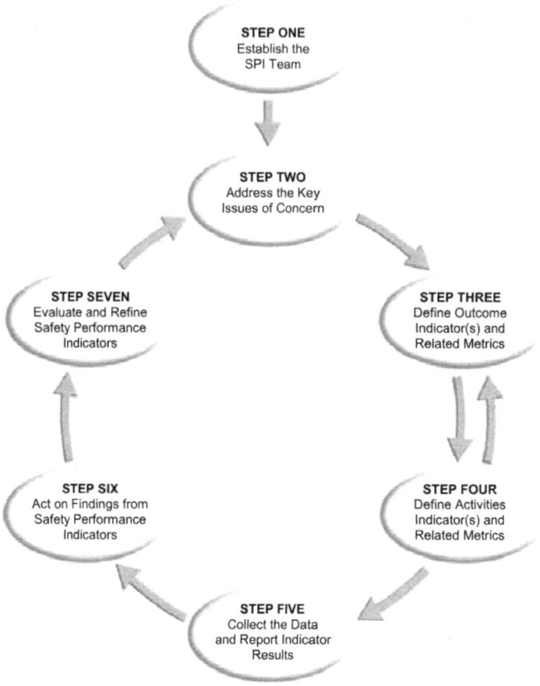

To facilitate understanding of this process, the Steps will be illustrated by using an example of a fictitious medium-size company that formulates small-batch specialty chemicals. In the example, the company has decided to develop an SPI Program using the *Guidance* to determine whether their health and safety program is effective and to improve their ability to set priorities for future safety-related activities.

Step One: Establish the SPI Team

The first step in beginning the process of developing an SPI Program is to establish the team that will be involved in the development of the SPI Program. This effort should start with identifying a leader – an individual or group – who will generally oversee the Program's development, effectively communicate the efforts of the team and promote the Program's implementation. It is also important that appropriate staff members, including technical experts and employees with hands-on knowledge, are involved in the process so the SPI Program reflects the hazards and safety measures at the facility.

Additionally, the involvement of management is critical to the success of the effort to ensure resources are committed to the Program and actions identified by the SPI Program, which are needed to improve chemical safety at the enterprise, are taken. Once the team is established, a plan should be developed which clearly identifies a timetable and milestones to ensure adequate progress is made in developing the SPI Program and appropriate resources are committed to the project.

SMALL SPECIALTY CHEMICAL FORMULATOR – EXAMPLE OF STEP ONE

The enterprise assigned the health and safety manager and one of its most experienced shift foreman to explore the use of SPIs. This team would develop a proposed SPI Program for review and approval by the company president. The company president stated that the team should consult with other relevant employees and that the proposals should include estimates of the cost of implementing the Program as well as estimated savings associated with the reduction in lost-time incidents.

Step Two: Identify the Key Issues of Concern

Once the SPI team and other arrangements are in place, the next step is to identify the scope of the SPI Program. Each enterprise will need to decide on its own priorities, in order to choose the appropriate indicators and the way they will be measured.

There are several ways to identify the programs, policies and/or procedures that are most critical and therefore should be part of the SPI Program. One way is to focus on each process in the enterprise and identify critical hazards, and the related safety policies, procedures and practices that are most critical for risk control or most vulnerable to deterioration over time. Alternatively, priorities can be identified by considering the potential consequences of failure of each of the policies, procedures and practices in place. Regardless of how this step is approached, the most recent process hazard analysis (PHA) can provide vital information to help get started.

The scope of the Program should begin with a limited number of indicators, which can be expanded as more experience is gained in implementation. In this step, enterprises should focus on what to monitor rather than how to monitor. Additionally, enterprises should work to avoid the pitfall of measuring what they *can* measure instead of what they *should* measure.

SMALL SPECIALTY CHEMICAL FORMULATOR – EXAMPLE OF STEP TWO

The SPI team started by reviewing its most recent HAZOP and considering possible accident scenarios. They realized that safety is an issue when there are new formulations or significant changes to existing formulations. Based on this analysis, the team concluded that their priority for indicators should be operator competence, process engineering or implementation of safety-related procedures. The team decided to consider the use of SPIs for each of these areas. For simplicity, the remainder of this example will focus on efforts to develop SPIs for safety-related procedures.

Step Three: Define Outcome Indicator(s) and Related Metrics

In this next step, the SPI team would define outcome indicator(s) addressing the key issues of concern. Outcome indicators can help measure the extent to which a targeted safety policy, procedure or practice is successful and is achieving the desired result. These indicators need to convey clear information on safety performance to those with the responsibility and authority to take action.

Outcome indicators are designed to collect data and provide results to help you answer the broad question of whether you have achieved a desired result but, unlike activities indicators, do not tell you why the result was achieved or why it was not.

Indicators – both outcome and activities – consist of two key components:

- A definition, which should clearly state what is being measured in terms that are meaningful to the intended audience; and
- A metric, which defines the unit of measurement. This should be precise enough to highlight trends in safety over time and/or highlight deviations from safety expectations that require action.

When defining a relevant outcome indicator, it is useful to ask "what would success look like?" and "can this successful outcome be detected?" The answers to these questions will assist the enterprise in defining the specific target trying to be achieved.

A metric must be assigned to each indicator. The metric is the approach by which safety data will be compiled and reported in the SPI Program and determine whether the program provides the insight necessary to assess and act on safety issues. Insights on how to choose metrics is provided in the OECD *Guidance*.

SMALL SPECIALTY CHEMICAL FORMULATOR – EXAMPLE OF STEP THREE

In response to the question, "what would success look like?" the SPI team agreed that a successful process for implementing procedures would result in a set of procedures that ensure that employees carry out their tasks safely. As a result, the team referred to sub-section B.3 ("Procedures") of Chapter 3 which has the target: "employees carry out their tasks safely and under conditions necessary to satisfy the design intent of the installation." In reviewing the possible outcome indicators, the team identified "extent incidents are attributed to procedures (due to, *e.g.*, procedures lacking, procedures inadequate and/or procedures not followed)" as one that addresses their needs.

The team noted that the percentage of incidents attributed to issues related to procedures might remain high even if the enterprise was successful in decreasing overall incidents (*i.e.*, if total incidents and number of incidents attributable to procedures decreased proportionally). Therefore, in considering possible metrics, they chose to track raw tallies of incidents attributed to issues related to procedures.

Step Four: Define Activities Indicator(s) and Related Metrics

The next step is to define activities indicators to monitor the key elements of the safety programs, procedures and policies identified in Step Two.

Activities indicators relate to outcome indicators and help to measure whether critical elements of safety programs, procedures and policies are in place to achieve the desired outcomes. Whereas outcome indicators are designed to provide answers about whether you have achieved a safety outcome, activities indicators are designed to provide information about why or why not the outcome was achieved. Therefore, well-designed activities indicators should provide the information needed to correct programs, procedures and policies when the desired outcome is not being achieved.

In identifying the appropriate activity indicators, the enterprise should consider those activities that are most important to achieving the intended target and most closely related to the outcome indicator(s) chosen in Step Three. If a change in the activity indicator does not result in a change of the outcome indicator, it may mean that the activity indicator is too far removed from the outcome indicator and a new activity indicator should be chosen. As with the outcome indicators, an appropriate measurement system needs to be established for the activities indicators.

SMALL SPECIALTY CHEMICAL FORMULATOR – EXAMPLE OF STEP FOUR

The team reviewed the possible parts of the procedures process that could result in safety issues and incidents. The incident investigations suggested that "new or changed formulations requiring additional or updated procedures" was a key issue. Therefore, the team defined the following activities indicators and related metrics adapting the suggestions presented in sub-section B.3 of the *Guidance*:

- "Is relevant information passed on from one stage to another and incorporated in procedures when developing or introducing new products, processes or equipment?" This would be measured by asking the degree to which each new or updated procedure incorporated safety information, using a three-point scale (*i.e.*, "complete," "somewhat complete" or "incomplete") based on periodic reviews.
- "Are users informed about changes in the procedures?" with relevant users periodically asked whether they were aware of specific changes in procedures. Their responses would be compiled by identifying the percentage of users aware of procedures relative to total users.
- "Are new procedures being implemented?" where periodic spot checks would be conducted to observe whether procedures were being followed. The percentage of spot checks where procedures were being followed would be periodically calculated and reported.

Step Five: Collect the Data and Report Indicator Results

Once the outcome and activities indicators and metrics have been determined, the next step is to identify how to collect and report the results. The initial step is a review of existing data sources (*e.g.*, information collected for quality control or other business purposes) to see what might be relevant and determine if they are of adequate quality to be used in the SPI Program.

Data collection procedures should address how frequently the data is to be collected and results reported. The data should be collected frequently enough to identify critical changes in the process in a timely manner so action can be taken to ensure safety at the facility. Reports should be provided in a timely manner to allow for appropriate actions to be taken. For indicators that use threshold metrics, the procedures should identify specific thresholds or tolerances, *i.e.,* the point at which deviations in performance should be flagged for action.

The presentation of the data should be simple and relevant to ensure understanding of any deviations from tolerances and to identify important trends and/or problems with the processes, policies and procedures. The presentation should also allow the reader to understand the links between outcome indicators and associated activities indicators.

SMALL SPECIALTY CHEMICAL FORMULATOR – EXAMPLE OF STEP FIVE

The team decided to use investigation reports as the source of information for the outcome indicator, "extent incidents are attributed to procedures..." For activities indicators, the team decided that they would work with shift foremen and monitor all new formulations and significant changes to existing formulations. They would: collect and review new and updated procedures for completeness; observe the degree to which the procedures were made available to users; and analyze the degree to which they were followed using the metrics defined in Step Four.

The team calculated that the cost of implementing this program would be low. It did not require any additional resources. If the program could identify and prevent even a few incidents, it would result in a net benefit. The president approved the approach.

Step Six: Act on Findings from Safety Performance Indicators

Results from SPIs, including tolerances being exceeded, disturbing trends over time and inconsistent results, must be acted upon. Otherwise, there is little point in implementing an SPI Program. Timely reports with important information highlighted should be provided to senior managers, safety management personnel, engineers, operators and other relevant employees. This information is paramount to ensuring quick follow-up action on adverse findings to fix problems in the associated processes, policies and procedures.

SMALL SPECIALTY CHEMICAL FORMULATOR – EXAMPLE OF STEP SIX

Over the first few months, the team noted that the number of incidents attributed to procedures did not change significantly. They also noted that when new and modified formulations were introduced, some but not all of the safety-related procedures were updated (*i.e.*, the majority were rated "somewhat complete"). For example, sometimes changes to procedures for filling and start-up were covered but not changes for shut-down and unloading. Some procedures addressed changes in normal operations but not changes associated with abnormal or emergency situations. The health and safety manager worked with process engineers to correct this problem.

The team noticed that without exception, operators were informed of new and modified procedures (*i.e.*, 100% of users were aware of the changes). However, the team also noticed that new and modified procedures were not always followed. Based on this, the health and safety manager interviewed several operators to determine the reason for these deviations. Operators described situations where new procedures conflicted with other existing procedures or relied on instrumentation that was not installed on all of the batch tanks. In order to meet production demands, the operators developed "workarounds" to deal with these issues. The team reviewed the incident reports for the period and noted that almost all of the incidents involved situations where workarounds were used.

Based on this, the health and safety manager implemented changes to the procedures process where operators would be consulted by engineers during the development of new or updated procedures. In addition, the introduction of new or significantly modified procedures would include a "shake-down" period where engineers would be on-call to receive feedback and address conflicting or unworkable procedures.

Step Seven: Evaluate and Refine Safety Performance Indicators

The SPI Program, including the indicators and metrics, should be periodically reviewed and evaluated. These reviews help to ensure that the indicators are well-defined, continue to address priority areas of concern and provide the information needed to monitor safety measures and to respond to potential safety issues. In addition, it will help to identify when specific indicators are no longer needed and allow adjustments to the SPI Program to focus on the most important issues and indicators.

For example, it may be discovered that some indicators do not provide useful measurements for the enterprise or that the metrics are not precise enough to recognize small but significant changes that require action. Also, the issues of concern can change over time due to improvements in safety or insights into previously unidentified issues. Based on experience, an enterprise might decide to expand the SPI Program to include a larger number of indicators in order to provide greater insights into a particular issue or address a larger scope of safety concerns. Finally, the SPI Program can be amended to take into account the experience of other enterprises (including, for example, other companies in the same industry or in the same community).

SMALL SPECIALTY CHEMICAL FORMULATOR – EXAMPLE OF STEP SEVEN

Based on their initial experience, the SPI team proposed the following approach for the procedures-focused SPI effort:

- Continue to use the outcome indicator, "extent of the number of incidents attributed to procedures (due to, *e.g.*, procedures lacking, procedures inadequate and/or procedures not followed)." This number had shown little change over the first six months. The team hoped that the new process for updating procedures would result in a decrease in this number.
- Continue to use the three activities indicators. The issues of addressing procedures associated with all stages of a new or modified formulation had been completed and the degree to which operators were informed of new procedures remained high. Despite this, the safety and health manager reasoned that measuring these activities would ensure that they continued to operate correctly. The team was hopeful that the new process for updating procedures would result in better implementation of written procedures.
- Implement two new activities indicators: 1) "is participation of employees built into the development of procedures?" and 2) "is there a means to ensure that procedures are corrected when conflicting with other procedures or if not working properly?" These indicators would help to monitor the degree to which the new process for updating procedures was implemented.

HOW TO CHOOSE ELEMENTS THAT ARE RELEVANT TO YOUR COMPANY?

Chapter 3 contains a long menu of possible indicators; the industry publication has six sections each divided into a number of sub-sections based on the various responsibilities of industry related to accident prevention, preparedness and response.

Figure 2 provides a list of the sections and sub-section in Chapter 3.

Figure 2. Table of Contents for Chapter 3

Section A.
Policies, Personnel and General Management of Safety

- A.1 Overall Policies
- A.2 Safety Goals and Objectives
- A.3 Safety Leadership
- A.4 Safety Management
- A.5 Personnel
  - A.5a Management of Human Resources (including training and education)
  - A.5b Internal Communication /Information
  - A.5c Working Environment
- A.6 Safety Performance Review and Evaluation

Section B.
General Procedures

- B.1 Hazard Identification and Risk Assessment
- B.2 Documentation
- B.3 Procedures (including work permit systems)
- B.4 Management of Change
- B.5 Contractor Safety
- B.6 Product Stewardship

Section C.
Technical Issues

- C.1 Research and Development
- C.2 Design and Engineering
- C.3 Inherently Safer Processes
- C.4 Industry Standards
- C.5 Storage of Hazardous Substances (special considerations)
- C.6 Maintaining Integrity/Maintenance

Section D.
External Co-operation

- D.1 Co-operation with Public Authorities
- D.2 Co-operation with the Public and Other Stakeholders (including academia)
- D.3 Co-operation with Other Enterprises

Section E.
Emergency Preparedness and Response

- E.1 Internal (on-site) Preparedness Planning
- E.2 Facilitating External (off-site) Preparedness Planning
- E.3 Co-operation Among Industrial Enterprises

Section F.
Accident/Near-Miss Reporting and Investigation

- F.1 Reporting of Accidents, Near-misses and Other "Learning Experiences"
- F.2 Investigations
- F.3 Follow-up (including application of lessons learned and sharing of information)

Each sub-section contains some general information about the subject matter and references the related text in the *Guiding Principles*. This is followed by the "Target" for the subsection and related "outcome indicators" and "activities indicators."

The outcome and activity indicators in Chapter 3 are not a complete list of all possible indicators; enterprises may decide to create different indicators to meet their needs. Additionally, enterprises will need to adapt these indicators to fit their specific circumstances.

Figure 3 on the following page illustrates a sub-section of Chapter 3.

Figure 3. Sub-Section of Chapter 3

B.4 Management of Change

See Guiding Principles document, paras.:

- 2.f.1 Formal procedures
- 2.f.2 Hazard analysis to be reviewed and process documentation to be supplemented
- 2.f.3 Provide procedures for startup after a modification
- 2.f.4 Employees to be informed and trained about the modifications
- 2.f.5 Contractors to be included in procedures

Based on historical evidence, inadequate reviews of changes in enterprises have resulted in accidents. The definition of what constitutes a change includes: modifications in equipment, technology or software; changes in personnel (including reducing and increasing staff size); and administrative/managerial adjustments, including temporary modifications. In this regard, it should be noted that changes in the organisational structure or in manning can be triggered by economic shifts.

TARGET

Change is managed to ensure that it does not increase, or create, risks.

POSSIBLE SAFETY PERFORMANCE INDICATORS:

Outcome Indicators

i) Extent technical modifications or other changes follow management of change procedures (or extent of non-compliance with management of change procedures).

ii) Number of incidents resulting from failure to manage change appropriately (*e.g.*, change in procedural process made without following the management of change policy).

iii) Percentage of change requests that are processed as "emergency changes" (*i.e.*, requiring immediate attention for safety reasons).

Activities Indicators

i) Is there a clear definition of a change (modification)?

ii) Are there procedures addressing the management of change, which cover all the necessary steps from planning to implementation and follow-up? Do the procedures address:
   - approval by the relevant responsible person before proceeding to the next step;
   - risk assessment, as appropriate;
   - clear allocations of roles and responsibilities;
   - a formal control form to steer and to keep track of the various steps in the procedure.

iii) Do the procedures apply to technical changes as well as changes of organisational or administrative character? Do they address modifications in the following areas:
   - technical, including changes in equipment and buildings (mechanical, instrumentation and control systems and other software, electrical, civil, etc.);
   - process parameters and recipes, including raw material and chemicals, utilities, etc. (*e.g.*, deviations from the approved "operating window");
   - organisation and management;
   - personnel (manning, working times, outsourcing, etc.).

iv) Do the procedures address permanent as well as temporary modifications (including pilot projects)?

v) Do the procedures provide for a risk assessment and/or other appropriate review including pre-startup review for relevant modifications? Does this address the need for competent personnel, independent from those directly responsible for the proposed change (recognising that depending on the complexity and risk level, external expertise may be needed)?

vi) Are there clear requirements related to the updating of technical and other documentation (*e.g.*, do they require updating before a modification is implemented)?

vii) Are there clear requirements for the updating of instructions/procedures and for information and training of employees before a modification is implemented?

WHAT HAPPENS NEXT?

The OECD and its member countries are pursuing three primary goals with respect to the SPI Guidance:

- distribution and promotion of the *Guidance*;
- translations; and
- development of an interactive Web site.

The OECD has distributed the *Guidance* to all OECD member countries and to a number of international organizations via the WGCA. The documents are also available on the OECD Web site at www.oecd.org/ehs under "chemical accidents." The OECD will be undertaking a number of additional activities to further promote the distribution and use of the *Guidance*.

The OECD is currently translating the documents into French and, it is expected that several member countries will undertake prepare translations for use in their countries. As those translations are completed, they will be made available on the OECD Web site.

Finally, the OECD, working with the US, is developing an interactive Web site for the *Guidance*. On this Web site, users will be able to access various topic areas and specific subjects as well as create a SPI Program specific to their organization. That Web site should be available in mid-2009, and will be also be accessible at www.oecd.org/ehs.

Industry and professional organizations like CCPS can have a critical role in the further dissemination, promotions, use, and improvement of the *Guidance*. For example, CPPS and its member organizations can:

- Distribute copies of the *Guidance* to your constituents;
- Make it available on your Web site, or create links to the OECD site;
- Use, promote and adapt the *Guidance* for use in your organization;
- Participate in development and testing of the web-based version;
- Establish systems for sharing experience within and among countries; and
- Provide feedback on the *Guidance* and its implementation.

Additionally, it would be useful to collaborate with member companies to exchange information on lessons learned in developing and implementing a SPI Program using this OECD *Guidance* or similar guidance such as that being developed by CCPS. Specifically, case studies added to the interactive Web site would provide users with real examples of the successes and challenges faced by stakeholders in implementing a SPI Program. The OECD Web site can be easily updated to include these case studies.

OECD would like to work with CCPS to determine how our two efforts in this area can be mutually supportive as we move forward in implementing SPI Programs to improve chemical safety.

49f CCPS Process Safety Metrics Status Report

Tim Overton
The Dow Chemical Company
Building B-101, 2301 North Brazosport Boulevard
Freeport, Texas 77541-3257 USA
Phone: (979) 238-1358
E-mail: toverton@dow.com

Abstract

In 2006, the Center for Chemical Process Safety launched a project committee to develop guidelines for Leading and Lagging Process Safety metrics. A few months later, the Baker Panel investigating the BP Texas City incident and the US Chemical Safety Board each released reports calling for improved industry process safety metrics.

The CCPS project has been conducted in two phases. The first phase was to develop new definitions and criteria for industry-wide process safety metrics. This was prepared by a committee of greater than 50 international stakeholders representing industry, academia, regulatory agencies, and labor. The second phase was the development of a more detailed guideline book on the most effective implementation and use of metrics for individual companies.

The presentation will address the implementation status of the numerous organization and individual companies who have implemented the CCPS metric definitions, as well as the status of additional discussions with an API committee and international trade associations to promote broad acceptance and implementation of those metric definitions. The presentation will also describe the current status and key concepts from the forthcoming CCPS Guideline book on developing metrics.

CCPS Guidelines for Chemical Transportation Safety, Security and Risk Management

Brad A. Fuller
AcuTech Consulting Group
2000 North Beauregard Street, Suite 500
Alexandria, VA 22314

Prepared for Presentation at
American Institute of Chemical Engineers
2009 Spring National Meeting
5th Global Congress on Process Safety
43rd Annual Loss Prevention Symposium
Tampa, Florida
April 26–30, 2009

UNPUBLISHED

INTRODUCTION

This paper presents and overview of the CCPS *Guidelines for Chemical Transportation Safety, Security and Risk Management*, 2008. This new book, deals with transportation safety, security and risk management on a broader basis and provides tools and methods to benefit a wider range of transportation professionals and stakeholders. In particular, it presents a comprehensive framework for managing transportation risk and introduces more qualitative and practical techniques for screening, identifying, and managing higher-level risks issues that balance both safety and security.

Due to the complexity of a transportation supply chain, risk management is a shared responsibility of all stakeholders. The transportation supply chain (Figure 1) should be considered as a single, complete system beginning with the raw material supplier, through the manufacturing and distribution process, to the final end-use customer, and any residuals management, including the transportation network in between. While the roles and responsibilities may differ for each stakeholder, and differ within the functional areas of an individual company, these groups need to understand how their activities and actions can impact the risk to the overall supply chain. The goal of this guideline is to assist in the development of transportation risk management (TRM) programs where information from these various groups can together support decision-making that balances risk with operational efficiencies, while optimizing the entire supply chain's activities.

Given the complexities of the chemical industry, the intended audience for this guideline is extensive and not limited to just shippers. To manage the risk across the entire supply chain, the following groups can benefit from the primary management systems, risk concepts, risk assessment tools and techniques, examples, and overarching risk management framework presented in this guideline:

- Logistics service providers and managers
- Shippers
- Chemical manufacturing companies
- Chemical distribution companies
- Carriers
- Business unit managers and owners
- Transportation managers
- Safety professionals
- Security professionals
- Risk professionals
- Regional and global government regulators
- Insurers
- Industry associations

Figure 1. Transportation Supply Chain

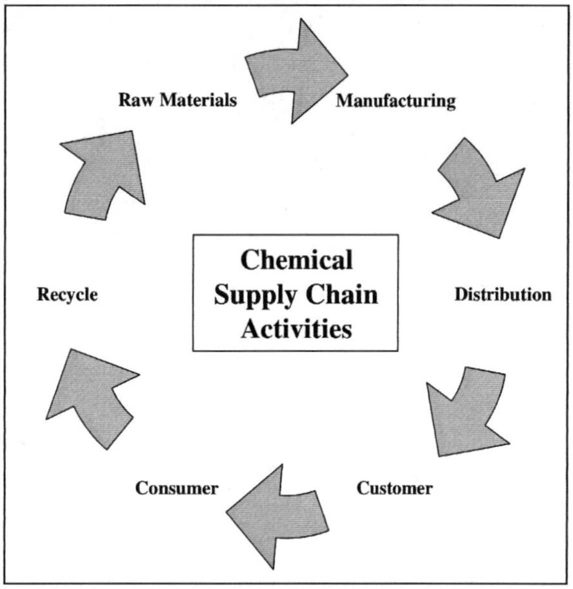

A major goal of the safe and secure transportation of hazardous materials is a reduction in incidents that could lead to a release or to misuse. To date, the achievements in safety are the result of regulations, industry standards, individual company initiatives, and emergency response preparedness, as well as investments in training, systems, and technology. Even with the foundation that these programs and activities supporting the day-to-day operating practices provide, the safe transportation of hazardous materials is complex due to a number of issues:

- Number of regulated hazardous materials (thousands are listed in regulations worldwide)
- Regulations that vary by mode, region, and country
- Different hazards classes including toxicity, flammability, corrosivity, and reactivity
- Various modes of transportation including road, rail, marine (including bulk vessels), pipeline, and air
- Multiple packaging types including bulk and non-bulk
- Use of more than one mode during a shipment (intermodal)
- Complexity of the supply chain including multiple parties and changes of custody during transit
- Overlapping and potentially unclear responsibility of various parties
- Transport routes where the risk profile can change depending on proximity to the public or other sensitive areas

The complexity of these activities, even with the current safety regulations and operating practices, is one of the reasons that accidents involving the transportation of hazardous materials occur.

In addition to safety considerations, security has become an increasingly greater concern for hazardous materials during transport. Certain hazardous materials could potentially be used as a weapon to spread fear, cause injuries or fatalities, and/or result in negative economic impacts. To address the potential for deliberate misuse, additional security regulations and industry standards continue to be developed.

This new publication provides a framework and common practices that will allow shippers and supply stakeholders to manage chemical transportation networks. The framework should embrace and utilize a company's existing systems, but must meet regulatory requirements in areas where shipping activities occur. It is written from an international perspective, recognizing that many companies transport materials across country borders and around the globe. Local customs, regulations, and working cultures may vary, but the common principles of safety, security, and risk management should always be recognized.

FRAMEWORK FOR MANAGING TRANSPORTATION RISK

To ensure the safety and security of chemicals in transit, this guideline promotes a comprehensive risk management program covering the entire supply chain. This holistic approach to risk management encompasses involvement from all stakeholders, and includes a general risk management framework highlighted in Figure 2.

The framework is designed to be flexible, enabling companies to tailor their TRM process to their individual circumstances and specific applications. The TRM framework is comprised of the following steps:

- *Primary Management System* is the foundation. Ensuring these elements are complete and current is the first step in the process. With the potential for cross-border shipments and international operations, understanding and keeping current with changes in regulations is critical.
- *Identification and Prioritization* consists of (1) cataloging the hazardous materials and modes of transportation, (2) identifying sensitive areas and potential points of failure along the transit route, and (3) understanding interactions with other stakeholders in the supply chain.
- *Risk Analysis* concepts presented in this guideline are applicable to people new to TRM, or those needing a refresher and range from simple, quick, and general to detailed, time-consuming, and quantitative assessments. Simpler techniques should be the starting point with the more complex and resource-intensive activities conducted on a risk-justified basis.

Figure 2. TRM Framework

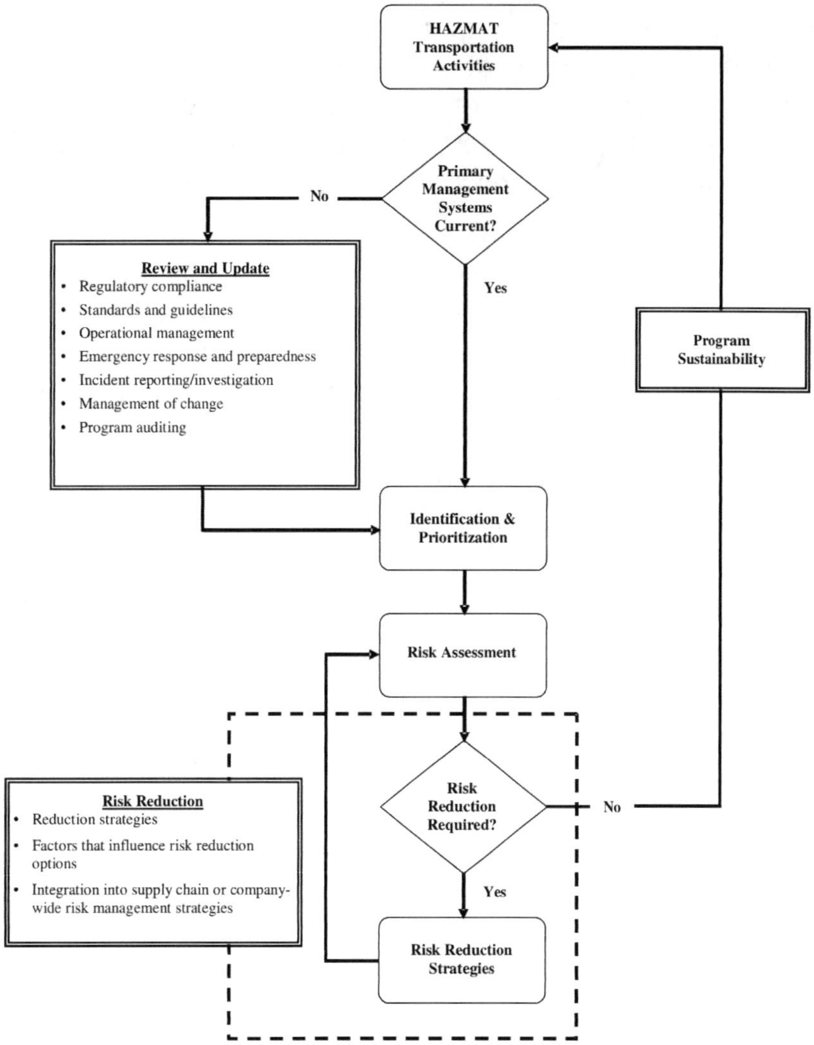

- *Risk Reduction* is the process of determining if the assessed risk is being managed adequately and evaluating/implementing options to reduce risk, as needed. If the risks merit further reduction, methods can be developed for the hazardous material operation, coordinated with the enterprise risk management philosophy of a single company, and/or developed in coordination with other supply chain partners.
- *Program Sustainability* is the final step in the TRM framework and is designed to ensure the risk management process remains current. Includes an ongoing commitment to managing risk, continuously improving, reacting to emerging trends, and keeping current on evolving transportation risk analysis practices

PRIMARY MANAGEMENT SYSTEM

The first step of the TRM framework is to review the baseline programs for safety and security management for all modes of hazardous material transport. These baseline or "primary" management systems are those things that need to be in place prior to an organization being ready to attempt a risk analysis. The first tier of risk management activities for all companies is to ensure that their baseline programs are in place, complete, and current.

All members of the supply chain should have primary management systems in place to address compliance with all applicable regulations and implementation of additional programs and practices to manage the safety and security of hazardous material transportation activities. An overall transportation safety and security management system is recommended to assist in:

- Ensuring regulatory compliance
- Establishing and meeting training requirements
- Establishing procedure and protocols
- Ensuring equipment integrity (containers, rolling stock, pipelines, etc.)
- Monitoring and improving practices for the supply chain
- Managing changes to transportation procedures and practices
- Preparing for incidents and emergencies
- Identifying and managing transportation risks

It is difficult to develop a comprehensive list of primary systems that will meet every need and suit all situations. However, it is important to recognize that the success of any initiative or program is dependent on an existing management system, the corporate culture, and the continued support from management. Figure 3 illustrates that one's daily practices should be built on company policies and standards, while ensuring that all applicable codes and regulations are followed.

Figure 3. Primary Management Structure

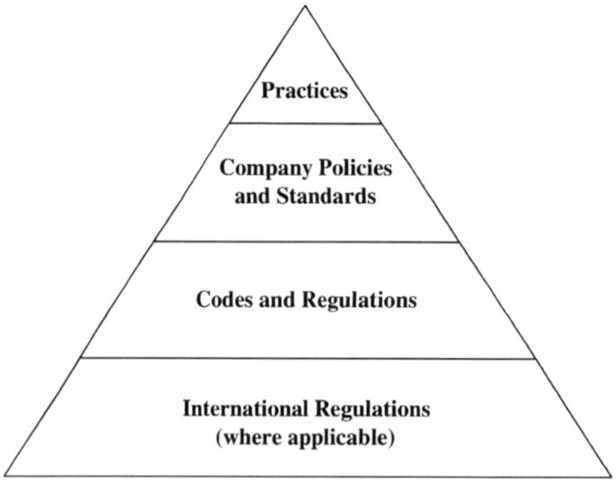

Where practicable, transportation risk management activities should be integrated into existing management systems including those for safety, process safety management, security, environmental stewardship, quality assurance, and business proficiency. Existing systems may provide a firm foundation for incorporating transportation risk management program elements and regulatory requirements. Many times, however, in the area of transportation activities, regulatory, safety, and security improvement programs are handled in different parts of the organization. Therefore, if integration is not feasible or practical, at least some level of connectivity should be established including an active and sustainable communication process.

In addition to a more discussion of the essential components of a transportation management system, this section of the guideline also includes a more detailed review of regulatory compliance, including:

- UN Model Framework
- International regulations
- Regional regulations
- Harmonization of transportation regulations

RISK ANALYSIS

Before a risk analysis is initiated, it is important to specify the scope and have a protocol in place for these activities. For a transportation network, the scope may need to consider:

- The entire supply chain
- A segment of the transportation network
- A single hazardous material or a mixed shipment
- A subset of all the materials shipped
- One mode or route of transport

The overall theme of this guideline is to put lower-level risks into the appropriate perspective and focus increased energy and resources on the most important or complex transportation issues. Figure 4 presents a risk assessment protocol that can be applied to accomplish this goal, with the first step being the identification and prioritization process.

For those risks that are escalated following the identification and prioritization process, many can be evaluated, and mitigated through the use of more detailed risk analysis techniques. In practice, the most risk analyses are qualitative/semi-quantitative in nature, and only lead to a more detailed quantitative risk analysis (QRA) when a simpler approach cannot provide a clear understanding of the risk and more information is required for proper management. Additionally, qualitative analysis typically precedes all other higher levels of analysis as it is an essential data-gathering step of the entire risk analysis process.

Figure 4. Risk Assessment Protocol

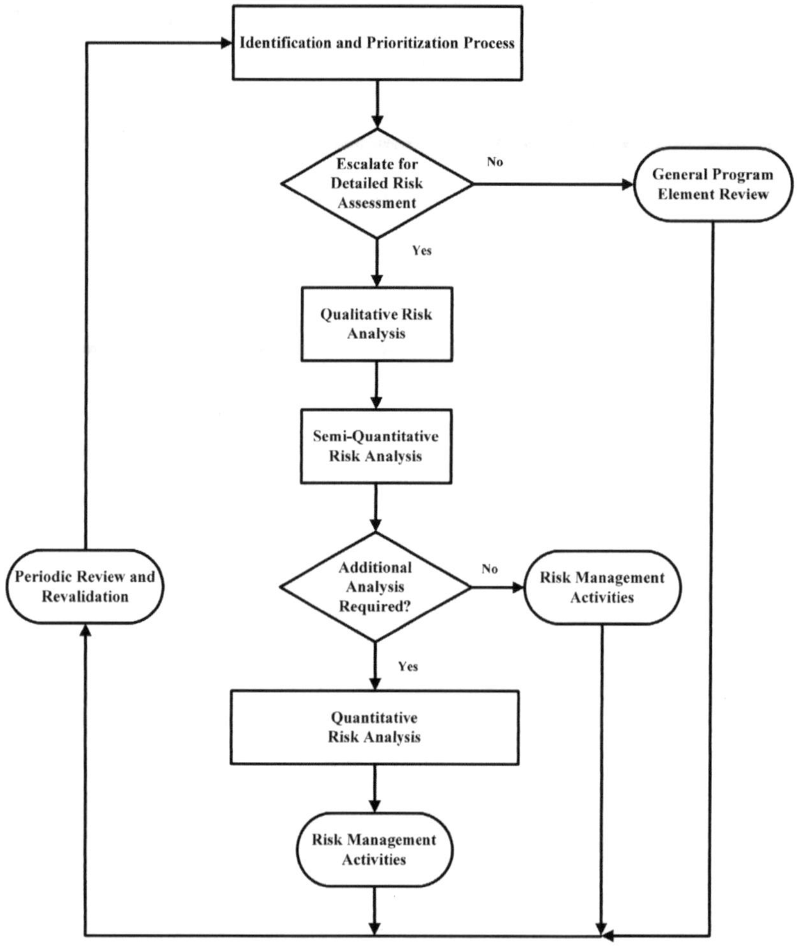

The results of a risk analysis are fed into the risk reduction step of the TRM framework, shown as risk management activities in Figure 4. As discussed later in this paper, the risk reduction step is where different options for managing risk are evaluated and compared. For all issues that go through qualitative, semi-quantitative, or quantitative analysis, periodic re-evaluation of the level of risk should be conducted, with changes implemented as required.

Identification and Prioritization Process

There are a number of transportation issues that can be reviewed alone or in combination to identify those requiring additional detailed analysis. The identification and prioritization step can be completed using a checklist to identify issues that may require escalation, including:

- Hazards of the materials being transported
- Specific initiating events that may be considered

- Potential consequences and impacts while in transit
- Accident history
- Potential gaps in safety practices
- Sensitive areas along the route for people, property, and the environment
- Complexity of the transportation activities and number of stakeholders in the supply chain

Once the issues along the supply chain are identified, there are a number of options available for identification and prioritization. As shown in Figure 5, these prioritization options include:

- *Hazard Based*—where all chemicals of a certain class are ranked the same (e.g., acute toxics, flammability rating, ecotoxics)
- *Consequence Based*—where potential high risk areas along the route are identified (e.g., population, environmentally sensitive, critical infrastructure)
- *Likelihood Based*—where recent history (internal and external accident performance of a specific carrier or of the industry as a whole) drives priority
- *Risk Based*—where the potential consequence and likelihood are used

As illustrated in Figure 5, the different techniques can be used alone or in combination to escalate an issue more detailed risk analysis. For example, a company may always require a detailed risk analysis for a specific chemical based solely on its hazard rating, whereas another company may only require additional analysis if the chemical is shipped by a specific mode and/or in large specified quantities.

The hazard-, consequence-, and likelihood-based approaches are founded on the potential issue or combination of issues. The risk-based approach discussed here is qualitative and could result in a subjective estimate of the level of risk (based on chemical hazards, potential impacts alone the route, and likelihood of an accident).

Regardless of the prioritization process utilized, the end result is a screening of transportation issues that can be limited to a general program element review, with a subset of issues escalated for more detailed analysis.

The complete guidelines book contains a number of industry tools that are used by companies to conduct this risk prioritization. The process illustrated in Figure 5 and the industry tools are intended to provide examples, and provided a basis for developing a tailored approach for your company or specific supply chain activities.

Figure 5. Identification and Prioritization Process

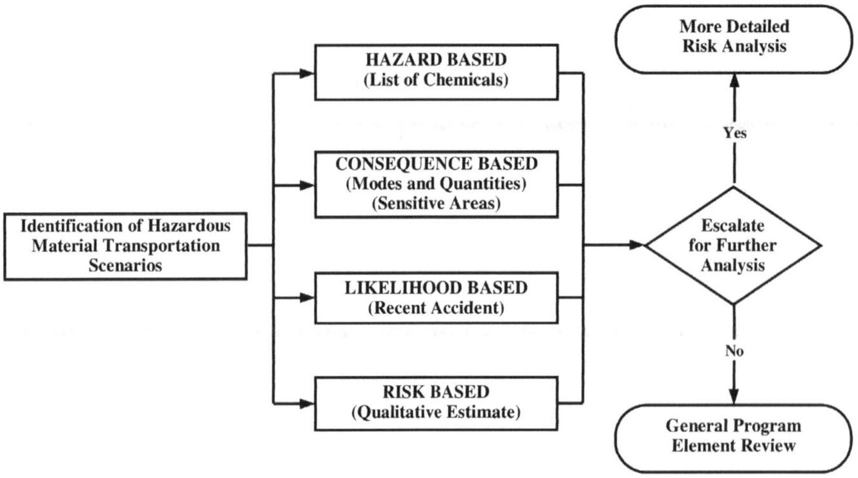

Qualitative and Semi-Quantitative Risk Analysis

Figure 6 illustrates the risk assessment process for qualitative and semi-quantitative techniques that can be applied to transportation safety scenarios. Typically, these will be a scenario and issues developed and escalated for more detailed analysis during the identification and prioritization process described above.

As shown in Figure 6, issues escalated from the identification and prioritization process are first evaluated using qualitative techniques. A qualitative risk analysis is based primarily on description and comparison using historical experience and engineering judgment, with little quantification of the hazards, consequences, likelihood, or level of risk. This first level of risk analysis should be typically conducted before further semi-quantitative or full quantitative evaluations are initiated. Even at this stage of analysis, a reasonable risk assessment decision can be reached. With the findings documented, the activities managed as part of the transportation system, and the focus moved to other more complex issues escalated through this process. The qualitative risk analysis step also collects additional information, building from the prioritization process, which will be necessary for the more detailed analyses. This data collection activity, and ability to better understand the operation, is another reason for not bypassing qualitative risk analysis in lieu of more detailed risk analysis approaches.

For those issues or transportation scenarios requiring more detail and insight than a qualitative approach offers, the next level of analysis is a semi-quantitative evaluation. A semi-quantitative risk analysis includes some degree of quantification of consequence, likelihood, and/or risk level. Once evaluated, through the combination of consequence and likelihood, a risk-based decision can be made from the analysis or additional refinement in consequence or likelihood may be necessary. This refinement of the risk analysis may rely on additional information included as part of a semi-quantitative analysis or, as warranted, on some minimal level of quantification.

Figure 6. Qualitative and Semi-Quantitative Risk Assessment Process

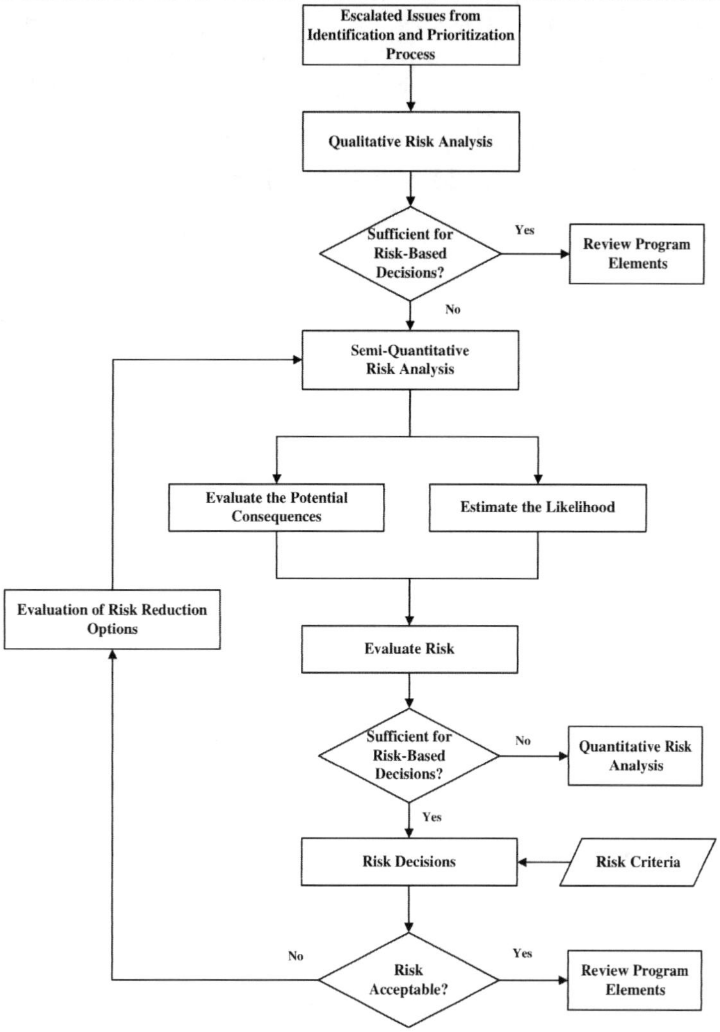

Regardless of the risk analysis technique utilized, risk decisions and evaluation of risk mitigation options can only be accomplished if evaluation criteria are defined. Evaluation criteria are the metrics used to determine if additional risk controls will be needed as well as the means to compare the benefits of competing risk mitigation options. While the selection of company-specific evaluation criteria is beyond the scope of this guideline, it is critical that risk analyses be evaluated against metrics consistent with the risk tolerance of the individual companies involved in the supply chain activity being reviewed.

Quantitative Risk Analysis

This section of the guideline provides an overview of QRA techniques for evaluating hazardous materials (or dangerous goods) transportation issues, including data requirements, analysis techniques, and the generation and interpretation of quantitative risk results. The intent is not to be the ultimate technical resource for transportation data, consequence modeling, frequency analysis, or calculation of quantitative risk results, but rather to serve as a resource for the transportation professional participating or supporting a QRA. *Guidelines for Chemical Process Quantitative Risk Analysis* (2000) is the technical resource on applying a QRA, and *Guidelines for Chemical Transportation Risk Analysis* (1995) is the technical resource for adapting these fixed-facility techniques to address the unique issues of hazardous material transportation.

This purpose of this section is not meant to suggest that rigorous analysis techniques should be employed in all cases, and is only intended for advanced risk analysis applications. The goal of this practical review of transportation QRA is to help the reader better understand:

- Basic QRA concepts and terminology
- QRA calculation techniques
- Data requirements
- Data limitations
- Results and presentation formats
- Common pitfalls

Similar to the process defined for semi-quantitative risk analysis, quantitative risk analysis is a function of consequence and likelihood. Decisions are made regarding the resulting risk level and the need for additional risk mitigation based on a comparison of the results to a set of criteria. The QRA process is illustrated in Figure 7.

The differences between a QRA and the other simpler techniques are:

- Scope is based on the issues escalated through the use of simpler techniques.
- Data requirements to complete the analysis can be significantly greater.
- Supplemental and better quality data are critical inputs to a more detailed analysis.
- Frequency and consequence are both analyzed in more detail.

Before a QRA is pursued, it is important to ensure that the simpler risk analysis techniques, and their analysis options, have been conducted to the fullest extent. A QRA will provide benefits to the decision process if:

- The risk estimate requires additional refinement of consequence and/or likelihood.
- Simpler risk analysis approaches have left too many options.
- Decisions cannot be made or there are unanswered questions/issues.
- Valuable insight into the "risk drivers" is needed to identify the selection of appropriate risk mitigation options.
- A cost-benefit analysis of the relative difference between options needs to be evaluated.
- Corporate management or a regulatory agency demands a defensible analysis and a clear rationale for estimating risk.

Figure 7. Quantitative Risk Analysis Process

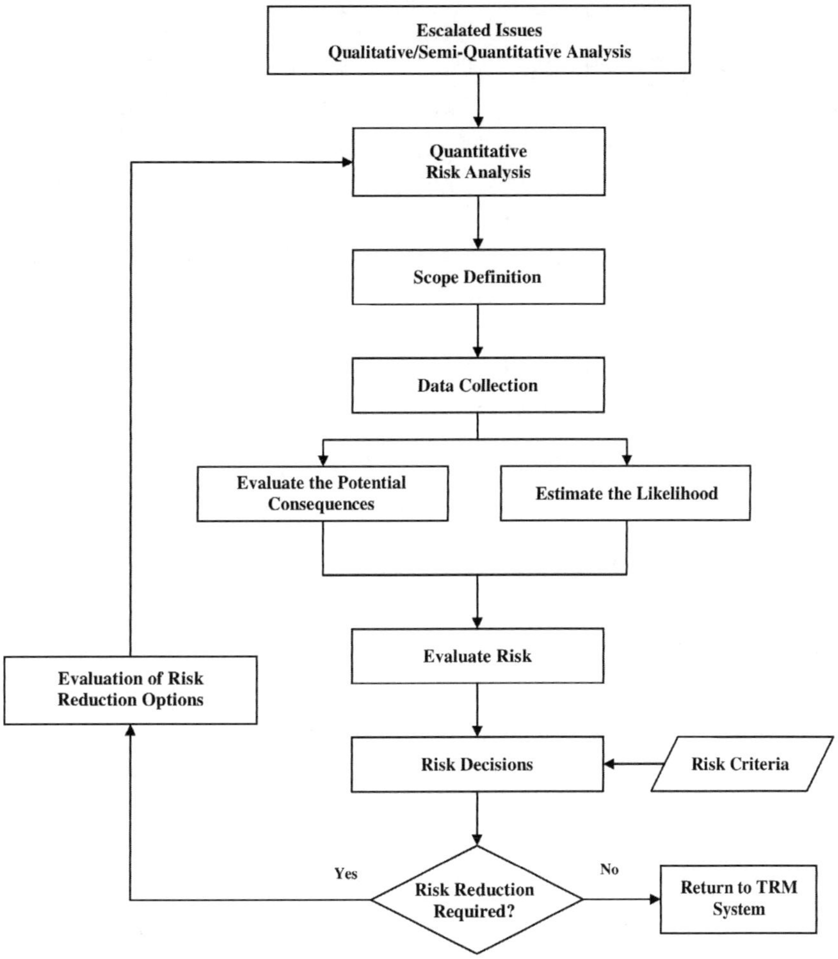

Transportation Security and Risk Analysis

Worldwide, there is an increased level of security awareness in all aspects of our daily activities. Specific to transportation security, we are faced with many challenges as a result of a global economy and supply chain. These challenges include not only terrorism, but also piracy, organized crime, illegal drug manufacturing and trafficking, and smuggling. It is important to consider all potential threats when evaluating chemical transportation security, since a chemical shipment can fall victim to a number of different types of attacks, not just terrorism. The good news is that many of the same security countermeasures can be effective in addressing these widely divergent security threats.

A number of terrorist-related attacks have highlighted the vulnerabilities of the transportation infrastructure. These types of attacks, combined with recent accidents, are changing the perception, tolerance, and means of managing the risk of hazardous material transportation operations. It is increasingly common that strong public reaction following a hazardous material transportation accident is not limited to safety, but also focuses on transportation security. Specific accidents in the United States include the following:

- *Graniteville, South Carolina (January 6. 2005)*—train derailment and chlorine release
- *Minot, North Dakota (January 18, 2002)*—train derailment and anhydrous ammonia release
- *San Antonio, Texas (June 28, 2004)*—train derailment and chlorine release

These incidents have highlighted the vulnerabilities of the transportation infrastructure and the potential consequences if a shipment was compromised en route. This concern has led to worldwide calls for legislation directed at increased security, rerouting of hazardous material shipments, as well as a push for the application of inherently safer technologies (commonly referred to as IST).

The reason for this increased focus on transportation security is the fear that hazardous materials could be attractive terrorist targets due to the potential consequences resulting from an intentional release, contamination of the material in transit, or theft of the material for use as a weapon against another target. Since the chemical industry is critical to the global economy, any successful attack, damage, or disruption could possibly cause negative local/regional/national impacts or rippling economic disruptions across numerous industry sectors. Therefore, the chemical industry and its value chain have been identified as a critical infrastructure requiring security measures in most countries to address today's threat environment.

This section of the guideline provides:

- Discussion of current security guidelines and regulations.
- Methodology for adapting and applying security vulnerability assessment (SVA) techniques designed for fixed chemical facilities to account for the differences specific to transportation.
- Practical transportation security elements.

Safety and Security Synergies and Tradeoffs

When developing or applying baseline security elements and proposing recommended countermeasures as part of a security review, one must understand compromises between safety, security, and convenience. The key difference between security and safety is that potential security incidents are the result of intentional acts. This is the basis for understanding the hazards, potential consequences, and the likelihood of an intentional malevolent event. While an increase in security may result in changes and inconveniences that everyone in the supply chain will need to understand, accept, and make appropriate adjustments, security decisions should not be blindly made without considering transportation safety implications. As such, it is recommended that security be integrated into the overall transportation risk management process. Examples of safety and security synergies and tradeoffs are presented in Table 1.

Table 1. Examples of Safety and Security Synergies and Tradeoffs

Issue	Safety	Security
Placards	Commodity information needed by emergency responders to react appropriately to an accident and minimize any impact.	Commodity information could be used by terrorists to target specific chemicals.
Rerouting	May result in more accidents if there are longer transits or the infrastructure along an alternate route may be less well maintained or contain undesirable features (uncontrolled intersections, no shoulders, etc.).	Eliminating a shipment near a specific location (most likely a highly populated or critical area) may inadvertently transfer the risk from one community to another.
Working with supply chain partners (implementing security countermeasures)	Technology can be used for both safety and security (e.g., GPS to indicate location en route, emergency response to accident, and monitoring time-sensitive chemicals/materials).	Technologies focused on security should not distract the main function of the carriers (e.g., the safe transport of chemicals from point A to B).
Risk Analysis Methods	• Rational and structured results lead to recommendations • Participation and engagement by individuals with different perspectives, roles, and backgrounds/skill sets for safety, security, and transportation • Similar methodology • Same decision metrics (guidelines)	

Security Prioritization Process

Similar to the safety risk analyses, a security screening process should be utilized to identify specific issues that require a full Transportation Security Vulnerability Analysis (TSVA), screening out those scenarios that justify only a general security review. As compared to the prioritization process focused on safety, security screening brings about new concerns; therefore, the chemicals and movements escalated through each distinct activity may be different. Some chemicals and movements that may be ranked as a lower priority, or even screened out for further detailed analysis based on safety issues, may rank *higher* in light of the security concerns. For example:

- A material transported in a specially designed container to minimize the likelihood of a release based on safety concerns (accidental release) may have ranked low as having a potential for a large release. However, since the goal of an intentional act is to create a maximum release, a larger security event may be possible than that anticipated from an accidental scenario.
- The transportation of a chemical in a limited number of small cylinders may have ranked low from a safety concern. However, this activity might be escalated for security reasons based on the potential for theft and the subsequent use of this material to produce a chemical weapon that could be mobilized and used against any number of potential targets.

- A food additive may have screened out of a safety analysis since there are no hazards associated with the product, but the potential for tampering with a loaded tank truck or rail car may raise the concern of potential contamination of the product. Therefore, a review of the measures in place to prevent and detect tampering at the origin, along the route, and at the destination may be warranted.
- A tank truck of flammables may have ranked low in a safety analysis as the consequence footprint of such materials can be much smaller than that of toxics. But, if an attack along this route has the ability to result in a consequence that could significantly damage a critical transportation route and result in an extended shutdown for cleanup and repair, then this segment of the route may warrant escalation in the security review. Also, the shipment could be hijacked, enabling the flammables to be diverted from the planned route.

Specifically for security, the following factors may be helpful in developing a company-specific security prioritization process:

- Chemical hazards
- Quantity transported per container
- Number of shipments or campaigns
- Mode of transit
- Proximity to people
- Proximity to sensitive environmental areas
- Interim (unsecured) storage
- Proximity to other critical assets or infrastructure
- Specific threat information

No one prioritization factor is intended to escalate an issue to a full TSVA; however, it is recommended that a multistep process be developed to identify the specific chemicals, movements, and segments that may require further evaluation. An example prioritization process is illustrated in Figure 8, consisting of the following steps:

- *Step 1—List Chemicals and Hazards*: The inherent physical properties (e.g., toxicity, flammability, reactivity) of a hazardous material in transit should be used to develop a list of chemicals that will be carried forward in the prioritization process. This list of chemicals may be different than that developed for accidental releases and may need to consider the following security issues:
  - Loss of significant containment of a hazardous material in transit from intentional damage of equipment or malicious release with the potential to impact people, property, or the environment
  - Theft, diversion, and misuse with the intent to cause harm along the route or an alternate location
  - Contamination with the intent to cause a reactive event while in transit, an event at the final destination, or other offsite impacts that could result from tampering with the material

Figure 8. Security Prioritization Process

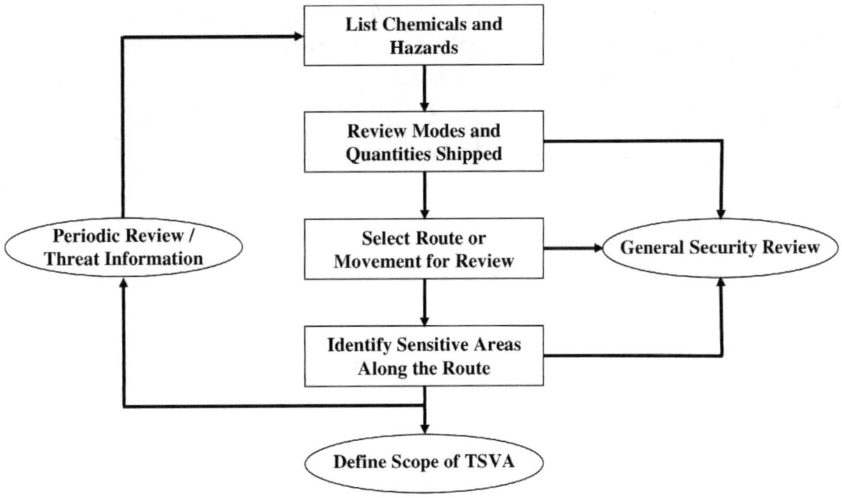

- *Step 2—Review Mode(s) and Quantities Shipped*: For all chemicals carried forward from Step 1, the mode and quantities should be reviewed to further refine a list of chemicals and modes of transportation and may be based on:
  - Minimum quantity or a threshold for toxics, flammables, and explosives that, if released in transit, could pose an impact to people, property, or the environment
  - Dedicated or contract carrier versus common carrier
  - Features of the mode related to the difficulty of attack or vulnerability during various stages of the operation or at specific segments along the route
- *Step 3—Select Route or Movement for Evaluation*: Depending upon the complexity of the supply chain network, it may first be desirable to divide the transportation activities into more manageable groups. These may include specific chemicals or chemical classes, modes of transportation, movements, regions, in-transit storage (yard or warehouse), etc.
- *Step 4—Identify Sensitive Areas Along the Route*: Except for chemicals that remain on the list because of potential theft and contamination issues, the chemicals and modes carried forward from Step 3 should be evaluated to focus on specific sensitive segments along the route, which may include:
  - Proximity to people
  - Proximity to sensitive environmental areas
  - Proximity to other critical assets or infrastructure
  - Attack along the route that could result in the degradation of critical infrastructure or other critical assets
- *Step 5—Security Evaluation*: Each of the chemical movements screened out through the prioritization process should be analyzed by a general security review before closing these items. All issues that make it through the prioritization process should be reviewed using a TSVA methodology.

- *Step 6—Periodic Review*: It is important to periodically review the security of the transportation network to ensure that the baseline program elements are still in place, recommended countermeasures have been implemented, and vulnerabilities reassessed. Additionally, this process should be reinitiated if there is specific intelligence or threat information pertaining to your supply chain, mode of transport, or specific chemical movement.

Following this prioritization process or a similar one, a transportation system can be evaluated to identify those issues that need to be escalated to a more detailed TSVA to better understand the potential consequences, vulnerabilities, and level of risk, or to compare potential security countermeasures.

RISK REDUCTION

Several analysis techniques are presented in this guideline for identifying and evaluating the safety and security risks associated with the transportation of hazardous materials over a route(s) and across different modes of transit. Any of these techniques can be used to determine where additional risk reduction options might be warranted and to identify and evaluate potential options. Additional risk reduction opportunities may include identified inconsistent operating practices, changes/advances in industry guidelines, or new regulatory requirements.

The value of a systematic approach to risk management is to help ensure that risk reduction resources are applied appropriately to address the highest priority risks. Although it may be tempting to bypass the prioritization and risk analysis steps, and to arbitrarily implement some of the commonly practiced risk reduction strategies, risk analysis provides valuable information and insights that can be incorporated into a comprehensive decision-making process. Whether the need to reduce risks has been warranted through analysis or through management or regulatory dictate, an action plan to identify, evaluate, compare, and select appropriate risk reduction options should be prepared.

Opportunities for reducing risk can be identified at any point in the transportation risk management process. A more detailed transportation risk analysis, including full quantification, is generally not required to identify and recommend operational improvements. As discussed, options for improvement may be recommended as part of a review of the primary management systems (e.g., regulatory, training, protocols and procedures, equipment integrity, monitoring, incident preparation), or during the prioritization and qualitative screening activities. As illustrated in the Figure 9, when risk reduction (or mitigation) options are identified, they need to be incorporated into the risk analysis process. This will help to ensure that the risk reduction options being considered are evaluated consistently, as compared with the current (or baseline level) operational level of risk, and that the desired level of risk reduction can be achieved.

As part of this section of the guideline specific risk reduction options are presented and discussed, including:

- Balancing safety and security
- Factors influencing risk reduction options
- Overall supply chain risk reduction options
- Pre-shipment risk reduction options
- Process for selecting risk reduction options

Figure 9. Risk Reduction Identification and Evaluation

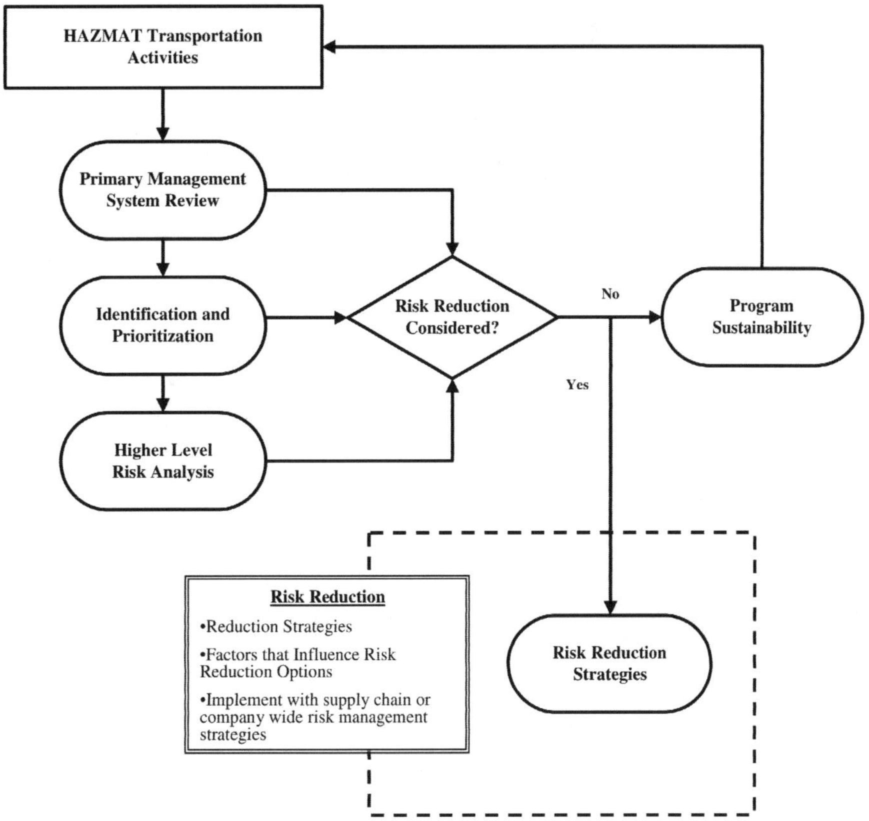

PROGRAM SUSTAINABILITY

As companies continue to expand globally, and supply chains continue to increase in complexity, the adaptability and sustainability of the transportation risk management (TRM) system will be a key competitive advantage. The risk management framework and safety and security analysis techniques presented in this guideline are flexible, adaptable, and appropriate for an uncertain environment of changing public acceptance of hazardous operations, evolving worldwide regulations, changing supply chain partners, and growing business opportunities and operational environments in developing countries. Even with the uncertainties, challenges, and opportunities that the future holds, the implementation of the TRM framework as a management system combined with the concepts, tools, and techniques presented in this book will position a company to be able to respond to changes and to manage its transportation risks in the future.

The TRM framework is not a one-time risk assessment activity, but rather builds upon the original safety management ideas process by the CCPS in the late 1980s, integrates lessons learned and operational changes, applies the management system principles of "plan, do, check, act," and organizes a system in a way that will be useful to all transportation operations—even with relatively lower hazards —throughout the life cycle of the operation. The TRM framework from Figure 1 fits the "plan, do, check, act" principles as shown in Figure 10.

Figure 10. Plan, Do, Check, Act—TRM Framework

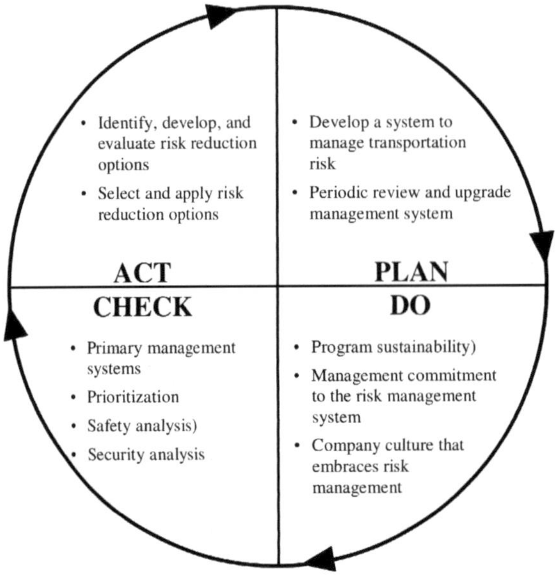

As detailed below:

- *Plan*—development of a system (e.g., TRM Framework) for managing transportation risks, and includes periodic review and update of the management system, as needed
- *Do*—sustainability piece of the management system, and includes management commitment and a company culture that implements risk management concepts in daily decision-making
- *Check*—primary management systems, identification and prioritization activities, and additional risk analysis of elevated safety and security issues
- *Act*—identification, development, evaluation, and selection of risk reduction options

This final step of the TRM Framework and this guideline reviews the following concepts that will help ensure the sustainability of a transportation risk management program:

- Ongoing commitment to risk management
- Continuous improvement
- Emerging safety and security trends
- Evolving transportation risk analysis practices

SUMMARY

In the decade since the CCPS *Guidelines for Chemical Transportation Risk Analysis* was published, it has received widespread recognition as an excellent resource on transportation risk analysis. It has been used extensively by major chemical companies and shipping organizations both domestically and internationally. In order to best serve the needs and interests of its clients and stakeholders, CCPS periodically reviews the use and effectiveness of its publications and issues updates or revisions. This new book, *Guidelines for Chemical Transportation Safety, Security and Risk Management* (2008), is not meant to replace the earlier 1995 edition, but rather to augment it. The new publication deals with transportation, security, and risk management on a broader basis and provides tools and methods to benefit a wider range of transportation professionals and associated stakeholders. In particular, it introduces more qualitative and practical techniques for screening, identifying, and managing higher-level risks issues that balance both safety and security. These two books should be used in combination to effectively manage and analyze transportation risk. For this reason, the original *Guidelines for Chemical Transportation Risk Analysis* is reproduced on an enclosed CD.

As highlighted, this paper has only provided an overview of the concepts presented in this new guideline. In addition to the further guidance, details, checklists, tools, and resources presented in the full guidance document, a comprehensive set of appendices, in CD format, is included. In addition to the original 1995 book, the appendices include tools shared by stakeholder companies, links to current and pending regulations and to additional resources current at the time of publication, and a complete example illustrating the transportation concepts that are developed throughout each chapter of this guidelines book.

A Model to Evaluate Pipeline Failure Frequencies based on Design and Operating Conditions

Vania De Stefani, BP International, Sunbury on Thames, UK.
Zoë Wattis, Advantica, Loughborough, UK.
Michael Acton, Advantica, Loughborough, UK.

ABSTRACT

Underground pipelines are widely used in the petrochemical industry to transport raw materials and products. Although the pipeline industry in general has a good safety record, historical evidence reveals the loss of mechanical integrity of such pipelines can be due to a variety of causes such as corrosion, external impact, defects and natural hazards. Such failures present a potential risk to people and the environment from the hazardous effects of the released material, which is often transported at high pressure.

In order to predict the level of risk to people and the environment, a means of predicting the likelihood of failures and material release is required.

Generic failure frequencies derived from local databases are largely used in pipeline risk assessments. However, the generic nature of these databases mean that they cannot be used to take pipeline-specific factors (including wall thickness, depth of cover, country of operation, or specific local issues) into account. As a result, risk assessments for pipelines may not accurately reflect operational experience for a specific pipeline or country of operation. In addition, it may not be possible to quantify the beneficial effect of any pipeline-specific risk reduction measures (such as physical protection, increased surveillance, etc.) that may be taken to reduce the likelihood of failure.

This paper describes a new model developed for use within BP to estimate the annual failure frequency of onshore cross-country pipelines. This model uses historical databases, such as CONCAWE (liquid hydrocarbons), EGIG (natural gas) and DOT, to develop failure frequencies as a function of several influences, including both design and operational parameters.

INTRODUCTION

In recent years, a great deal of effort has been dedicated to accident prevention. Statistics indicate that pipelines are inherently safer for the public than other modes of hydrocarbon transportation. However the pipeline industry is increasingly requested to demonstrate this safety performance in response to requests from regulating authorities or from the public with regard to new or existing pipelines.

Quantitative Risk Assessment (QRA) is an invaluable method for making informed risk-based process safety and environmental impact planning decisions, as well as being fundamental to any facility-siting decision-making. Information regarding failure frequency and failure mechanisms is of increasing importance in carrying out meaningful risk assessments and is required by both regulatory bodies and pipeline operating companies.

Government authorities and pipeline operating companies collect data on pipeline incident and their causes. These "incident databases" are, where available, essential for the authorities, operating companies and engineering companies as well as for the general public to both demonstrate and ensure that pipelines remain a safe and reliable means of transporting gas.

However, pipeline incident frequencies derived from different databases may lead to inaccurate conclusions as inherent differences in the databases are not always taken into account (Ref. 1). With the aim of ensuring consistency in risk assessments, BP initiated a study to review the most frequently used pipeline incident databases and to develop a model which estimates the annual failure frequency of onshore cross-country pipelines. In addition, the influences of the major operational and design parameters on the failure frequency have been examined. The model has been developed for use within BP, and the views informing its development are for BP's purposes. This paper describes and explains the model, in order to enable others to consider whether something similar may be of benefit in their own operations.

PUBLIC DATA SOURCES

There are a range of databases available for comparing pipeline failure frequencies. The information stored within these databases varies widely depending on the reporting criteria. For instance, the reporting criteria for some pipeline databases only capture failures to the pipeline itself, whereas others also record failures to fittings and other equipment. The databases listed below are compared in Table 1 of this report:

- European Gas Incident Group (EGIG), Ref. 2: Natural gas pipeline transmission incident data have been collected by EGIG since 1970 from pipeline systems for twelve operators in Europe. Incident data is summarized for the period 1970 to 2004 for all releases that result in a loss of gas from onshore, steel, transmission pipelines located outside the fences of installations. Within this database, a rupture is defined as a hole with a diameter greater than the diameter of the pipe. It should be noted that the work in this paper was carried out prior to the latest EGIG report (Ref 3) becoming available.
- The Oil Companies' European Organization for Environment and Health and Safety (CONCAWE), Ref. 4: CONCAWE produces an annual statistical summary of performance of oil industry cross-country pipelines in Western Europe since 1970. The latest report contains a summary of all incident data to pipelines and associated equipment from 1971 to 2005 with an oil spillage size of $1m^3$ or greater. Within the

database, 176 releases can be categorized by hole size allowing further breakdown of the failure frequencies. It should be noted that, within this database, a rupture is defined as a hole greater than 75mm long and with a width at least 10% of the pipeline diameter.

- US Department of Transportation (US DOT), Ref. 5: Onshore transmission incident data have been collected by US DOT/Office of Pipeline Safety under federal requirement for all gas and hazardous liquid pipeline companies in the US since 1970. The raw data is publicly available and this has been further analyzed to obtain the rupture frequency for gas transmission lines.
- UKOPA (United Kingdom Onshore Pipeline Association), Ref. 6: UKOPA is a forum for UK gas and oil pipeline operators. In the latest report, pipeline and product loss incident data from onshore Major Accident Hazard Pipelines (MAHPs) has been collected to determine exposure and associated frequencies. The vast majority of the pipelines included are high pressure natural gas transmission pipelines.
- Australian Pipeline Industry Association (APIA), Ref. 7: A study was carried out to compare available pipeline statistics for the State of Victoria in Australia with both national and international data. Within this report, reference is made to the Australian and New Zealand database which records exposure and incident data resulting in rupture, leak and no damage for gas transmission pipelines between 1971 and 1995.
- National Energy Board (NEB), Ref. 8: The NEB requires all ruptures on Canada's major pipeline systems to be reported in order to assess the safety of the network. This report defines rupture as a loss of containment event that immediately impairs the operation of the pipeline and includes all liquid spills greater than 1.5 m3. The report provides information ruptures in both liquid and gas lines as well as exposure data over a twenty year period up to 2003. Summary tables are provided which break down the number of ruptures by cause; however these are not further subdivided by fuel type.
- Alberta Energy and Utilities Board (EUB), Ref. 9: The EUB report summarizes the inventory of all energy-related pipelines in Alberta including gas and crude oil lines. As part of this summary, the length of pipeline is given by year and fuel type. The type and frequency of all incidents, including part-wall damage, and failures between 1990 and 2005 is discussed. The reporting criteria require failures to both steel and non-steel pipelines and fittings/other equipment to be recorded. Tables and charts are provided which break down incidents by fuel type, cause, spill size and pipe diameter.
- World Bank, Ref. 10: The characteristics of oil pipelines and the causes of ruptures occurring within Russia and states of the former Soviet Union (FSU), between 1986 and 1996, were assessed on behalf of the World Bank.

- OSSA-1 pipeline, Ref. 11: A paper was presented at the International Pipeline Conference (IPC) in 2004, which summarizes incident data by cause for the OSSA-1 pipeline over a twenty year period between 1983 and 2003. It is unclear from the paper whether this incident data is for ruptures only or all leak sizes. It should be noted, according to Ref. 11, that for this pipeline geohazards contribute 52.5% of incidents and sabotage/theft make up a further 10%.

Table 1 presents a summary of the results in each of the data sources described above.

Table 1 Pipeline Accident Databases

Database Name	Region	Fuel Type	From	To	Exposure (km.yr)	Failure Frequency (per km.yr)	Rupture Frequency (per km.yr)
EGIG	EUROPE (incl UK)	Gas	1970	2004	2,770,000	4.05×10^{-4}	6.06×10^{-5}
CONCAWE	EUROPE (incl UK)	Oil	1971	2005	858,705	5.08×10^{-4} (1)	5.12×10^{-5}
DOT (PHMSA)	US	Gas	1986	2006	9,300,060	8.18×10^{-5} (2)	3.03×10^{-5}
		Oil	1986	2006	5,056,591	5.69×10^{-4}	N/A
APIA	Australia	Gas	1971	1995	348,000	4.89×10^{-5}	2.87×10^{-5}
NEB	Canada	Gas	1984	2003	556,000	N/A	4.86×10^{-5}
		Oil	1984	2003	304,000	N/A	6.25×10^{-5}
World Bank	Russia & FSU	Oil	1990	1996	588,379	N/A	1.90×10^{-4}
OSSA-1	Bolivia	Oil	1983	2003	8,300	N/A	4.82×10^{-3}
UKOPA	UK	Gas	1952	2006	700,463	2.48×10^{-4}	9.99×10^{-6}
EUB	Alberta, Canada	Oil	1990	2005	231,250	1.69×10^{-3}	1.05×10^{-4}
		Gas	1990	2005	2,313,942	1.57×10^{-3}	9.75×10^{-5}
		Sour	1990	2005	171,233	1.98×10^{-3}	1.23×10^{-4}

1 CONCAWE records 436 incidents, 176 of which can be categorised by hole size.

2 DOT website records 764 incidents; however, the raw data used in the analysis for the rupture frequency only records 761.

3 Data taken from the EGIG 5th Report

Figure 1 below depicts the total failure frequency for Crude Oil releases and Figure 2 presents a similar set of results for Natural Gas releases. Figure 1 demonstrates that the historical overall failure frequency (per km.yr) for liquids is comparable for Europe and the US. When the rupture failure frequency data for hazardous liquids is plotted, failures to pipelines in South America are shown to dominate. The high rupture failure frequencies in South America are considered to be largely due to land instability.

Figure 2 shows that the overall failure frequency (per km.yr) for Natural Gas is low in Australia and the US when compared with the European and UK data. The gas pipelines in Alberta, Canada (EUB) have much higher failure frequencies both overall and for ruptures specifically. However, the EUB data includes failures to non-steel pipelines and fittings and other equipment in addition to those that fit the criteria used in the UK and European databases.

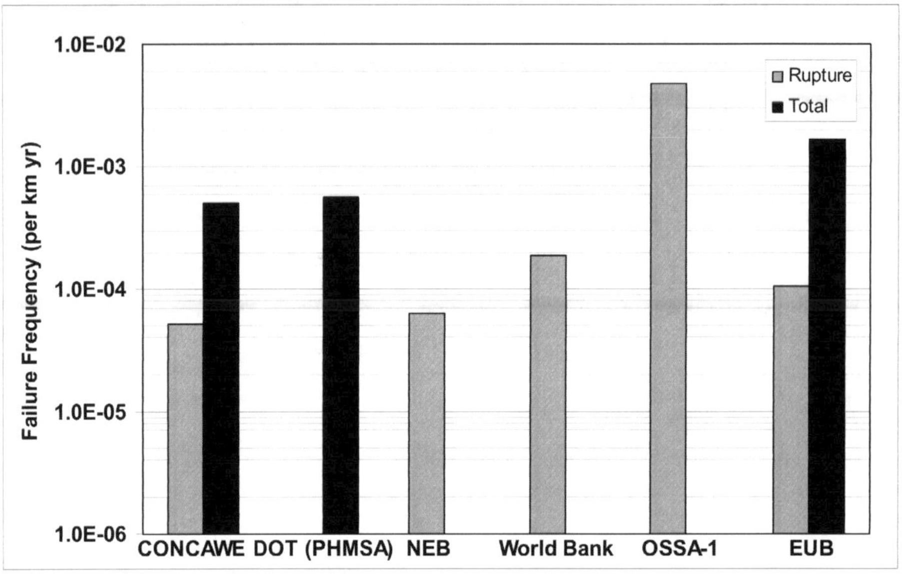

Figure 1: Failure Frequency for Hazardous Liquid Releases

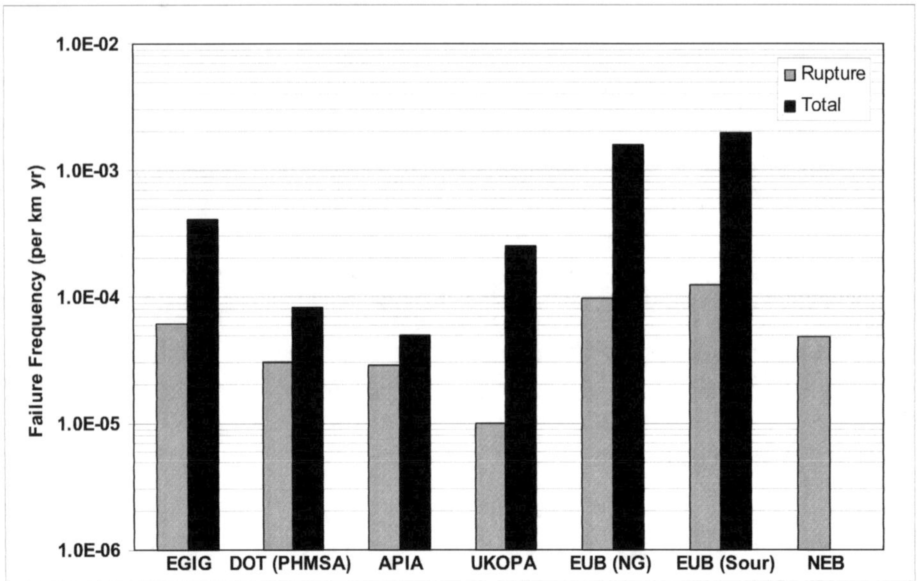

Figure 2: Failure Frequency for Natural Gas (including Sour Gas) Releases

Influencing Parameters

A detailed breakdown of failure mechanisms can be found in individual databases depending on the extent of accident analysis performed and the available data. They are most frequently classified in the following six cause categories:

- **External interference or third party activity:** External interference, mostly third party activity involving interference using machinery, is generally the dominant failure mechanism both for gas and oil pipelines.
- **Corrosion:** Corrosion, in all forms, is another major cause for incidents and is increasingly prevalent in ageing pipelines.
- **Construction defect and mechanical or material failure:** Construction and material defects (caused during processing or fabrication) are often connected with welds and equipment associated with the pipeline.
- **Natural hazards:** For the majority of pipelines, there is little or no risk of failure due to the occurrence of natural hazards. However, under certain circumstances, pipelines may need to be routed through difficult terrain or hydrological conditions where the risk from such hazards is significant.
- **Sabotage and Pilferage:** these include deliberate attacks aimed to pilfer fuel from pipelines for personal use or to sell on the black market. In addition, terrorism may lead to damage of the transportation and distribution systems connecting petroleum fields to the markets. Sabotage and pilferage are deliberate acts and highly specific to particular pipelines or locations.
- **Other or unknown causes**.

An initial list of parameters, considered to have a significant influence on the failure frequencies for these causes, was derived. These are summarized in Table 2 below.

Probabilistic models that assess the theoretical probability of failure due to third party interference are widely available and these are used in this report rather than extrapolating the failure frequency from the historical data. For corrosion and material defects, failure frequency modifiers have been derived from historical databases. For other pipeline-specific threats, such as natural hazards and sabotage, appropriate frequency contributions can be added to the overall frequency.

Table 2 Relevant Influencing Parameters to Failure Mechanisms

Parameter	Comment
External Interference (Third Party Damage)	
Diameter	Limit state (a measure of the resistance of the pipeline to mechanical damage) is a function of all the parameters but may be calculated by fracture mechanics models.
Pressure	
Wall thickness	
Material grade and toughness	
Geography (country)	Influence the hit rate (and possibly damage distribution) and may be derived from historical data where records exist.
Location (Rural or Suburban)	
Depth of Cover	

Parameter	Comment
Protective measures	Measures may be taken to reduce hit rate, including physical protection and surveillance, which can be quantified, and public awareness campaigns, which is more difficult to quantify.
Material and construction defects	
Quality of construction	Quality of construction has improved through modern construction and inspection methods and quality control of steel manufacturing. Year of Construction may be an indicator of quality, although construction practices in a given period may vary between countries.
Hydrostatic testing	Hydrostatic testing to levels exceeding the maximum operating pressure of a pipeline plays a significant part in controlling the possibility of mechanical failure. The greater the margin of hydrotest pressure over operating pressure, the greater the benefit. Historical databases don't record this information therefore this parameter is not included in the final model.
Corrosion	
Diameter, pressure, wall thickness and material properties	General corrosion typically leads to small leaks (usually detected) before a rupture occurs, and so the contribution to the frequency of ruptures and large leaks is expected to be small.
Corrosion Management	For pipelines transporting aggressive fluids, potential for corrosion can be managed by introducing corrosion inhibitors and use of "pigging" to clean out pipelines. Historical databases don't record the type of corrosion management in place therefore this parameter is not included in the final model.
Fluid type	Some fluids (e.g. crude oil) generate higher rates of internal corrosion than others (e.g. dry natural gas). It should be noted that the presence of water within the fluid will influence the potential for corrosion.
In-line inspection (ILI)	Inspection and repair plays a major part in controlling the corrosion threat.
Coating Quality	Pipeline coating technology has improved significantly in recent years, with older pipelines being more vulnerable to external corrosion. However, the quality of coating may vary between pipelines of a similar age, and so should be taken into account.
Cathodic Protection	The vast majority of pipelines are fitted with cathodic protection.
Stress Corrosion Cracking	Stress Corrosion Cracking (SCC) appears to be specific to certain pipelines, and not a general threat. Where a pipeline has a history of SCC, or where there is reason to believe that SCC is an issue, it may be appropriate to include a contribution from SCC.
Natural Hazards	
Terrain	Mountainous, hilly or flat. Includes mobile terrain where relevant.
Number and type of river crossings	Washout may influence failure frequency under specific local conditions where pipelines may be exposed. Not included in the final model.
Level of geotechnical input to design	Evidence that modern methods of designing pipelines to withstand ground movement reduce failure frequency.

Parameter	Comment
Sabotage and pilferage	
History of attacks on pipelines or similar infrastructure within the same country.	Very specific to particular countries and areas.
Fluid type	Liquid lines (e.g. gasoline) more vulnerable to pilferage than other fluids (e.g. high pressure gas pipelines).

The frequency modification model typically used in BP is discussed in the following paragraphs for each of the failure modes. It should be noted that many of the parameters used in this model are dependent on hole size and so separate modified failure frequencies can be generated for punctures and ruptures. In this work, punctures are defined as failures from 50mm holes.

EXTERNAL INTERFERENCE MODIFICATION FACTORS

Limit State Model

External interference is generally the dominant threat for most pipelines and has been studied extensively. Approaches adopted in pipeline risk assessments range from simple statistical analysis of historical data (taking no account of the specific conditions for a particular pipeline) to sophisticated limit state models, requiring detailed information on pipeline properties. A highly sophisticated approach is not necessarily appropriate; however, predictions of failure frequencies must reflect pipeline-specific parameters as far as practical in a simple tool. Therefore, within this paper, a well-established limit state model, FFREQ (Ref. 12), was used for external interference in combination with modification factors derived from historical data to take account of other parameters. FFREQ is the failure frequency prediction tool for external interference recommended by UKOPA, (Refs. 13, 14).

By adopting FFREQ, many of the parameters identified as important in estimating pipeline failure frequencies due to external interference can be taken into account relatively simply, without the need to derive modification factors. Important parameters handled by the model include:

1. Pipeline-specific parameters, such as diameter, pressure, wall thickness and material properties.
2. Location, i.e. land use type – Rural or Suburban as defined in IGEM/TD/1, Ref. 15. The definition of Rural given IGEM/TD/1 (i.e. < 2.5 persons per hectare) is broadly equivalent to Class Locations 1 and 2 in the ASME pipeline code, ASME B31.8.
3. Depth of Cover. The FFREQ model takes into account the mitigating effect of depth of cover on the likelihood of occurrence of external interference damage. The factor that is applied, within the FFREQ model is based on measured

damage observed on UK gas transmission pipelines, as a function of depth. UK gas transmission pipelines are laid to a minimum depth of 1.1m in accordance with IGEM/TD/1.

The remaining parameters that were considered to be important were:

4. Geography (i.e. country)
5. Protective measures (where applied)

The hit rate will vary between countries. In principle, damage distributions used to determine the probability of a failure given an impact could also vary if the machinery causing damage to pipelines differs between countries. In this study, it has been assumed that the main variation between countries will be the hit rate, and that the type of machinery capable of causing damage at the depth to which cross-country pipelines are buried will be similar in different countries.

The FFREQ model has been developed using hit rates appropriate to the UK. The hit rate for external interference is a factor that is believed to vary significantly between different parts of the world for a variety of reasons including regulatory regime, rate of new construction in that country and cultural aspects.

Some databases record incidents that result in damage as well as those that lead to failure. These hit rates are not widely available world wide and so other factors should be used to understand a particular country and its attitude to risk and compliance with the regulatory environment. In this paper, the road traffic accident rate was chosen as a measure of the attitude to risk in a particular country, and hence the likelihood that people will comply with safety rules such as contacting a pipeline company prior to any excavation work. Another important factor is that countries undergoing rapid growth are likely to have many more construction activities taking place which could damage a pipeline.

For those countries where sufficient data is available, BP would consider that the hit rate is only affected by country and location. Therefore the proportion of pipeline operating in rural or suburban areas was also used to generate hit rate modification factors, taking the UK hit rate for normalisation purposes. It was concluded that the location factor included in the FFREQ model should be appropriate for the countries where data is available, and because most other countries follow similar pipeline design and routeing requirements to the UK and US, it would be reasonable to extrapolate this conclusion to all other countries as well.

The hit rate modification factors were then compared with modification factors derived using the fatal accident rate (FAR – Ref. 16) and the Gross Domestic Product (GDP – Ref. 17). Comparison of these results suggested good agreement and a simple system proposed, whereby countries are placed on a grid of FAR (per 100,000 inhabitants) versus growth in GDP. Each country is assigned a hit rate modification factor based on these two variables, in accordance with the chart shown in Figure 3.

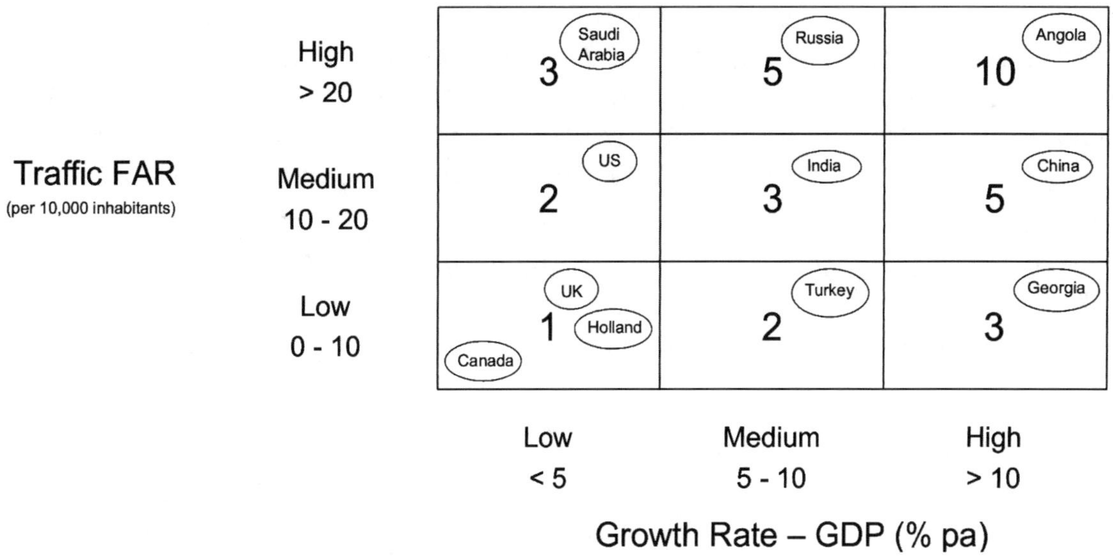

Figure 3: Country Chart for Hit Rate Modification Factors

Protective Measures

A number of experiments were performed by Advantica (then British Gas Research and Technology) in the 1990's that investigated the effectiveness of different types of barriers in preventing pipeline damage (Ref. 18). Contractors were asked to excavate in the region of a buried pipeline that was protected by a concrete or steel barrier to determine if the pipeline would be damaged. Warning tape was also used either alone or in addition to the physical protection in some of the experiments. Based on the results of these experiments, "Damage Reduction Factors" were derived depending on the combination of warning tapes and physical protection. These experiments were undertaken using a specific design of concrete slab in use by British Gas at that time.

Subsequent research has been carried out to study the effectiveness of such risk reduction measures, including consideration of how the design of the slab (width, thickness, type of reinforcement, etc) influences the risk reduction achieved. Work by UKOPA considered this additional information and other references in open literature, and recommended the values given below (Refs. 13, 14).

These factors can be directly applied to determine the "protective measures modification factor" for use in estimating failure frequencies provided that the design and construction of the concrete slabs are comparable, in particular the 3m width.

Table 3: Failure Frequency Reduction Factors for Pipeline Protection

Type of Protection	Modification Factor
No Protection	1
Concrete slab	0.16
Concrete slab plus visible warning	0.05

In addition to providing physical barriers many companies use surveillance techniques to reduce the potential for external interference on the pipeline network. This surveillance may take the form of a walking, vehicle or helicopter survey. The aim of these surveys is to detect and prevent any unusual activity on or near the pipeline. A survey is considered to be effective if it can detect, intervene and prevent the activity from continuing.

UK pipelines are usually surveyed once every 14 days in accordance with IGEM/TD/1, and therefore the assumptions in FFREQ are based on this surveillance regime. However, surveillance intervals may vary between country and region and so a simple model was developed to quantify the effect of surveillance. The observed frequency of excavation activities near pipelines and the duration of those activities were used to estimate the probability that surveillance will coincide with those activities in order that further work may be prevented. It should be noted that the effectiveness of this surveillance regime may vary between countries and/or safety cultures. Therefore, if there is reason to believe that any concerns raised by a survey are likely to be ignored, then in the model under discussion BP would not expect to give any credit for increased surveillance activity.

Where both physical protection has been applied *and* the surveillance regime differs from the 14 day surveillance interval assumed in the model, the two factors could be multiplied together to give an overall modification factor for protective measures (P_{PM}).

MECHANICAL FAILURE MODIFICATION FACTORS

From CONCAWE (Ref. 4), the 5-year average (2001-2005) failure frequency due to all causes is 2.58×10^{-4} per km.year. Within this database, 22% of failures in that five year period were due to mechanical defects giving a failure frequency of 5.73×10^{-5} per km.year. The UKOPA report (Ref. 6) indicates a failure frequency of 7.1×10^{-5} per km.year for mechanical faults between 1962 and 2004. These values are also comparable to frequencies obtained from EGIG for material and construction defects. The failure frequencies from EGIG can be broken down by year of construction and so these were used within the BP model as the baseline failure frequencies and are summarised below.

It should be noted that the frequencies given in EGIG assumed that the pipeline under consideration had been built to meet the standards appropriate at the time of construction. Typical construction standards include hydrotesting, girth weld Non Destructive Testing (NDT) and low number of defects detected. If there is evidence to suggest that the pipelines have not been built to meet the appropriate standards, then the baseline failure frequency may be modified further.

Table 4: Baseline Mechanical Failure Frequency (per km.year), (F_{yr}) Ref. 2

Year of Construction	Total Mechanical Failure Frequency (per km.year)
<1954	2.00×10^{-4}
1954-1963	1.70×10^{-4}
1964-1973	7.00×10^{-5}
1974-1983	4.00×10^{-5}
>1984	7.00×10^{-6}

A series of modification factors were also developed for use in the BP model to adjust the failure frequencies to take into account the impact of pipeline age, using the five year average and other data within CONCAWE.

For the modified failure frequency due to mechanical failure, a split between punctures and ruptures is generally required for risk assessments. The following contributions were obtained from the CONCAWE and EGIG databases.

Table 5: Contribution of Hole Size to Mechanical Failure

Database	Percentage Contribution			Total
	Pinhole	Puncture	Rupture	
CONCAWE (Liquid)	45%	38%	17%	**100%**
EGIG (Gas)	71%	21%	9%	**100%**
Average	**58%**	**29%**	**13%**	**100%**

There may be physical reasons for the differences observed in the fatigue behaviour of material and construction defects of pipelines transporting gas and liquids, because liquid pipelines typically see far greater variations in pressure. Because gas is compressible, the pressure swings tend to be smaller and less frequent. The CONCAWE database appears suitable to use for liquid pipelines and the EGIG database is used for gas pipelines.

This is likely to be conservative; particularly for pipelines constructed after 1984, where there is no recent history of ruptures due to mechanical failure.

CORROSION MODIFICATION FACTORS

Corrosion is a significant failure cause for transmission pipelines. In general, corrosion typically results in a small leak which would grow over time before eventually resulting in a rupture. However, the introduction of corrosion management schemes and the use of modern inspection and repair techniques such as "pigging" allow the majority of corrosion to be detected and repaired whilst still small. Thus, the contribution of corrosion to rupture and puncture frequency is generally small.

The key parameters considered to influence the failure frequency associated with corrosion are listed in Table 2:

Baseline corrosion failure frequencies for the BP model were derived for natural gas and liquid pipelines using EGIG and CONCAWE respectively. For example, EGIG (Ref 2) gives a corrosion failure frequency of 3.53 x 10^{-5} per km.year for natural gas lines. Similar results can be derived for liquid lines from CONCAWE. A modification factor can then be applied to adjust the baseline failure frequencies to take into account the impact of wall thickness.

For pipelines transporting aggressive fluids, the potential for corrosion can be managed by the introduction of corrosion inhibitors and the use of "pigging" to clean out the pipelines. However, there is insufficient evidence within the public historical data to establish a link between the use of corrosion inhibitors and failure frequency. However, a modification factor was obtained to take into account in-line inspection from CONCAWE (P_{pig}). This results in in-line inspection factors for pipelines with or without an in-line inspection regime in place.

CONCAWE and EGIG give the proportion of internal and external corrosion events for liquid lines and natural gas lines respectively. These are presented in Table 6 below; with corrosion events in CONCAWE split between those in cold pipelines (oil transported at ambient temperature) and hot pipelines (such as hot crude oil, lubricating oils and heavy fuel oils transported at elevated temperatures).

Table 6: Proportion of External and Internal Corrosion Events (P_{IC} and P_{EC})

Database	Fluid	External	Internal	SCC	Other	Total
CONCAWE (cold)	Liquid	65%	29%	6%	0%	100%
CONCAWE (hot)	Liquid	98%	2%	0%	0%	100%
EGIG	Natural Gas	79%	16%	1%	4%	100%

The modification factors discussed above for fuel type, wall thickness and pigging apply to both internal and external corrosion mechanisms. However, the potential for external corrosion may also be affected by the coating in place around the pipeline. Therefore, a modification factor (P_{CT}) was derived from the UKOPA historical data (Ref 6) to be applied to account for coating type (coal tar, bitumen, polyethylene, epoxy resin and "other").

BP would consider that all pipelines are fitted with Cathodic Protection. However, in a small number of cases this may not be an appropriate assumption. Under these circumstances a modification factor has been derived from the PRCI report L51830 (Ref 19) and is applied to the external corrosion failure frequency for pipelines without Cathodic Protection (P_{CP}).

For the modified failure frequency due to corrosion, a split between punctures and ruptures is generally required for risk assessment. The following contributions were obtained from the CONCAWE and EGIG databases and are used in the model for liquid and gas pipelines respectively.

Table 7: Contribution of Hole Size to Corrosion (P_{fl})

Database	Percentage Contribution			Total
	Pinhole	Puncture	Rupture	
CONCAWE (Liquid)	59%	34%	7%	**100%**
EGIG (Gas)	97%	2%	2%	**100%**

Stress Corrosion Cracking (SCC) appears to be a very specific problem for certain pipelines rather than a general threat. Therefore, BP would only include a contribution from SCC if there is reason to believe it is an issue. In particular, SCC initiates if three conditions are present: potent environment at pipe surface, susceptible pipe material and tensile stress. Thus, when required, the SCC failure frequency should be added to the contributions from external and internal corrosion using local information.

For example, a report was published by NEB (Ref. 20) on near-neutral pH SCC in Canadian oil and gas pipelines. This report, records 22 SCC related failures over the 20 year period between 1977 and 1996. Using the exposure data, the overall SCC failure frequencies can be derived for this location.

NATURAL HAZARDS MODIFICATION FACTORS

Geohazards (geological hazards) are geological phenomena or conditions, either natural or man-made that are dangerous (or potentially dangerous) to the environment or its inhabitants. For the majority of pipelines, there is little or no risk of failure due to the occurrence of one of these natural hazards. However, under certain circumstances, pipelines may be required to be constructed through difficult terrain or hydrological conditions where the risk from such hazards is significant.

There is limited historical data available to allow factors to be defined against a baseline frequency. Typical landslide failure rates are demonstrated in Figure 4.

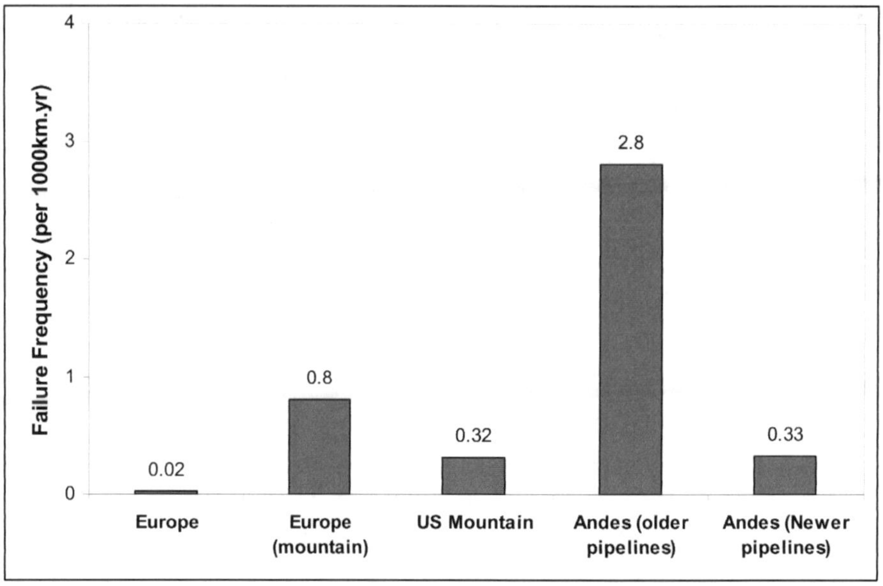

Figure 4: Landslide failure frequency, Ref 21

In European and US databases, landslides contribute about 3% of spills with CONCAWE reporting a failure frequency of 2.0 x 10^{-5} per km.yr. This low frequency of events is due to the benign terrain for the majority (95 to 99%) of the areas reported in the database. In mountainous terrain, landslide rupture for traditional pipelines overwhelms all other risks by several orders of magnitude. This is illustrated in Figure 4 with a rupture failure frequency of 8 x 10^{-4} per km.year in Alpine regions compared to a rupture failure frequency of 2.8 x 10^{-3} per km.year in the Andes.

In the BP model failure frequencies for natural hazards were developed by applying general guidance from UKOPA (Refs. 13, 14) where there is no known history of ground movement and the information presented above for terrain with a known history of landslide.

SABOTAGE AND PILFERAGE MODIFICATION FACTORS

The threat from sabotage and pilferage is generally negligible for buried cross-country pipelines, and it would not be appropriate to include a contribution by default to the pipeline failure frequencies used in risk assessment. However, there may be countries where pipelines are vulnerable to deliberate attacks, and then the threat from sabotage and pilferage can dominate. The factors below are based on case-specific examples of pipelines which have been subjected to sabotage and/or pilferage.

Where a pipeline is known to be at risk of sabotage or pilferage, local information is helpful to provide a best estimate of the contribution to the failure frequency, either directly from experience for that particular pipeline or similar pipelines in the area. Where local information is not available, but there are reasons to believe that a significant threat of sabotage or pilferage exists, then results derived from regions with similar threats may be used to give guidance on values that could be used.

In the BP model the threat levels defined in Table 8 were used to classify the pipeline. For each of the proposed categories, BP then developed appropriate failure frequencies using local information.

The failure frequency for sabotage and pilferage was then added to the failure frequencies for the other causes estimated as described in the previous sections.

Table 8: Threat Level for Sabotage and Pilferage (F_{SP})

Threat Level
High Specific, hard intelligence of imminent attack and ongoing, sustained attacks against industry, assets or personnel.
Moderate History of terrorist attacks in the region **AND** frequent illegal tapping.
Low History of terrorist attacks in the region **OR** risk of pilferage or illegal tapping.

FAILURE FREQUENCY MODEL

Combining the analyses presented in the previous sections gives the proposed frequency modification model,

$$F = F_{EI} + F_{MF} + F_{CO} + F_{NH} + F_{SP}$$

The modified failure frequencies for external interference, material and construction defects and corrosion defects are given by,

$$F_{EI} = F_{ffreq} \times P_{hit} \times P_{PM}$$

$$F_{MF} = F_{yr} \times P_{age}$$

$$F_{CO} = F_{BC} \times P_{fl} \times P_{wt} \times P_{pig} \times (P_{IC} + P_{EC} \times P_{CP} \times P_{CT}) + F_{SCC}$$

Each of these parameters is defined in Table 9 below. It should be noted that many of the parameters used in this model are dependent on hole size and so separate modified failure frequencies can be generated for punctures and ruptures.

Table 9: Parameters for Estimation of Pipeline Failure Frequencies

Parameter	Description
F	Calculated failure frequency – from all causes
F_{EI}	Calculated External Interference failure frequency
F_{MF}	Modified Mechanical Failure frequency
F_{CO}	Modified Corrosion failure frequency
F_{NH}	Modified Natural Hazards failure frequency
F_{SP}	Modified Sabotage and Pilferage failure frequency
F_{ffreq}	External Interference failure frequency generated using FFREQ – incorporates effects of pipeline-specific parameters, depth of cover, and location
P_{hit}	Hit rate modification factor – e.g. country
P_{PM}	Protective measures modification factor – e.g. physical protection, surveillance
F_{yr}	Mechanical failure frequency – dependent on year of construction
P_{age}	Pipeline age modification factor
F_{BC}	Corrosion failure frequency
P_{fl}	Fluid type modification factor
P_{wt}	Wall thickness modification factor
P_{pig}	In-line inspection modification
P_{IC}	Proportion of internal corrosion events – dependent on fluid type
P_{EC}	Proportion of external corrosion events – dependent on fluid type
P_{CT}	Coating type modification factor – applicable only to external corrosion
P_{CP}	Cathodic Protection modification factor
F_{SCC}	Stress Corrosion Cracking frequency – added if pipeline is vulnerable to such events

CONCLUSIONS

A study was carried out to derive a range of modification factors that could be applied to generic failure frequencies to take pipeline-specific factors (including wall thickness, depth of cover, country of operation, or specific local issues) into account. The study included a review of existing knowledge in order to derive generic baseline failure frequencies from publicly available databases, a proposed methodology for the application of risk modification factors, and the identification of historical data sources or predictive models that could be used to derive risk modification factors depending on the robustness of available information.

Baseline failure frequencies and modification parameters have been derived for each of the major failure causes identified, in particular: External Interference; Mechanical Failure; Corrosion; Natural Hazards and Sabotage and Pilferage. The model has been coded into a software tool (PipeFFRAC) for use within BP.

Future work includes the extension of the model to above ground pipelines and offshore pipelines, which are characterised by different failure mechanisms.

REFERENCES

1. "Guideline: Using or Creating Incident Databases for Natural Gas Transmission Pipelines," Report of Study Group 3.4, 23rd World Gas Conference, Amsterdam, The Netherlands, June 2006
2. 6th EGIG Report 1970-2004. European Gas Pipeline Incident Data Group (EGIG). EGIG 05.R.0002. December 2005
3. 7th EGIG Report 1970-2007. European Gas Pipeline Incident Data Group (EGIG). EGIG 08.TV.0502. December 2008
4. Performance of European Cross-Country Oil Pipelines; Statistical Summary of Reported Spillages in 2005 and since 1971. CONCAWE. Report No. 4/07. May 2007
5. http://primis.phmsa.dot.gov/comm/reports/safety/PSI.html
6. UKOPA Pipeline Product Loss Incidents - 5th Report (1962 – 2006), Advantica report 6957, August 2007, available from www.ukopa.co.uk
7. Overseas and Australian Statistics for Gas Transmission and Distribution Incidents. Risk & Reliability Associates Pty Ltd. Report No. 293-3. January 2004
8. Jeglic, F. (2004), "Analysis of Ruptures and Trends on Major Canadian Pipeline Systems", National Energy Board (NEB)
9. Pipeline Performance in Alberta, 1990-2005. Alberta Energy and Utilities Board (EUB). April 2007
10. Russia Pipeline Oil Spill Study. Joint UNDP/World Bank Energy Sector Management Assistance Programme (ESMAP). 1998
11. Esford, F., Porter, M., Savigny, K.W. and Muhlbauer, K.W. (2004), "A Risk Assessment Model for Pipelines Exposed to Geohazards" 5th International Pipeline Conference, ASME (IPC04-0327)
12. Corder, I. and Fearnehough, G.D. (1987), "Prediction of Pipeline Failure Frequencies", Second International Conference on Pipes, Pipelines and Pipeline Systems, Utrecht

13. IGEM/TD/2 "Application of pipeline risk assessment to proposed developments in the vicinity of high pressure Natural Gas pipelines", Institution of Gas Engineers and Managers, 2008

14. BS PD 8010-3:2009, Code of practice for pipelines. Steel pipelines on land. Guide to the application of pipeline risk assessment to proposed developments in the vicinity of major accident hazard pipelines containing flammables. Supplement to PD 8010-1:2004

15. IGEM/TD/1 Edition 5 "Steel pipelines and associated installations for high pressure gas transmission", Institution of Gas Engineers and Managers; 2008

16. http://www.who.int/healthinfo/statistics/bodgbddeathdalyestimates.xls

17. https://www.cia.gov/library/publications/the-world-factbook/index.html

18. Corder, I., "The Application of Risk Techniques to the Design and Operation of Pipelines", C502/016, I.Mech.E. 1995

19. Kiefner, J.F., Mesloh, R.E. and Kiefner, B.A. (2001) "Analysis of DOT Reportable Incidents for Gas Transmission and Gathering System Pipelines, 1985 Through 1997". PRCI. L51830

20. Stress Corrosion Cracking on Canadian Oil and Gas Pipelines. National Energy Board. MH-2-95. November 1996

21. Sweeney, M., Gasca, A.H., Garcia Lopez, M., and Palmer, A.C. (2004), "Pipelines and landslides in rugged terrain: A database, historic risks and pipeline vulnerability," Proc. Terrain and Geohazard Challenges Facing Onshore Pipelines, Thomas Telford, London

Analysis of the explosion linked to the formation of an explosive atmosphere in a shop carrying fuel products (gasoline, kerosene) through the accident involving the tanker M/T Chassiron that occurred on June 13, 2003

Sébastien EVANNO[a], Christian PELLIGAND[b], Germain VERLET[c], Daniel DREVET[c]
[a] INERIS
Parc Technologique Alata, 60550 Verneuil-en-Halatte, France,
sebastien.evanno@ineris.fr
[b] Ministry of Economy, Industry and Job, Office of the Security of Industrial Plants,
139, rue de Bercy 75572 Paris Cedex 12, christian.pelligand@industrie.gouv.fr
[c] Ministry of Ecology, Energy, lasting Development and town and country Planning,
Office of Investigation into Events at Sea, Tour Pascal B, 92055 Paris La Défense,
France, bea-mer@developpement-durable.gouv.fr

ABSTRACT

The rise in the flows of dangerous cargo shipped in bulk by sea has in some cases led to a rise in the explosion hazards onboard the ships that carry products that may release flammable vapors. This hazard is especially accentuated during loading, unloading and cleaning operations which may encourage the formation of an explosive atmosphere (ATEX) in the storage tanks.

In parallel with the safety of such vessels, aspects relating to crew protection and safety must be taken into account. The new Maritime Labor Convention adopted in 2006 opens the way to social standardization for the seafaring world.

In 2003, while conducting a joint technical and organizational investigation into the explosion onboard the petroleum product (gasoline and kerosene) tanker M/T Chassiron, INERIS and the Office of Investigation into Events at Sea identified the advantage of ensuring that this kind of oil tanker displacing less than 20,000 tons be inerted as part of the prevention of the formation of an ATEX environment in line with European Directive ATEX 1999/92/CE[1] relating to the minimum stipulations aimed at improving the safety and welfare protection of workers who may be exposed to explosive atmosphere hazards.

Keywords: ATEX, Explosion, inerting, flammable liquid.

1. Introduction

The Office of Investigation into Events at Sea (BEAmer), a part of France's Ministry of Ecology and the INERIS Accidental Hazards Department performed an analysis of the explosion in 2003 onboard M/T Chassiron, a tanker carrying petroleum products.

M/T Chassiron is a French tanker on the TAAF register (French Austral and Antarctic Territories) belonging to a company called Petromarine. This tanker of 10,000 metric tons heavy, mainly plies the coastal trade carrying light refined petroleum products.

The accident affecting the M/T Chassiron on June 13, 2003 highlighted the need to ensure improved prevention against the explosion hazard affecting petroleum product carriers displacing less than 20,000 tons during loading, unloading and washing operations.

The main recommendation made by the Office of Investigation into Event at Sea[2] report relates to the inerting of compartments during loading, unloading and washing operations onboard vessels with a capacity of less than 20,000 tons. Until now, this safety measure was not required by international maritime regulations for vessels of limited capacity.

As part of Office of Investigation into Event at Sea's international action, a presentation of recommendations (especially inerting vessels of less than 20,000 tons) was previewed at the IMO (International Maritime Organization) so as to include this safety measure in the context of international maritime regulations.

The feedback linked to the explosion of tanks that were not degassed during washing operations is well known. These accidents come from an INERIS analysis (collected from the "Hazardous Cargo Bulletin"). These accidents took place in port (or close by) during ship loading/unloading operations. The results of this research is presented in Table 1.

Table 1: Summary of feedback on explosions of vessels containing flammable liquids

Date	Location	Ship's name	Product	Brief description of the event
20 Jan 98	Ulsan South Korea	Dae Won	Vapors	During cleaning operations, the chemical carrier (1,200 tons, built in 1986) was shaken by an explosion followed by fire in tank #1. Three crew members were killed, one disappeared. No pollution reported.
30 July 96	Sungei Gerong Indonesia	Lido	Petroleum residue	Explosion and fire while cleaning the ship. Six crew members were killed. Tanks 3 and 4 were damaged.
17 June 96	Guayaquil Ecuador	Anne Sif	Ethanol	Explosion in tank #3 onboard this mixed oil/chemicals carrier during water cleaning. Two persons injured.
15 Aug 94	Baton Rouge Louisiana USA	Tank barge	Toluene	Explosion and fire in a tank onboard a barge during cleaning. Ignition was probably caused by static electricity along a flexible that was part of the vacuum system and that was not grounded. Four persons injured.
18 Sept 92	Piraeus Greece	Chemstar	Flammable vapors	Explosion onboard the chemical carrier during cleaning operations. The ship had previously carried hexane and white spirit. Presumed source of ignition: a portable electric torch. Two persons died.

Date	Location	Ship's name	Product	Brief description of the event
07 Mar 87	Gulf of Mexico	Golden Queen	Benzene vapors	The explosion in empty benzene tanks during cleaning caused two deaths among the crew.
20 Mar 84	Syracuse Italy	Andreas Z	Petroleum vapors	Six persons injured and one missing following the explosion caused by degassing the tanker's #5 tank. The ship which was waiting to load up with gasoline probably had soiled tanks.

More recently, feedback from events that INERIS and Office of Investigation into Event at Sea were involved in during these last few years included:

An explosion on board M/T Chassiron on June 13, 2003: air/gasoline ATEX explosion during a tank washing operation,

Difficulty encountered on board the Adamandas in September 2003: An air / H_2 ATEX situation was generated linked to iron oxidation obtained by direct reduction (DRI) through water and the correlative rise in product temperature in the holds (INERIS CASU was called in for this accident in September 2003),

Explosion onboard the Bow Mariner on February 28, 2004: Fire started which washing a tank triggering an Air / Methyl Tertio Butyl Ether (MTBE) ATEX. Three crew died and 18 others went missing following the explosion and wrecking of a chemical carrier carrying 11,000 tons of ethanol, 80 km off the shores of Virginia, as announced by the Coast Guard,

Explosion onboard the Methanol, a chemical carrier on April 30, 2005: A tank explosion during operations to unload methanol,

Explosion onboard a bulk carrier called the Thalassa, on October 4, 2005: Fire broke out during unloading triggering an air / H_2 ATEX linked to zinc oxidizing with water and the rise in product temperature in the holds.

2. Description and Methods

2.1 Description of the M/T Chassiron

The ship comprised 12 tanks for petroleum and chemical products. These 12 tanks are laid out in six sets of two tanks, port and starboard. Sets 5 and 6 were primarily affected by the explosion.

The characteristics of the ship's cargo prior to unloading were as follows:

FOD: total capacity of 795.642 tons (23.6°C ; 938.378 m^3),

Diesel: total capacity of 6,429.281 tons (29.5°C ; 7,729.932 m^3),

Lead-free premium 98 octane: total capacity of 907.135 tons (25.2°C ; 1,228.234 m^3).

Figure 1 presents the layout of tanks 5 and 6:

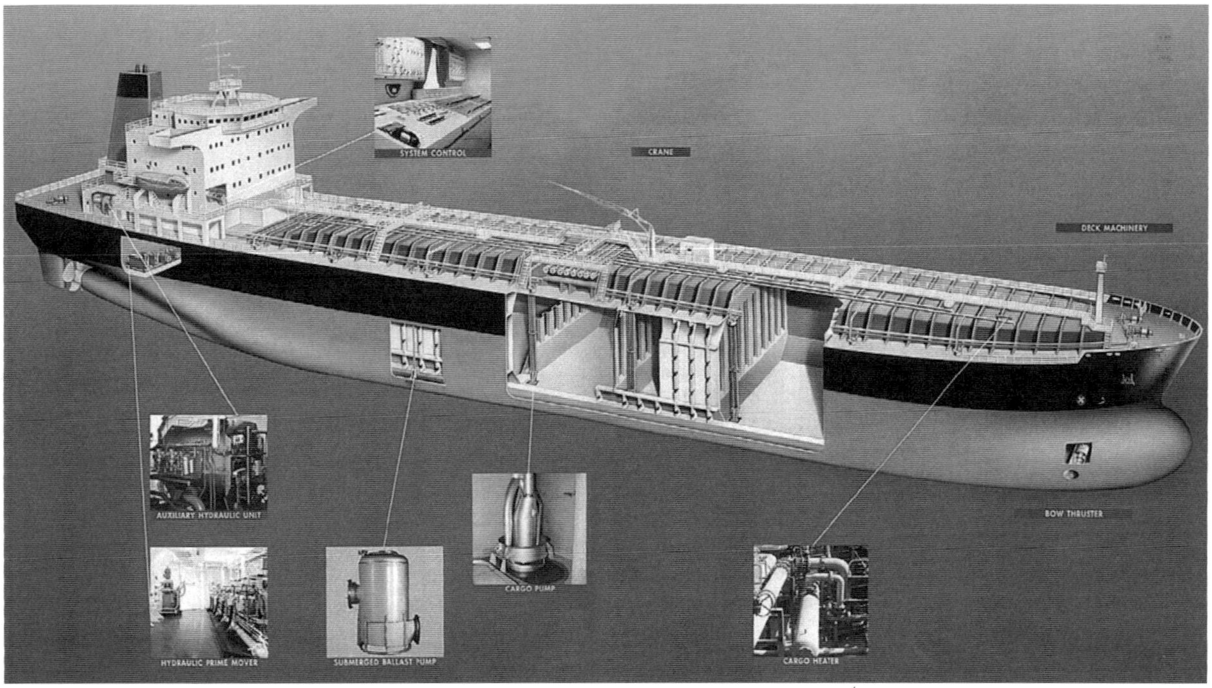

2.2 Presentation of the Hazard Analysis Method

The method retained for this determination was the one used by Office of Investigation into Event at Sea for all of their investigations in application of IMO resolution A849(20) modified.

The analysis of the prevention and protection measures is based on facts from the technical analysis and on a reconstitution of the chronology as put together from an observation of the damage and by collecting statements from the crew onboard.

The principle applied when analyzing preventive measures comprises identifying the system's defense barriers (from a causal tree type illustration) and querying whether the latter are suited to preventing the formation of an air/gasoline ATEX as well as finding the presence of a sufficient energy source for setting it alight.

The first question raised is the quality, generally speaking, of all of the preventive measures implemented to avoid the presence of the ATEX and to manage sources of inflammation during tank cleaning operations.

The second aspect relates to the seriousness of the event, i.e. the match between safety measures in relation to the human presence during some of the work entailed (the cleaning activity). One person died during the explosion onboard the ship for they were present on top of the tank during the tank washing operation.

2.3 Statements Collected and Chronology of Facts Before and After the Accident

Experts from INERIS and Office of Investigation into Event at Sea led personal interviews with the ship's crew so as to collect their statements on the accident and the facts that took place ahead of the explosion.

From these interviews, a chronology of the operations that preceded the moment of the explosion was determined.

The explosion occurred towards 07:10, during a tank cleaning operation, while 14 crew were onboard with the ship heading for the refinery at Donges, after having first unloaded its cargo at the Boucau terminal.

A sailor died following the explosion that caused major damage: part of the deck was ripped open causing an opening in the port side plating of the ship's double hull above the waterline.

Below we relate a summary of the chronology of operations that preceded the moment when the explosion took place (taken from the Office of Investigation into Event at Sea report and the statements collected).

The ship left the port of Bayonne, France, on June 13, 2003.

At 05:00, after completely unloading, M/T Chassiron sailed from the oil dock at the Raffinerie du Midi facility at Boucau heading for the oil refinery in Donges, France.

At around 06:40, washing out tanks 1 (port and starboard) was completed and the pump operator was to awake the crew chief to help him with his work.

At around 07:00, the tank washing system for tanks 6 port and starboard was in service, along with each of the pumps for tank 6. The ship was at position 43° 53',9 N – 001° 30',8 W.

At around 07:10, a violent explosion along with flames occurred at the level of tanks 6, which contained vapors from the unleaded 98 octane gasoline after unloading. Immediately the general alarm signal is triggered.

At 07:10, the Master is on the bridge. The ship is slowed to the mark zero, the second steering motor is started and maximum rudder angle selected.

At 07:11, the Chief Engineer is in the engine room. The engine room appears to be intact. He proceeds to start the fire fighting foam circuit. At 07:12, the foam circuit enters service.

At 07:15, a fire nozzle is installed on the poop deck but it cannot be used as the main fire manifold has been cutoff at the front of the superstructure by the explosion. As a result, the Chief Engineer decided to install a solid seal on the main manifold to the machine so as to recover fire manifold integrity for the engine room section and protect the compartment.

At 07:20, the crew assembles at the stern of the ship. After a number of roll calls and recounts, it is confirmed that the pump operator is missing.

At 07:24, the second mate tasked with communications calls the Socoa semaphore station to request assistance from the Aquitaine Explorer. The relief captain reports on the apparent integrity of the ship's hull.

At 07:30, observing that the fire is receding, the Master decides to stop the foam pump to retain a reserve supply. At this time the fire has reached the manger plate level on the port side manifold, around the main deck crane and on the destroyed part of the catwalk. The Master then decides to plot a fire route to first of all move closer to the coast then another fire route towards Bayonne.

At 07:39, contact is made with the CROSS emergency center and nearby fishing vessels to look for a possible man overboard.

At 07:50, the CROSS emergency center calls to indicate that the Aquitaine Explorer has been sent out from Bayonne as well as a large tug from Bilbao, Spain.

At 07:56, the SNSM 79 lifeboat from Bayonne and the SNSM 243 lifeboat from Cap Breton set out.

At 08:00, the fire is almost out. The fire pump is stopped to avoid any consequences on the ship's stability and structure. The ship proceeds on course and her speed is increased to eight knots.

At 08:08, the ship receives permission from the CROSS emergency center to continue along its course to the port of Bayonne.

3. Results and Interpretation

The results of the explosion hazard analysis which was used to process the accident analysis are presented below.

3.1 Evaluation of Product Flammability and Explosibility Characteristics

The first phase of the hazard analysis comprises determining the possibilities of an air / gasoline ATEX occurrence and an explosive air / diesel fuel ATEX after the tank unloading operations.

The temperature of the 98 octane unleaded gasoline was approximately 25°C and that of the diesel fuel some 32°C before the unloading operations.

The flashpoint of the unleaded gasoline is –40°C and that of the diesel fuel in excess of 55°C.

Tests to characterize the electrical conductivity of the unleaded and the diesel were performed by the INERIS Static Electricity Laboratory.

According to the test results, the tested samples show an electrical conductivity that is well below 50 pS/m. Furthermore, the tested liquid presents a discharge time of a few minutes to dissipate 90% of the initial electrical charge. This inflammable liquid therefore presents an aspect that is strongly insulated from carrying electrical charges.

3.2 Evaluating Explosible Atmospheres

Given the characteristics of the unleaded gasoline's flashpoint and temperature (25.2°C) before unloading, the climatic conditions did not have any effect on the formation of an air/unleaded gasoline ATEX while unloading the tanks.

Table 2 presents a summary of the hazards that lead to the formation of an Explosible Atmosphere (ATEX).

Table 2: ATEX presence after loading and unloading tanks 5 and 6

	Volume (V1) of the gaseous cap after loading the tanks (m3)	ATEX presence in the cap after loading the tanks	ATEX presence in the cap after unloading the tanks
Tank 6 Port	16	NO	YES
Tank 6 Starboard	20	As above	As above
Tank 5 Port	20	NO but diesel fuel vapors can self-ignite in the presence of a high temperature flame (self-ignition)	NO but diesel fuel vapors can self-ignite in the presence of a high temperature flame (self-ignition)
Tank 5 Starboard	24	As above	As above

Gasoline vapors are denser than air: they will therefore accumulate towards the base of the tank and this difference in density creates a gradient of lead-free gasoline vapor concentrations. The concentration will therefore progressively fall from the bottom to the top of the tank: it then creates an air/gasoline ATEX in tanks 6 port and starboard, where the richness of fuel is variable (more concentrated towards the bottom of the tank) but within the realm of flammability.

3.2 Evaluating Sources of Inflammation

The second phase of the hazard analysis comprises determining the chances that the inflammation sources that may have appeared during the tank cleaning operation may occur, and especially those which caused the explosions on June 13, 2003.

Among all of the possible inflammation sources, three of them were retained during the hazard analysis:

The source with a mechanical origin that is linked to the mechanical failure of either the pump or the washing gun. The two parts that were in motion during the tank cleaning operations were the after washing gun and the pump.

These turning elements may, in the event of a failure, create mechanical friction and trigger localized hot points or sparks of a mechanical origin with sufficient energy to set alight an air/gasoline ATEX. These malfunctions may be explained by:

The introduction of foreign objects into the pump (nuts, bolts...) or into the washing gun (e.g. shells, stones drawn in by the seawater pumping mechanism...),

Mechanical failures: an unbalance of the pump turbine, material wear, washing canon unbalance.

The source with an electrostatic origin that is linked to charge generation by the spraying of water from the washing gun[3] and through an equipotent level fault affecting a devices in the in service washing gun.

An electrostatic spark likely to inflame an air/gasoline ATEX may occur between conductive metal surfaces if the electrically charged washing gun was to neutralize itself on conductive equipment not connected to the same electrical potential level. This conductive equipment that is not connected to the same electrical potential level must be sought out within a short distance around the washing gun. It may for example, represent an equipotentiality breakdown that is linked to the washing gun's rotary nozzle. It appears improbable that a spark could have been triggered from father away (tank wall or between the washing guns).

The electrostatic source origin linked to an equipotentiality fault affecting one pump element during operation. We retain as probable, the source of inflammation coming from the pump following a pump equipotentiality breakdown and due to the presence of an air/gasoline ATEX within the inner pump body after setting it into service.

4. Conclusion: Most Plausible Event Scenarios

From the accident analysis, it appears that the series of explosions observed is due to a domino effect that is the result of three successive explosions:

- First air/gasoline ATEX explosion, continues in the sky over tank 6 starboard given the severity and the damage observed. As plausible inflammation sources, those with a mechanical origin as well as an electrostatic origin combined with the mechanical failure or the pump or the washing gun, were retained. This is because the air/gasoline ATEX found in starboard line 6 requires an energy level of just some few millijoules so as to catch fire.

- Explosion propagation towards tank 6 port.

- Second air/gasoline ATEX explosion contained in the tank by the heat brought in by the first explosion.

- Explosion propagation towards tank 5 port.

- Third explosion due to self-inflammation of diesel fuel vapors contained in tank 5 port. The blast of the explosion projected the transverse bulkhead between tank 5 port and tank 6 port towards the aft of tank 6 port (superstructure end).

This sequence of events oriented the recommendations towards a reduction in the air/inflammable vapor ATEX (especially that of the unleaded gasoline) by inerting tanks containing flammable liquid during the washing operation as well as during loading and unloading.

5. Recommendations

The analysis of preventive and protection measures is based on facts collected from the accident and by rebuilding the chronology, put together from an observation of the damage. According to the analysis of the accident[4], a series of adequate barriers have been proposed and are intended to reduce, as far as possible, the probability (based on the assumptions retained in this report) of the accident happening and its severity.

The first series of barriers put forward are intended to diminish the probability of any presence of an air / flammable vapor ATEX. The first barrier comprises inerting those segregations containing any flammable liquid during the loading, unloading and washing operations. If it is not possible to inert these tanks during the loading, unloading and washing operations, eliminating washing operations at sea is recommended; it this case the appearance of sources of inflammation should be limited.

The second series of barriers proposed is intended to reduce the probability that there will be any presence of a source of inflammation due to an electrostatic and/or mechanical cause.

When inerting tanks that contain flammable liquids during the loading, unloading and washing operations, ensure that the washing pumps and guns used meet the requirements of category II 3 G for an ATEX zoning of tanks in zone 2.

When not inerting tanks that contain flammable liquids during loading, unloading and washing operations, ensure that the washing pumps and guns used cannot be a source of inflammation. To this end, this equipment must meet the requirements of category II 1 G for an ATEX zoning of tanks in zone 0.

It is necessary to ensure that the installation is permanently brought to the same electrical potential level and to ensure electrostatic continuity between all conductive parts of the equipment used.

We recommend using cold seawater (not soiled with washing products or with insulating liquid products) during washing operations while still retaining the current flow rates and spray pressures used for the washing water.

It is necessary to instigate a regular schedule of service and maintenance for the pumps, washing guns and the seawater feed circuit.

Fire permits for all grinding, cutting or welding work must be subject to surveillance requirements that must be scrupulously complied with by the maintenance teams and/contractors. These fire permits must come with effective surveillance of the affected locations after the work is completed.

The third set of barriers proposed are intended to reduce the seriousness, in terms of human losses, of an explosion.

It is necessary to limit the presence of crew and staff above the segregations that contain unleaded gasoline when inerting is not used during the loading, unloading and washing operations, to ban the opening of manholes and gangway panels used to enter the tanks during these loading, unloading and washing operations. It is also necessary to train the crew on electrostatic hazards and on hazards links to the formation of flammable solvent vapor ATEX situations.

The fourth of the barriers proposed is intended to reduce the seriousness, in terms of material losses and especially to protect against tank explosions in those segregations that contain flammable liquids (explosion venting system).

6. References

1. European 1999/92/CE ATEX directory covering the minimum stipulations aimed at improving protection in terms of the safety and welfare of workers who may be exposed to explosive atmosphere hazards (Official Journal of the European Community).

2. Explosion (one victim) onboard M/T Chassiron on June 13, 2003 off Bayonne, France. Technical Investigation Report. French Ministry of Ecology, Energy, lasting Development and town and country Planning, Office of Investigation into Events at Sea.

3. The avoidance of ignition hazards due to electrostatic charges occurring during the spraying of liquids under high pressure. Journal of Electrostatics, 23 (1989) 99 – 109.

4. International Safety Guide for Oil Tanker and Terminals (ISGOTT, 4th Edition), International chamber of shipping, Oil companies international marine forum, International Association of ports and harbours.

7. Acknowledgments

We wish to thank the French Ministry of Ecology, Energy, lasting Development and town and country Planning, Office of Investigation into Events at Sea for their help during this accident analysis.

Evaluation of Permanent Buildings at an Existing Upstream Oil and Gas Business

Mark Aspengren, Tracy Whipple, David Vralsted, BP Exploration, (Alaska) Inc.
Vijay Raghunathan, Robin Pitblado DNV Energy, Houston TX 77084

ABSTRACT

Operators, both upstream and downstream, are evaluating protection for employees in permanent buildings against multiple process threats – vapor cloud explosion, jet fires, pool fires and toxic releases. This paper describes an approach, consistent with the near final update to API752 and internal standards for upstream facilities in Alaska operated by BP.

The paper describes the methodology to select events for analysis, calculation assumptions, and shows typical outputs. A unique aspect of the study is that most process facilities on the North Slope are located inside large enclosed modules for weather protection. This is unusual when compared to many other process plants around the world. A typical facility might encompass 20 to 40 of these process modules. This leads to some differences for consequence analysis compared with typical facilities with processes located outside. Consequences were predicted using PHAST (Process Hazards Analysis Software Tool) and this was followed by full geometrical analysis to establish peak pressure loads and thermal impacts in spreadsheet with extensive embedded visual basic coding.

The final results were then accumulated and prioritization was made for the business to take forward into a project.

INTRODUCTION

BP Alaska has recently embarked on a hazard review of occupied permanent buildings at their facilities. This review is part of a broader comprehensive approach towards risk management and focuses on the process safety hazards to occupants of buildings. The goal of the program was to identify the buildings that were occupied, estimate the consequences to the occupants, and prioritize the buildings that may require mitigation. The methodology of the program is presented in this paper, including a discussion of the reasoning behind the methodology development.

A risk assessment process takes place in multiple stages. The data gathering for the analysis as well as the tools to perform the analysis can vary in level of detail and method. The following sections discuss data gathering methods and application of the data to assess the facility siting risks on the North Slope of Alaska and at a GTL (Gas-to-Liquids) pilot facility in South Central Alaska.

The risk assessment process included the following phases:

1. Hazard Identification
2. Scenario Development
3. Consequence Analysis
4. Frequency Analysis
5. Risk Analysis
6. Using the Outputs

Figure 1 Risk Assessment Procedure

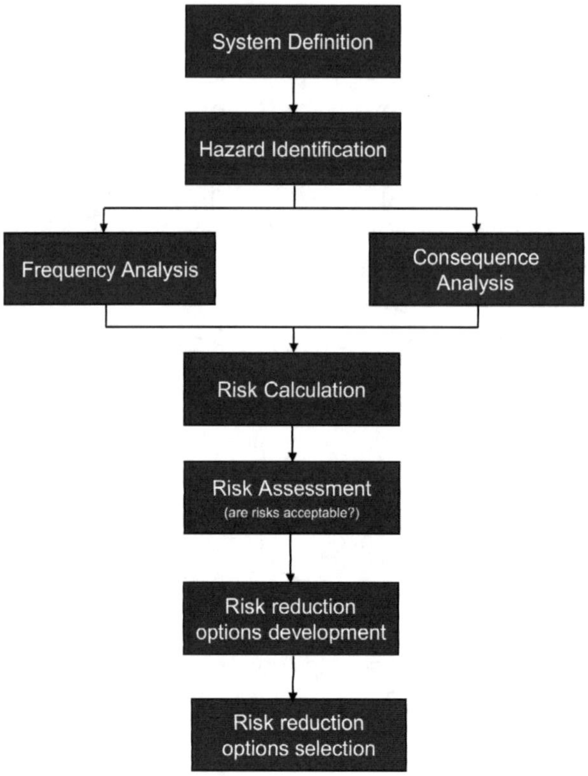

ALASKA NORTH SLOPE EXPLORATION AND PRODUCTION

BP Alaska is an upstream business unit of BP. The Alaska business is the second largest resource base for BP worldwide and the largest known undeveloped gas resource in the United States. The North Slope is a flat, treeless plain, which extends 88,000 square miles from the foothills of the Brooks Mountain Range to the Arctic Ocean and west from the Canadian border to the Chukchi Sea. The developed area of the North Slope encompasses about 312 square miles.

Prudhoe Bay is about 600 air miles north of Anchorage and about 1,200 miles south of the North Pole. It is about 250 miles north of the Arctic Circle.

Winter temperatures across the North Slope frequently dip to minus 30 degrees Fahrenheit with winds to 30 to 40 miles per hour, resulting in severe chill factors. Temperatures average in the 50s (Fahrenheit) in the summer months.

There are 408 total miles of roads. The field is approximately 80 miles from east to west and 30 miles from north to south. This includes open and roadless areas and open ocean to two island facilities.

BP operates several types of facilities on the North Slope. These facilities include over 70 well pads and 13 hydrocarbon processing facilities as well as additional facilities for living quarters, power generation, and sea water injection.

Oil, water, and gas flow from well pads to gathering centers for separation of the phases. The water is re-injected at the well pads, the gas is sent to a central gas facility and the oil is sent to Pump Station 1 of the Trans Alaska Pipeline System. Central gas facilities recover natural gas liquids and compress gas for re-injection.

The North Slope of Alaska is a unique environment in that almost all process equipment is located inside of skids or modules. The large modules typically range in volume from 50,000 to 300,000 ft^3. The modules are of steel frame construction with insulated panels enclosing the process. The interiors of the modules are heated and well ventilated to shelter process equipment and controls from the harsh weather and allow for improved access. The modules are raised off the ground on piers to minimize the transfer of heat to the permafrost layer and reduce snow drifting around the facilities.

A separation facility layout is divided into water handling, oil separation, gas compression systems, and pipeline distribution system. Multiple modules supporting a common processing function may be arranged together on the site.

Figure 2. Example Facility Map

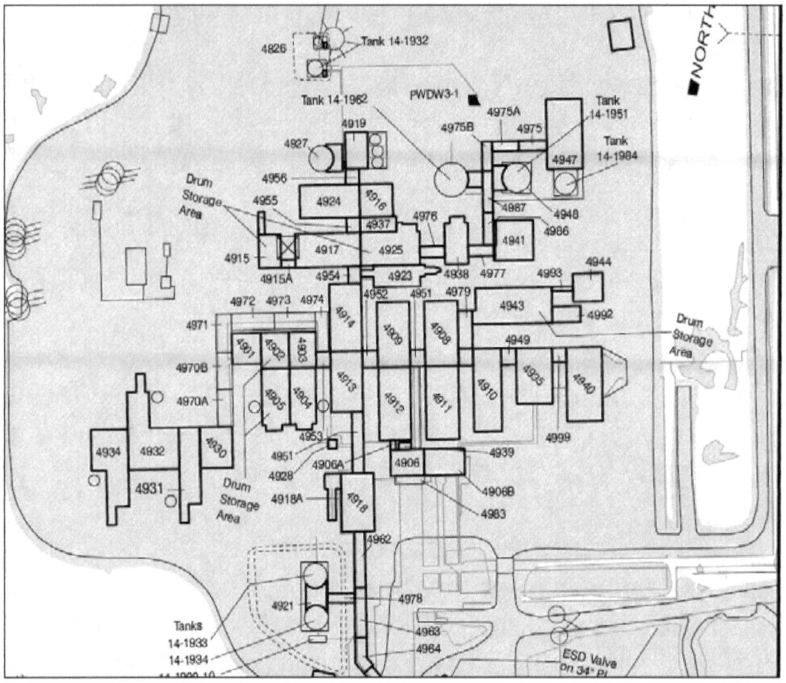

Figure 3. Example Facility Photo

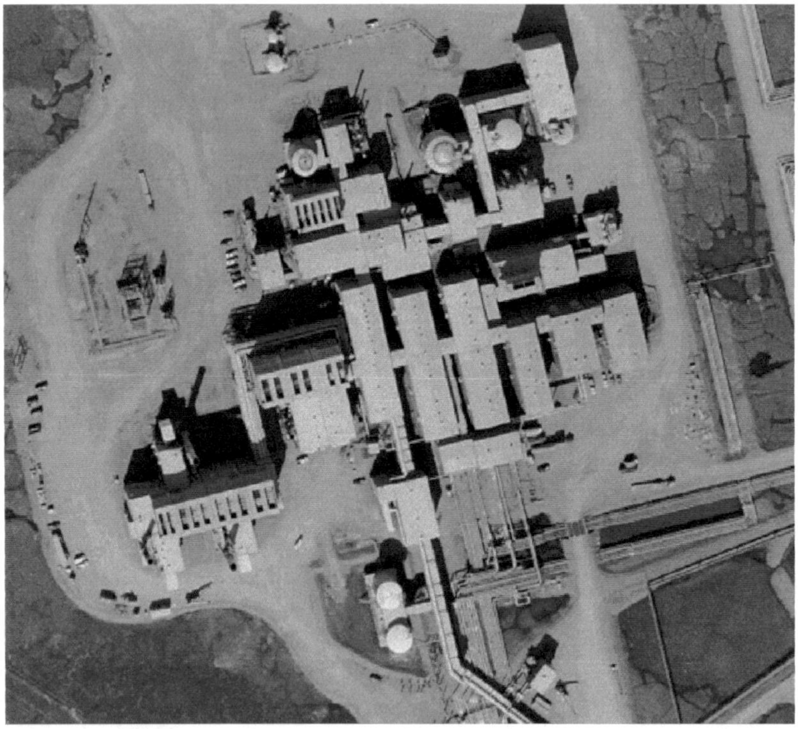

Siting challenges include consideration of facility footprint and size. Minimizing impacts to tundra is an environmental priority. This presents challenges to facility siting seeking separation distance from potential process hazards. Increase in separation distance may increase other risks such as transportation and weather related risks.

THE STUDY

The purpose of the facility siting study was to generate a list of all occupied buildings sorted high to low based on potential risk. This included the development of the methodology. There were a large number of occupied buildings and potential hazards to be reviewed. The exact number and location of the occupied buildings and the potential sources was yet to be determined. These factors influenced the level of detail set for the analysis and the phrase "screening level" study best described the objective. A "screening level" methodology was designed to be intentionally conservative to capture possible occupied area and potential hazard.

Follow up work could be prioritized based on level of risk to review the higher risk locations in greater detail. Further analysis would be required to understand the risk, generate options, and make risk informed decisions.

1.) Hazard Identification

The first phase in a program to manage risk is to identify the potential hazards that exist in a process facility. For this facility siting study, hazard identification focused on determining the worst case credible events for each process module. This study considered the energy stored in a process and the worst case credible release of that energy, only giving consideration to large releases that may result in further reaching consequences.

Site visits were necessary to ensure accuracy and understanding of the potential hazards to be identified. Prior to the site visits, information was gathered related to processes and how the facilities were laid out. Data with respect to what process occurred in which buildings (oil separation, water, gas), equipment lists, P&IDs, and process data from control systems was collected.

During site visits, data was gathered by a team and confirmed with operations lead technicians with respect to processes at the facilities. A number of study methodology issues were also agreed upon during the meeting between DNV and BPXA team to assure consistency with other assessments underway for BPXA and to comply with BP guidance. The hazard identification team was made up of DNV experienced risk analysts, a dedicated facility operations representative knowledgeable in the facility operation, and a BP Technical Lead with Process Engineering background.

A map of the facility was a starting point for the hazard identification. A facility is divided into modules containing pieces of process equipment. All modules listed on the map were put into a table. Tables of major equipment and processes, gas or liquid state, typical operating pressures and temperatures, stream compositions, and inventory estimation data were created for each module. Pipelines leading to, from, or past the facility were included in the table with information about pipeline diameter, compositions, temperatures, pressures, flow rates, line length, and automated isolation points.

The hazard identification process reviewed a facility on a module by module basis. The presence of hydrocarbon piping even passing through the module was identified as a potential hazard.

Along with the collection of the process details, permanent building occupancy data for each facility was recorded. A rule set was developed to collect data in a consistent manner and was also validated by facility operations.

2.) Scenario Development

Another phase in a risk management program is the scenario development. This work involves identifying a mechanical failure, a release, and the potential consequence outcomes that could evolve. The potential hazards scenarios given most consideration included vapor cloud explosions, jet fires, pool fires, toxic and flammable gas dispersions, and BLEVE.

During the facility visits, notes were taken to identify credible sources for potential hazardous events for each module and for pipeline scenarios near the facility. Data collected included module dimensions (volume), estimated level of congestion, HVAC functions (building ventilation, dividing walls, pressurized electrical rooms), containment systems, and site grading. Notes were made if modules were designed with specific intent to mitigate a hazard such as specialized fire or blast designs.

Scenarios were built out for each module and pipeline at the facility. The scenarios developed were verified with the facility operations the next day.

Scenarios considered attempted to determine a worst case single event and would account for escalation from other smaller scenarios. Only a single event for each hazard type was identified for each module. For example: In Module 1 a jet fire case was identified for a gas leak in the compressor discharge, a pool fire case was identified for a separator, a VCE was identified for the compressor room, and a gas dispersion case would be considered from the same source as the jet fire.

3.) Consequence Analysis

Once the hazards have been identified and the scenarios developed the next phase is to evaluate the potential consequence of those hazards. Consequence analysis involves a detailed analysis for each release dispersion, thermal radiation, and blast overpressure calculations.

The parameters for modeling the potential hazard scenarios come from the data collected in the Hazard Identification Phase and the Scenario Development Phase. Information related to the numbers of people vulnerable to a hazard scenario is considered in this phase.

During the site visits, data was gathered to aid in the consequence analysis of the potential scenarios defined. This data included building ventilation controls (i.e., HVAC with gas detection and shutdown), fire resistant structures, area electrical classification, isolation and blow down valve locations.

A Vapor Cloud Explosion (VCE) was predicted to be a high contributor to a risk profile due to the large number of enclosed modules. The effect of a VCE on an occupied building is a complex analysis requiring a detailed structural review as well as details on the impulse characteristics. BakerRisk was hired to conduct an independent site visit to the Alaska facilities and gathered structural data required for structural analysis of occupied buildings. This data included construction type, beam size, purlin and girt sizes and spacing, window dimension and thickness, and door construction. Structural response and component level analysis to blast loads provided a more detailed estimate of building damage and percent occupant vulnerability.

A typical output from a consequence analysis would indicate the potential occupant vulnerability for each occupied building based on the scenarios developed. For the screening level approach, all consequences were analyzed as if free field. No risk reduction benefit was assumed for structures between the hazard and the occupied area. In follow-up studies, the team intends to evaluate the benefit of shielding effects for buildings acting as jet fire and VCE shielding.

The number of occupants strongly influences the societal risk calculation. Time was spent ensuring these numbers were as accurate as possible. The determination of occupancy was done during the facility visits. Draft API RP 752 and internal BP guidance do not allow screening based on the number of man-hours. An area is identified as occupied if there were people assigned to the location (control rooms, offices, etc.) or if there were signs of occupancy, such as a desk with a chair.

Additional occupancy clarifications were made for the Alaska North Slope facilities. The function of the building was considered. Since the primary function of a process modules is to provide protection for equipment and workers will only be in the process module for normal operator rounds, these modules were not considered occupied. In some cases, interior rooms within process modules were identified as occupied.

The duration of occupancy was included in the calculation of risk and can be evaluated at various levels of detail. For this screening level study, average typical occupancy rates for day and night were generated. The higher of the day time or night time rate was used in vulnerability calculations.

Consequence Modeling Study Basis

The BP Engineering Technical Practice (ETP) (Ref 1) was used as the main reference document for this siting analysis. Failure cases were selected in sufficient detail to identify possible consequence loads on occupied buildings. The ETP nominates the BP Internal CIRRUS consequence modeling tool but allows alternative models. In view of the very large number of cases and the short time span, BP chose to use the DNV PHAST model which generally produces results similar to CIRRUS. Both the CIRRUS and PHAST models have been verified by Dr Rex Britter, Cambridge University, using the EU SMEDIS model evaluation protocol.

Failure Cases and Hole Sizes

Generally, DNV followed the ETP method which specifies both a two-inch hole size and Full Bore Rupture (FBR). The key source events are as follows: Pool fire, Vapor Cloud Explosion (VCE), Fireball, Boiling Liquid Expanding Vapor Explosion (BLEVE), and Jet fire. The failure cases / scenarios modeled as part of the study were grouped based on the different areas of the facility such as:

- Process Areas Inside Modules (Process equipment/piping inside a confined building)
- Process Areas Outside Modules (Process equipment/piping adjoining any confined module but not covered by the building)
- Pipelines (All pipe racks and pipelines running across the facility)
- Wellheads (Confined well houses)

In practice, due to the large equipment size and the consequence approach, there is no actual difference in result for two-inch hole and FBR for many releases inside the modules. For example, both a two-inch hole and FBR can, in theory, fill a module with flammable vapor and lead to a VCE. Similarly, most liquid spills can cover the floor up to the sills whether two-inch or FBR. The only difference is speed of the event – obviously FBR releases are much greater than two-inch holes, but the outcome is the same.

Jet fires were a little different and DNV modeled a two-inch hole as a jet fire and an FBR as a fireball. It was noted that isolation in many cases is automatic as part of the Fire and Gas detection system. Some pressure drops also cause equipment to trip. Two-inch jet fires were simple to model as most line sizes can sustain this leak rate for a long duration. FBR events were likely to be short lasting due to rapid depressurization. These have been modeled as fireballs based on the total loss of hydrocarbon in the first 30 seconds of release starting with the initial FBR discharge rate, declining to 10% within 30 seconds. Jet fires were assumed conservatively to be aligned horizontally; these jet fires will quickly cut through the light steel walls and potentially impact occupied spaces. FBR for the pig launcher/receiver fed by a long pipeline was modeled as a horizontal jet fire and aligned in the direction of the launcher only.

Pool fire and VCE have similar outcomes for both two-inch and FBR as most modules will quickly reach size limits of the module and the only difference between the two leak sizes is the time duration required. This was checked for a variety of leaks and including the effects of HVAC ventilation rates (before and after flammable alarms might sound). Pool fires are limited to the width of buildings due to internal low height dike and drainage preventing complete floor coverage. Pool fires are treated as free field even though they are inside buildings. This does mean that some flame drag beyond the building dimension does occur. However, flame impact to the light steel wall may mean early wall collapse .

VCEs were modeled as stoichiometric mixtures filling the entire space. Where the space was congested the VCE was modeled using TNO Curve 7 with congestion conservatively assigned to 150% of the measured congestion (subject to not exceeding 100% of the space). To account for cases where internal congestion is low, all modules are also modeled using TNO Curve 4 and using 100% of the space. The higher of these two values was used for subsequent load estimation.

BLEVE events were only relevant indoors for vessels with volatile contents (usually natural gas liquids [NGLs]). Three-phase oil vessels were screened out as these contain large quantities of water and this limits the temperature rise of the oil to below a significant flash potential. Fire protection does not usually include water deluge for vessels inside modules. However, outside BLEVEs were discounted due to installed fire protection systems and response of fire fighting crews.

Outdoor events (mainly pipelines) used slightly different modeling. Pool fires of oil were modeled indoors as temperatures inside are warm; an outside pool fire of stabilized crude was not considered credible in arctic conditions. VCEs are not considered credible outdoors due to minimum congestion and thus these are modeled as flash fires, with ignition assumed at the maximum cloud extent.

Outputs have been calculated to several end-points of interest:

- Overpressures to 2.9psi, 0.7psi and 0.4psi (along with pulse duration in seconds)
- Thermal flux to 35, 12.5, 6.3 and 1.6 kW/m2
- Dispersion to LFL and 3.3 LFL
- Fireball / BLEVE to 1% fatality
- Toxicity to ERPG 2 for relevant cases

To summarize, the following hole sizes/failure cases modeled in this study are shown in Table 1.

Table 1. Failure Cases on the North Slope

Scenario	Event Case	Modeling Basis
Process Areas Inside Modules		
Pool Fires	2" case - superseded by FBR	The 2" case is always less than the FBR
Pool Fires	FBR case	Modeled as pool equal to width of module - as modules have internal dikes and drainage and this limits pool dimension to the width, not the length
Jet Fires	2" case	Modeled as a standard sustained 2" case - although the leak will diminish over time once isolation occurs, the main damage will have occurred quickly. Jet modeled horizontally as this will maximize outcome on adjacent modules.
Jet Fires	FBR module case - not relevant	The FBR will not be sustainable for more than a few seconds due to lack of inventory behind the leak. This could be considered to fill the module with gas above the UFL and the point where a VCE could occur that is equivalent to a Fireball event.
Jet Fires	FBR case for Pig Launchers	Pig launchers are a special case as these incorporate a designed Full Bore hole. They also have large inventories behind them which can sustain a FBR for some time - sufficient for jet fire impacts to do damage to occupied spaces. By design these are horizontal leaks but are oriented in one direction only.
VCE Case	2" case - superseded by FBR	The 2" case is always less than the FBR; in terms of VCE it only means it takes longer to reach stoichiometric concentration in the module. But it could still happen. For all large cases DNV examined, the 8-12x/hr maximum HVAC rate is insufficient to prevent a 2" vapor case from reaching flammable concentrations.
VCE Case	FBR case	Modeled as a module full of stoichiometric concentration vapor (average of UFL and LFL concentrations). Modeled as TNO7 based on 150% of the congested volume subject to 100% module volume. TNO 4 case using the empty module volume has been examined but for all typical module sizes and congestions in GPB, TNO 7 outcomes were larger.
Fireball	Not relevant	Fireballs tend to cause only local damage inside the module and outside - only thermal radiation for 30+ sec and this is unlikely to impact occupied spaces in adjacent module.
Flash Fire	Not relevant	Like a fireball - causes fatalities within the module but not in external modules

Scenario	Event Case	Modeling Basis
BLEVE	Possible - where justified	BLEVEs inside modules were possible due to lack of deluge and smoke impeding firewater cooling from hose streams. Most situations do not favor BLEVE due to heavy oils, or oils mixed with water limiting the oil temperature and thus potential to flash. The main case will be NGL liquids.
VCE Case	FBR case	Modeled as a module full of stoichiometric concentration vapor (average of UFL and LFL concentrations). Model as TNO7 based on 150% of the congested volume subject to 100% module volume. TNO 4 case using the empty module volume has been examined but for all typical module sizes and congestions in GPB, TNO 7 outcomes are larger.
Process Areas Outside Modules		
Pool Fires	2" case - superseded by FBR	See FBR. 2" case is less likely to ignite due to smaller thermal mass than an FBR.
Pool Fires	FBR case	Pool fires for an outside release will spill on the pad. The average temperature is 13 deg F, so ignition is very unlikely, but the release may be warm and we should calculate the outcome of a FBR on an infinite flat plate with no drainage. Inventory to lose is unclear - so assume largest outside vessel (often 10 min of holdup) and add same again for additional oil in pipe work.
Jet Fires	2" case	Model as a 2" jet fire in horizontal orientation.
Jet Fires	FBR process case - not relevant	Not relevant - same basis as for Module case. Also, there are no Pig Launchers outside.
VCE Case	2" case - not relevant	Insufficient congestion outside
VCE Case	FBR case - not relevant	Insufficient congestion outside
Fireball	Not relevant	Possible but no impact to occupied buildings due to thermal impact only for a short duration, not sufficient to fail steel structures.
Flash Fire	2" case	Superseded by FBR case
Flash Fire	FBR case	Model dispersion to LFL and 3.3xLFL. Only 3.3xLFL affects occupied buildings. Assume release and target heights are both 1.5m.
BLEVE	Highly unlikely but possible - where justified	Fire protection for external vessels is good and is unlikely to be hampered by dense smoke as inside a module. This effective cooling should occur and BLEVE events can be screened out.
Pipelines		
Pool Fires	2" case - superseded by FBR	See FBR. 2" case is less likely to ignite due to smaller thermal mass than an FBR.
Pool Fires	FBR case	Pool fires for an outside release will spill on the pad. The average temperature is 13F, so ignition is very unlikely, but the release may be warm and it will sit on a much larger volume of warm water. Calculated the outcome of a FBR on an infinite flat plate with no drainage.
Jet Fires	2" case - Aboveground	2" case superseded by FBR case, model as horizontal
Jet Fires	2" case - Belowground	2" case superseded by FBR case, model at 45 degrees
Jet Fires	FBR case -	Release assumed to be possible with horizontal orientation, large

Scenario	Event Case	Modeling Basis
	Aboveground Pipe	inventory behind pipe means initial rate can be sustained for sufficient time to do maximum damage predicted.
Jet Fires	FBR case - Belowground Pipe	Release assumed to be possible with 45 degree orientation, large inventory behind pipe means initial rate can be sustained for sufficient time to do maximum damage predicted.
VCE Case	2" case - not relevant	Insufficient congestion outside
VCE Case	FBR case - not relevant	Insufficient congestion outside
Fireball	Not relevant	Possible but no impact to occupied buildings due to thermal impact only for a short duration, not sufficient to fail steel structures.
Flash Fire	2" case	2" case superseded by FBR case
Flash Fire	FBR case	Modeled dispersion to LFL and 3.3xLFL. Only 3.3xLFL affects occupied buildings. Aboveground pipes: assume release and target heights are both 1.5m. Belowground pipes assume 45 degree orientation from zero height.
BLEVE	Not relevant	Pipelines do not BLEVE
Well heads		
Pool Fires	2" case - superseded by FBR	Exactly same as FBR event
Pool Fires	FBR case	Well head module completely filled with liquid
Jet Fires	2" case	Horizontal event Jet Fire
Jet Fires	FBR module case - not relevant	Blowout event
VCE Case	2" case - superseded by FBR	Exactly same as FBR event
VCE Case	FBR case	Completely filled well head module, modeled as TNO 7 with 150% of congested volume in well head building
Fireball	Not relevant	
Flash Fire	Not relevant	
BLEVE	N/A	No vessels in well head

Consequence Modeling of Fire Thermal Radiation

Torch / Jet fires

FBR releases were not modeled as a jet fire but as a less directional turbulent fire, fireball, or pool fire. A two-inch hole was used to calculate jet fire scenarios. These were modeled as a horizontal release (except using 45 degree upward for underground lines) with a wind speed of F-2 m/s and D-5 m/s. Thermal radiation contours were calculated at 1.6, 6.3, 12.5, and 35kW/m^2. The contours were calculated based on the heat flux at a height of 1.5 m above ground. Flame dimensions were not required as the upper thermal radiation boundary of 35 kW/m^2 is assumed to have the same effect on people and buildings as direct flame impingement. (35 kW/m^2 is assumed to result in 100% impact within 30 seconds.)

Locations of sources were considered at the center of a module. For pipeline jet fire scenarios, the most conservative location (closest to the facility) were chosen. Thermal flux contours were drawn for length of pipelines around and through the facility.

Pool fires

For pool fires, thermal radiation contours were calculated at 1.6, 6.3, 12.5, and 35kW/m^2. The pool burning rate will match the discharge rate. Stabilized crude (after 3rd stage separator) was not be modeled as a pool fire outside due to the very low ignition probability in arctic conditions. Crude Pool fire was modeled as an inside module (same for arctic diesel). Appropriate smokiness was used with heavier liquids (e.g. Crude) burning with a smoky flame and lighter liquids with a bright flame (e.g. NGL).

Fireball / BLEVE

Fireballs and BLEVE are similar instantaneous events. Fireball is the more general term embracing both vapor and liquid releases. BLEVE specifically addresses liquid releases, normally of a pressure vessel containing hydrocarbon liquids at elevated pressure and with a significant flash fraction on release to atmospheric conditions.

Fireball calculations were based on the normal maximum inventory within the pressure vessel being studied. The output from fireball modeling gave distance from the source to fireball radius and 1% lethality contour.

A BLEVE from a Knock-out drum was not modeled in this study as the knock-out drum is typically near empty. BLEVE fragments (missiles) were also not considered as the range can be far and the direction unpredictable. Normal operating level in vessels was used to model BLEVE volume. No outside BLEVEs were modeled as fire protection and response time is considered adequate.

Location of source was at the center of a module location or at the most conservative location.

Modeling of Vapor Cloud Explosion Overpressure

Most units containing flammable gasses were assumed to be capable of suffering a VCE. An exception would be a release of materials which are above their auto-ignition temperature; these releases should be assumed to catch fire before a flammable cloud could develop.

Overpressure modeling was done using the more conservative of two options:

- The total volume of the building is calculated. If the building is of heavy construction, e.g. concrete or brick, then this volume was used in the TNO multi-energy model to determine blast overpressure using coefficient 5. If the building is of light construction, e.g. steel cladding, this volume was used in the TNO multi-energy model to determine blast overpressure using coefficient 4. Essentially all buildings on the North Slope are of light steel or wooden construction. There were no observed brick or concrete buildings.

- The volume of the obstructed (congested) areas within the building was calculated. An additional 50% was added to this (but limited to the maximum volume of the building). Using this enhanced volume in the TNO multi-energy model the blast overpressure was determined using co-efficient 7. The method to determine congested volume was by estimating the percentage of clear, unobstructed volume and subtracting this from 100%.

Overpressure Load Calculations

The overpressure load calculations for the different receptor buildings was based on the American Society of Civil Engineers (ASCE) methodology but simplified from the full possible analysis (reflecting the screening study level of detail) as follows:

- For normal incidence (90°) and up to 65°, the shock wave parallel to facing surface was calculated
  - Free field ground level propagation of the shock wave was assumed (i.e. no allowance for source building structure or intervening structures)
  - Peak Side on overpressure and Duration at the leading edge / facing wall
  - Peak Side on overpressure and Duration at the center of the roof
  - No reflected overpressures were calculated.
- For oblique incidence, with shock wave impacting at an angle to the facing surface
  - Peak Side on overpressure and Duration at the leading corner or surface closest to source.
  - The angle of incidence for the wave was estimated and for this oblique incidence case, the 65° criterion for angle of incidence was used. For angles between 45° and 65°, DNV reported overpressures at the corner. For angles less than 45°, then face considered was changed to the larger angle face and the above rule set was applied.
  - No reflected overpressures were calculated.

Modeling in the Facility Siting Spreadsheet

The PHAST consequence results were summarized in a facility siting spreadsheet listing all the event consequences.

The facility siting spreadsheet was also input with all the geographic information of sources and occupied buildings (targets). This data was used to determine the distance to every target and the geometric relationship – straight side-on or angled. This determines the reported load results as affecting one or two facing walls as well as the angle of incidence. The spreadsheet calculates all the thermal and explosion loads with a visual basic code using the PHAST consequences and the geographic input.

4.) Frequency Analysis

The calculation of risk is based on a consequence paired with a likelihood of that occurrence. Up until this point in the risk analysis the focus has been on determining the effects of the hazard scenarios. The likelihood of the event is calculated as the product of event frequencies and probabilities in an event tree.

For the screening level study the factors applied were relatively simple to ensure consistent application across the entire North Slope. Due to the volume of data being collected, and the time frame to complete all locations, a simplified method was chosen. The priority was placed on using a consistent methodology applicable to all the facilities on the Alaska North Slope. This simplified method was expected to affect the results by less than an order of magnitude. This assured that a relative ranking could be applied to all facilities for risk prioritization.

Calculation Methodology

A North Sea database provided leak frequency rates for defined systems of equipment. A gas compression system leak rate is defined for a number of flanges, valves, pumps, heat exchangers, turbines, pressure vessels, instruments expected in a typical gas compression system. An assumption was made that Alaska processes are designed and operated in a similar fashion to the North Sea or at least within the same order of magnitude with respect to leak rates.

Some conditional probabilities were considered in the analysis. For example, not all leaks would be large enough to contribute to the potential hazard scenarios defined in the consequence modeled scenarios. A 25% factor was uniformly applied to the leak frequency to calculate the large leak frequency.

Ignition probabilities for this study were assumed to be 10% for controlled ignition environments (electrical classified areas), 30% ignition rates were used for outside and other uncontrolled ignition areas. These rates were considered conservative based on actual experience.

A 25% directional probability was applied to account for the directional nature of dispersions. A detailed analysis of typical wind speed and direction was done but was not used in the screening level analysis. 5 mph and 11 mph wind speeds were considered to determine the worst case dispersion scenarios.

For the screening level study being conducted no conditional probability was considered for protective systems such as gas detection, fire detection, fire suppression, emergency response plans, or fire monitors.

5.) Risk Analysis

The combination of likelihood with a consequence analysis produces a risk analysis. The metric used in risk analysis for this study was an 8x8 risk matrix. This matrix provides 15 levels of risk referred to as Risk Scores. The BP matrix derives from the ISO17776 standard but has a greater degree of resolution by virtue of the 8x8 categorization.

Severity Level	BP OMS 8x8 Risk Matrix							
A	8	9	10	11	12	13	14	15
B	7	8	9	10	11	12	13	14
C	6	7	8	9	10	11	12	13
D	5	6	7	8	9	10	11	12
E	4	5	6	7	8	9	10	11
F	3	4	5	6	7	8	9	10
G	2	3	4	5	6	7	8	9
H	1	2	3	4	5	6	7	8
	>10-6/yr	>10-6 to >10-5 /yr	>10-5 to >10-4 /yr	>10-4 to >10-3 /yr	>10-3 to >10-2 /yr	>10-2 to >10-1 /yr	>10-1 to 1/yr	>1 yr
	1	2	3	4	5	6	7	8
	Frequency							

A risk score is determined by table look up of severity level and likelihood. The severity level for this siting study considered the range of potential vulnerabilities. The frequency band is subdivided by events per year categories.

An overall risk score for a location considers all the events that contribute to the risk. This can be found by looking up the sum of event frequencies and a frequency weighted average of potential vulnerabilities to get an overall risk score.

Equation 1

$$f_{overall} := \sum_i f_{event}$$

$$n_{ave} := \frac{\sum_i \left[f_{i,event} \cdot \left(n_{i,event}\right)^S \right]}{\left(\sum_i f_{i,event} \right)}$$

Greater resolution than the risk score may be desired. Due to the large numbers of buildings reviewed, multiple overall risk scores in the same category were expected. This caused difficulty in ranking two buildings of the same risk score relative to one another. A societal risk index calculation was created by summing the product of frequency and occupant vulnerability. The risk index may be used to prioritize occupied buildings relative to one another within the same risk score level.

Equation 2

$$R_{Index} := f_{event} \cdot n^{S}$$

f_{event} = frequency of event [yr^{-1}]
n = potential vulnerabilities associated with event
S= Societal risk weight (range likely from 1 to 2)

The risk score and risk index were calculated for a location by summing all the event frequencies affecting the occupied area divided by a weighted average of the potential vulnerabilities.

For a location considering multiple hazard scenarios (e.g. jet fires and VCEs), the overall risk index can be calculated as the sum of the individual risk indexes.

Equation 3

$$R_{Index} := \sum_{i} \left(f_{event} \cdot n^{S} \right)$$

The occupant vulnerability may be raised to a power (S) to correlate with the magnitude of the risk score based on the societal weighting. For example a risk score of 10 may be equivalent to a risk index in the 1.0E-3 range and a risk score of 11 may be equivalent to a risk index in the 1E-2 range, differing in risk index by an order of magnitude. Both the risk score (generated from a risk matrix) and the risk index account for a societal risk component.

6.) Using the Outputs

The screening level risk analysis goal was to gain an understanding of the risk profile of the North Slope of Alaska as a whole. Outputs from the study indicate some occupied areas at potentially higher risk relative to other areas. The data collected and scenarios developed help to understand the hazards associated with these risks. The screening level study, based on conservative methodologies, resulted in a comprehensive list of all potential hazard source locations.

The output data generated from a siting study can be useful in prioritizing any mitigation activities. The outputs from this screening level study allowed BP to risk rank all the occupied buildings at hydrocarbon processing facilities in Alaska. Over 200 occupied areas were reviewed and ranked by risk score and risk index calculations in this study. Planned analysis takes a closer look at the higher priority occupied areas based on their risk rank. A more detailed risk assessment methodology may be developed to ensure an understanding of the hazards as well as mitigation systems in place. Emergency response plans will be reviewed to ensure they are consistent with the potential hazards identified.

There are other projects ongoing in Alaska with respect to North Slope infrastructure renewal. Information from the siting study is used to make informed decisions when considering prioritization. For example, analysis of the risk calculation data could determine which process module presents the greatest risk and therefore should receive priority for preventative maintenance, inspections, system upgrades, and other risk reduction actions.

REFERENCES

1. BP ETP 04-30 Guidance on Practice for Design and Location of Occupied Permanent Buildings subject to Blast Loading.

Basic Tools for a QRA of an Oilfield Cold Vent

Pablo R. Zunana
Karina I. Peñaflor
Juan I. Farina
Pan American Energy LLC, Av. L. N. Alem 1180 Buenos Aires, Argentina

BACKGROUND

Pan American Energy LLC is an oil and gas exploration company with operations in South America.

In the past the Company Management had decided to develop in–house expertise in major accident risk analysis as part of its Risk Based Process Safety program.

This paper presents the risk analysis of a cold vent in an oilfield in Patagonia, Southern Argentina. The purpose of the study is to make recommendations about an exception request to allow emergency gas venting by design in oil and gas production gathering stations.

While some aspects of the analysis may be original, we feel that the main contribution of this paper lies in the lessons learned from a corporate point of view, beyond the technical issues.

OILFIELD CONTEXT

Cerro Dragón is an on-shore oilfield area comprising over 2500 production wells. Their average productivity is relatively low at 5.7 cubic meters of oil per day. The area covers 3375 km^2 and the production gathering system is extensive. There are more than 70 telemetry-monitored, unattended gathering stations ("Batteries"), where gas-liquid separators send gas production to gas gathering lines and oil and water are sent to atmospheric tanks and pumped into oilfield trunklines. Harsh climate adds to the complexity of the operation. Upon emergency upsets, shutdown of wells is firstly avoided to reduce risks such as vehicular accidents in frozen roads, breaches of system integrity due to freezing, operational errors and personal safety accidents. This is achieved by diverting production to an emergency pit where gas is vented.

STUDY SETUP

The study setup takes into account the following considerations:

Separate risk analysis models have to be prepared taking into consideration the fact that venting could affect people located in the battery as well as traffic in nearby internal roads.

Venting scenarios have to be considered separately due to different gas composition, duration and frequency.

Consequence analyses are needed to obtain safe distances between battery – venting point and venting point – roads.

The frequencies of single and multiple fatalities have to be computed separately to gauge risk according to the corporate risk matrix.

An inherently safer design approach is used: firstly, safe distances covering different venting scenarios are considered, then the risks of lower distances are assessed to recommend a design and layout.

Figure 1 below illustrates the case.

Figure 1. Study Setup

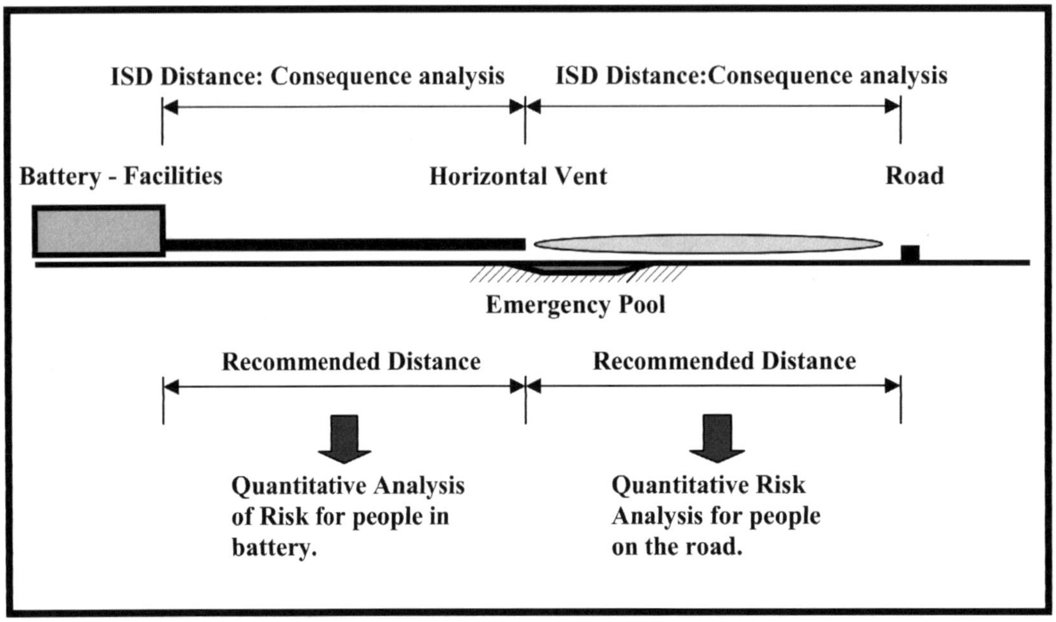

ISSUES

Geometry not Covered by Consequence Analysis Software

The direction of the flow in the vent line is from the battery to the discharge point, while the dispersion cloud that poses a risk to people at the battery heads in the opposite direction. A dispersion model for this geometry was not available in the commercial consequence analysis software package used by the company. Simulations were performed considering a minimum gas speed at the line outlet, as shown in Figure 2.

Figure 2 – Simulation diagram of the battery

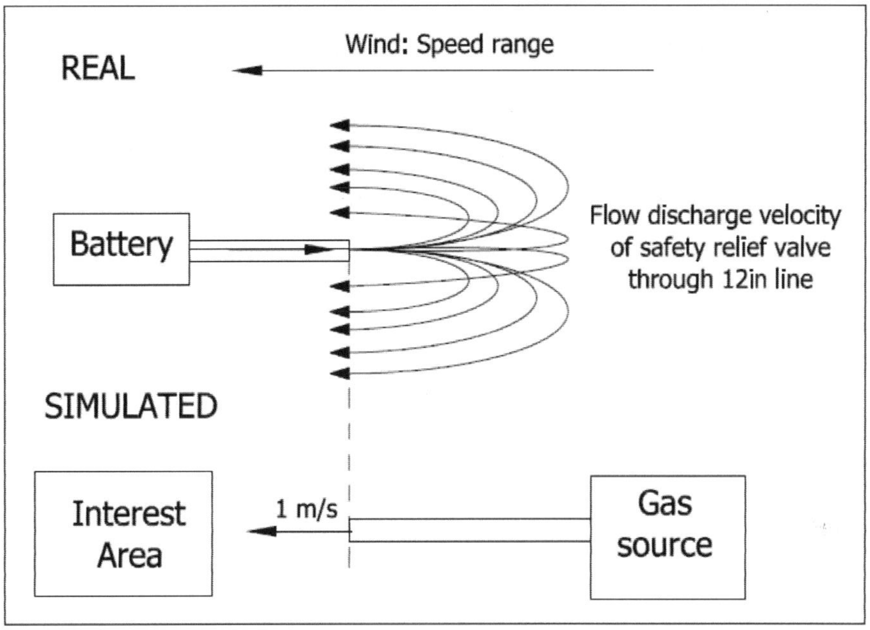

Venting Dynamics due to Multiphase Flow

The flow is multiphase in the production manifold as well as through its safety relief valve. The vented gas flowrate during short periods of time can be significantly higher than nominal.

In the study case, with a nominal gas flowrate of 30,000 SCMD (total nominal liquid rate of 6,000 CMD and a pressure safety valve set pressure of 12 kg/cm^2 gage), if the selected PSV opened when only gas is present at its inlet, the instantaneous gas release rate would be as high as 760,000 SCMD. Part of the increase was due to the selection of a nominal orifice with an area of 4.34 in^2, the required area being 3.52 in^2. Such a high flowrate was deemed feasible just during up to 0.1 second. For the nominal flowrate of each phase, 95% of the section of the PSV orifice was occupied by liquid. A release of 210,000 SCMD happens when 72% of this section is occupied by liquid. This variation is deemed to be quite probable and frequent, and the dispersion cloud of a continuous release of 210,000 SCMD was found to have a reach that is similar to the 760,000 SCMD - 0.1 sec extreme case. Other flowrates in between were deemed to be limited in time and have similar reach. So 210,000 SCMD was used for risk calculations.

Frequency and Duration of Initiating Events

It was deemed critical to provide an accurate estimate of the vents frequency and to understand the causes in order to estimate their durations properly. The available HAZOP study report was used to identify the initiating events. Once all of these events were listed, the mitigated venting events frequency was calculated using a commercial HAZOP/LOPA software package.

Results are summarized in Table 1.

Table 1. Mitigated Events Frequency

Events	**Duration** [hours]	**Quantity**	**Mitigated Events Frequency (MEF)** [1/year]
Total venting events	0.1 - 6	28	6.40 E-01
Venting events through manifold PSV	0.1 - 6	13	2.89 E-01
Venting events through separator PSV	0.1 - 6	15	3.51 E-01
Aggregate of venting events grouped by duration	6	13	3.09 E-01
	1	11	2.56 E-01
	0.1	4	7.50 E-02

RISK ANALYSIS MODELS AND RESULTS

Risk analysis models are described next.

Models were programmed in Excel and Visual Basic and result samples are shown in Tables 3 to 6. Conditional formatting of cells was used to visualize the level of risk according to the corporate risk matrix.

Battery Risk Analysis– Other than Fire

Battery-fatality frequency is calculated as the venting events frequency multiplied by several factors that take into account the multiple conditions that have to occur simultaneously ("conditional probabilities"). These probabilities are presented below and defined further under sub-sub-heading *Risk Analysis Model.*

Coverage Conditional Probability: For the dispersion cloud to be dangerous for people located in the battery it should reach the battery location. This depends on the Battery-Vent cardinal point relative orientation. For the flammable mixture to reach the battery, the wind should blow from the venting point to the battery with certain threshold intensity. The wind direction and intensity probabilities depend on that orientation. Battery-Vent orientation and wind intensity are expressed through a coverage conditional probability, based on the following observation: there is a Transition Speed u_t above which the dispersion cloud covers distances in the range of interest and below which, the cloud covers a shorter distance due to the difference between gas and air densities that acts as the predominant driving force over the wind. This is illustrated in Figure 3.

Figure 3. Transition Velocity

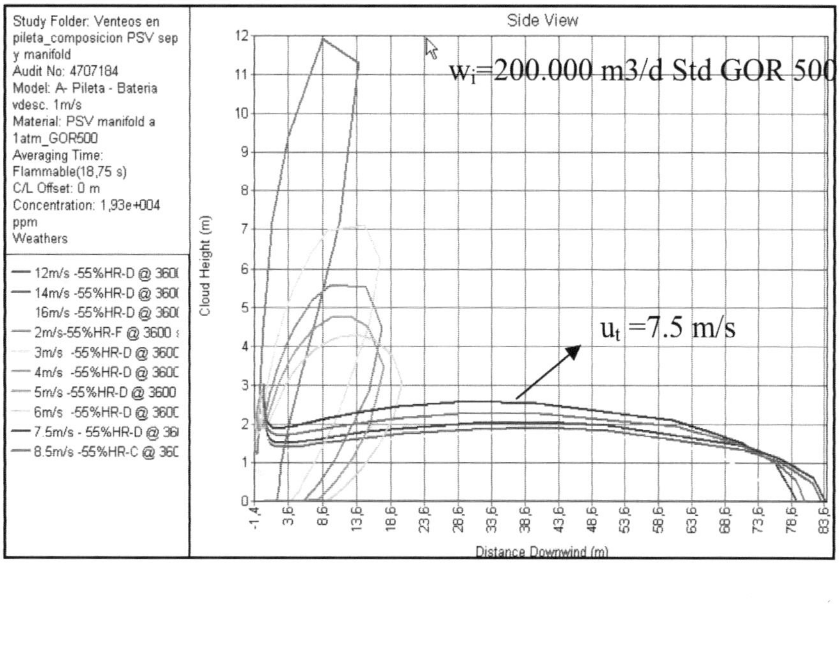

This behavior allows the conservative simplification of using the largest cloud size and associating it with the conditional probability of wind speed higher than or equal to u_t in a given direction.

Probability of Flash Fire and VCE: Defined as the probability that the cloud finds an ignition point (late ignition) and causes a FF or a VCE.

Probability of people present in the battery: Related to the number of persons located anywhere within the battery, considering the frequency and duration of visits.

Impact probability: Defined as the probability of people being inside the area affected by a Flash Fire or a Vapor Cloud Explosion. It is calculated as the ratio of the battery area covered by fire or explosion to the whole battery area, the underlying assumption being that any location of people in the battery is as probable as any other location. Figure 4 shows the geometry involved and how the battery-vent distance affects this conditional probability. This figure also shows that the area A_{FFi} was approximated conservatively to a rectangular area, as wide as the widest dispersion cloud.

Figure 4. Areas A_{BAT}, A_{FFi} y A_{VCEi}

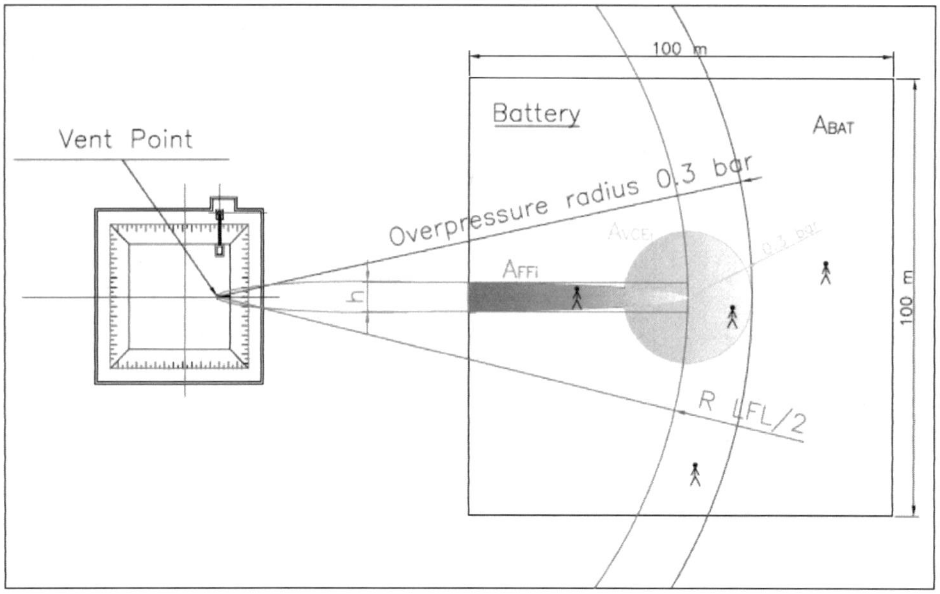

Vulnerability: Defined as the probability of people not surviving a Flash Fire or a Vapor Cloud Explosion.

Risk Analysis Model

Formulae of the risk analysis model follow.

Subscripts/Superscripts (Parentheses correspond to study case)

i	Nominal flowrates (i=1,...,5)
j	Battery-Vent distances (j=1,...,5)
n	Wind directions (n=1,...,12)
N	Battery-vent relative location (N=1,...,12).
S	Severity level (S = A o B)
q	Number of persons per activity. (Personnel visits the battery in groups of 1,2,4 or 10 people).

General expression for frequency of fatalities , F^S:

$$F^S = \left(\sum F_{me}\right).P_c.\left(P_{FF}.P^S_{iFF}.V_{FF} + P_{VCE}.P^S_{iVCE}.V_{VCE}\right)$$

(1) (2) (3) (4) (5) (6) (7) (8)

where

(1) $\sum F_{me}$ is the sum of frequencies of mitigated events.

(2) P_c , Takes into account the wind effects on the dispersion cloud (direction and intensity, as a function of the relative battery-vent location),

$$P_c\Big|_{D>20m} = \sum_{N-1}^{N+1} P_n . P_u$$

$$P_c\Big|_{D=20m} = \sum_{N-2}^{N+2} P_n . P_u$$

Figure 5 shows the relationship between D and the number of sectors.

Figure 5. Geometry shown for D = 20m

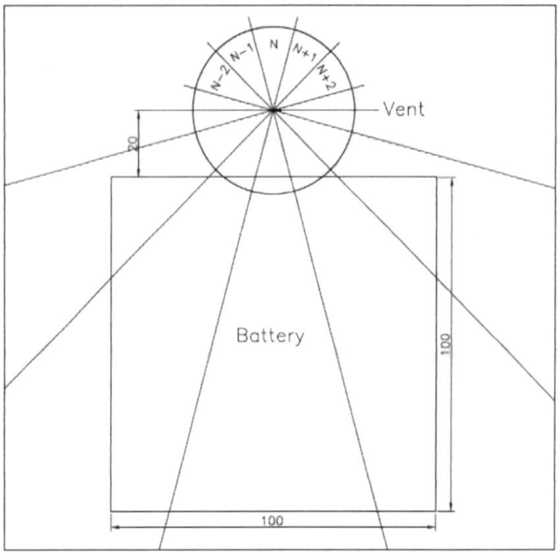

In previous expressions:

- P_n , Probability of wind orientation n (1 to 12), obtained from a Wind Measuring Station located in the oilfield.

- $P_u = e^{-\left(\frac{u_t}{A_n}\right)^{k_n}}$ Probability that wind speed is greater than or equal to transition velocity u_t for wind orientation n.

In this formula the term to the right corresponds to Weibull´s cumulative distribution function complement. k_n (shape parameter) and A_n (scale parameter), for each n wind orientation, from best-fitting the actual wind speed distribution, as measured in the Wind Measuring Station. Figure 6 below shows the Weibull´s speed general distribution shape and a *Pu* value as an example.

Figure 6. Probability of a wind speed higher than the transition speed

$$P(U)=\frac{k}{C}\left(\frac{U}{C}\right)^{k-1} e^{-\left(\frac{U}{C}\right)^{k}} \quad \Rightarrow P\{U<U_1\}=\int_0^{U_1} P(U)\,dU \Rightarrow P\{U<U_1\}=1-e^{-\left(\frac{U_1}{C}\right)^{k}}$$

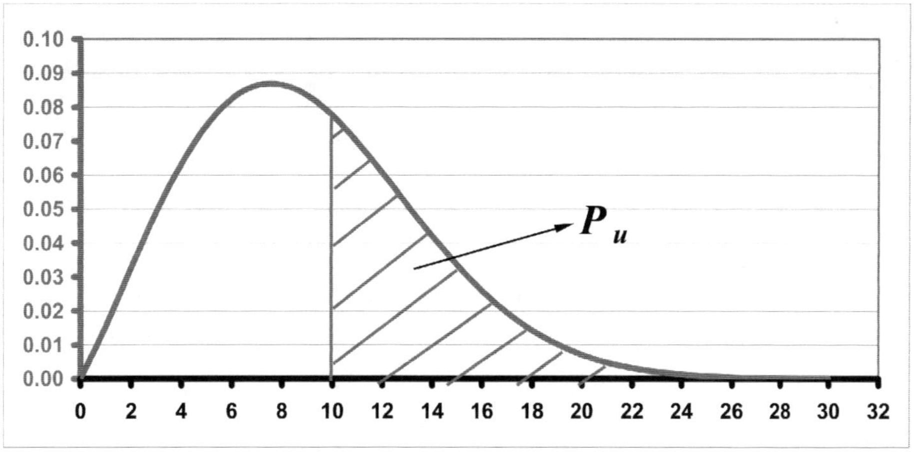

(3) P_{FF} Flash Fire probability.= 0.4 x 0.9 = 0.36.

These conditional probabilities are estimated as shown in the following events tree (Figure 7).

Figure 7. Events Tree

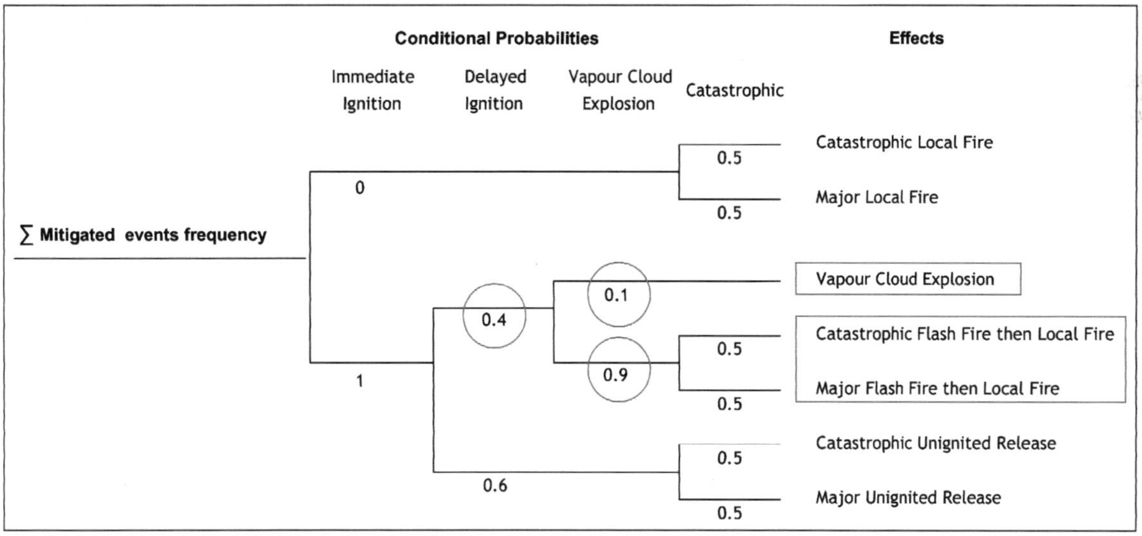

Vapour Cloud Explosion and Flash Fire are considered as possible final events. Both result from a late ignition after the explosive mixture cloud reaches the battery and finds an ignition point. As there are no ignition points near the venting point that could cause immediate ignition, the possibility of a jet fire event is disregarded.

(4)

$$P_{iFF}^{B} = \sum_{q} P_q \cdot \binom{q}{1} . a_{FF}{}^{1} . (1 - a_{FF})^{q-1} = \sum_{q} P_q . q . a_{FF}{}^{1} . (1 - a_{FF})^{q-1}$$

$$P_{iFF}^{A} = 1 - P_{iFF}^{0} - P_{iFF}^{B}$$

$$P_{iFF}^{A} = \sum_{q} P_q . \left(1 - (1 - a_{FF})^{q} - q . a_{FF}{}^{1} . (1 - a_{FF})^{q-1}\right) \quad q = 1,2,4,10$$

P^{B}_{iFF} and P^{A}_{iFF} calculate the probability of any group of q persons being present in the battery (Pq), and affected by a *Flash Fire*. P^{B}_{iFF}, is calculated by using the accumulated binomial distribution function, as the probability that just one person is affected out of a group of q persons, with an individual probability a_{FF}.

P^{A}_{iFF} is similar, but it considers the probability of 2 or more persons being impacted. So, it can be calculated on the basis of zero or only one person being impacted.

The probability that 2 or more groups of people visit the battery at the same time is disregarded (it is numerically insignificant, although it might happen).

Also in formulas above:

- P_q Conditional probability of a group of q persons being in the battery. Table 2 shows the study case values.

Table 2. Conditional Probability of People in the Battery

Activity	Q	Duration		Frequency		P_q
Normal Operation	1	0.7	hours	3	times/day	8.3E-02
Maintenance crews	4	2	hours	2	times/week	2.4E-02
Changes in the operation	2	4	hours	3	times/year	1.4E-03
Major works	10	8	hours	0.25	times/year	2.3E-04

- $a_{FF} = A_{FF} / A_{bat}$ Battery Area section covered by a dispersion cloud of concentration higher than LFL/2.

In order to simplify calculations while still obtaining safe results, the cloud was taken to be a rectangle and the rectangle corresponding to sector N was considered to represent any sector.

(5) V_{FF} Flash Fire vulnerability = 1

(6) P_{VCE} Vapour cloud explosion probability = 0.4 x 0.1 = 0.04. (See Figure 7 above)

(7)

$$P_{iVCE}^{B} = \sum_{q} P_q . \binom{q}{1} . a_{VCE}{}^{1} . (1 - a_{VCE})^{q-1} = \sum_{q} P_q . q . a_{VCE}{}^{1} . (1 - a_{VCE})^{q-1}$$

$$P_{iVCE}^{A} = 1 - P_{iVCE}^{0} - P_{iVCE}^{B}$$

$$P_{iVCE}^{A} = \sum_{q} P_q . \left(1 - (1 - a_{VCE})^{q} - q . a_{VCE}{}^{1} . (1 - a_{VCE})^{q-1}\right) \quad q = 1,2,4,10$$

where $a_{VCE} = A_{VCE} / A_{bat}$ Section of the battery area impacted by an overpressure equal to or greater than 300 mbar, as a result of a Vapor Cloud Explosion.

(8) V_{VCE} Vulnerability to Vapour Cloud Explosion´s overpressure of 300 mbar: 0.3.

The frequency of fatalities should be estimated as the sum of frequencies for each scenario, calculated on the basis of the frequency of the initiating events.

The following variables of separator and manifold venting scenarios were compared: dispersion cloud reach, P_u for the whole wind rose (i. e. not just for a specific sector) and event frequencies. Based on this comparison and in order to simplify calculations, separator scenarios were computed as if they were manifold scenarios, thus obtaining safe results.

Results

Frequencies obtained for lower-than-recommended distances between battery and venting point for any venting scenario except fire are shown in Table 3.

Table 3. Battery Risk other than Fire – Severity B Frequency [1/year]

Relative location 1

w_i [m3STD/d]	Distance Battery - Vent Dj [m]				
	60	50	40	30	20
42063	0.0E+00	0.0E+00	3.3E-06	1.5E-05	5.5E-05
54682	0.0E+00	4.1E-06	1.4E-05	2.7E-05	8.4E-05
210315	2.4E-05	3.6E-05	4.7E-05	5.8E-05	1.6E-04

Relative location 2

w_i [m3STD/d]	Distance Battery - Vent Dj [m]				
	60	50	40	30	20
42063	0.0E+00	0.0E+00	3.1E-06	1.4E-05	3.4E-05
54682	0.0E+00	3.9E-06	1.3E-05	2.5E-05	5.2E-05
210315	2.5E-05	3.6E-05	4.8E-05	5.9E-05	9.6E-05

Relative location 3

w_i [m3STD/d]	Distance Battery - Vent Dj [m]				
	60	50	40	30	20
42063	0.0E+00	0.0E+00	2.6E-06	1.2E-05	2.9E-05
54682	0.0E+00	3.3E-06	1.1E-05	2.2E-05	4.5E-05
210315	2.3E-05	3.4E-05	4.5E-05	5.5E-05	8.8E-05

Relative location 4

w_i [m3STD/d]	Distance Battery - Vent Dj [m]				
	60	50	40	30	20
42063	0.0E+00	0.0E+00	1.4E-06	6.3E-06	2.7E-05
54682	0.0E+00	1.8E-06	6.2E-06	1.2E-05	4.2E-05
210315	1.4E-05	2.0E-05	2.6E-05	3.2E-05	8.9E-05

Relative location 5

w_i [m3STD/d]	Distance Battery - Vent Dj [m]				
	60	50	40	30	20
42063	0.0E+00	0.0E+00	1.1E-06	4.8E-06	1.9E-05
54682	0.0E+00	1.4E-06	4.7E-06	9.0E-06	2.9E-05
210315	1.0E-05	1.5E-05	2.0E-05	2.5E-05	6.2E-05

Relative location 6

w_i [m3STD/d]	Distance Battery - Vent Dj [m]				
	60	50	40	30	20
42063	0.0E+00	0.0E+00	1.4E-06	6.4E-06	2.8E-05
54682	0.0E+00	1.8E-06	6.3E-06	1.2E-05	4.4E-05
210315	1.3E-05	1.9E-05	2.4E-05	3.0E-05	8.6E-05

Relative location 7

w_i [m3STD/d]	Distance Battery - Vent Dj [m]				
	60	50	40	30	20
42063	0.0E+00	0.0E+00	3.0E-06	1.3E-05	6.7E-05
54682	0.0E+00	3.7E-06	1.3E-05	2.4E-05	1.0E-04
210315	2.4E-05	3.5E-05	4.6E-05	5.7E-05	2.3E-04

Relative location 8

w_i [m3STD/d]	Distance Battery - Vent Dj [m]				
	60	50	40	30	20
42063	0.0E+00	0.0E+00	7.6E-06	3.4E-05	1.0E-04
54682	0.0E+00	9.6E-06	3.3E-05	6.3E-05	1.6E-04
210315	7.3E-05	1.1E-04	1.4E-04	1.7E-04	3.6E-04

Relative location 9

w_i [m3STD/d]	Distance Battery - Vent Dj [m]				
	60	50	40	30	20
42063	0.0E+00	0.0E+00	1.2E-05	5.2E-05	1.2E-04
54682	0.0E+00	1.5E-05	5.2E-05	9.8E-05	1.9E-04
210315	1.2E-04	1.7E-04	2.3E-04	2.8E-04	4.2E-04

Relative location 10

w_i [m3STD/d]	Distance Battery - Vent Dj [m]				
	60	50	40	30	20
42063	0.0E+00	0.0E+00	1.3E-05	5.7E-05	1.2E-04
54682	0.0E+00	1.6E-05	5.6E-05	1.1E-04	1.9E-04
210315	1.3E-04	1.9E-04	2.5E-04	3.0E-04	4.2E-04

Relative location 11

w_i [m3STD/d]	Distance Battery - Vent Dj [m]				
	60	50	40	30	20
42063	0.0E+00	0.0E+00	8.6E-06	3.8E-05	1.1E-04
54682	0.0E+00	1.1E-05	3.8E-05	7.1E-05	1.8E-04
210315	8.2E-05	1.2E-04	1.6E-04	1.9E-04	3.9E-04

Relative location 12

w_i [m3STD/d]	Distance Battery - Vent Dj [m]				
	60	50	40	30	20
42063	0.0E+00	0.0E+00	4.6E-06	2.0E-05	8.6E-05
54682	0.0E+00	5.7E-06	2.0E-05	3.8E-05	1.3E-04
210315	3.6E-05	5.2E-05	6.8E-05	8.5E-05	2.8E-04

A similar table was prepared for Severity A, showing low (green) risk in all cells.

Battery Fire Risk on People in the Battery

Table 4 below describes the events tree, frequencies and conditional probabilities for process fires and fires external to the battery process, on the battery. Fatalities caused by a fire escalation affecting people present at the battery are evaluated when vented gas, due to shut down, is blown back towards the battery location. This scenario is feasible when the distance between battery and venting point is shorter than the safe distance calculated by the consequence analysis.

Table 4. Risk Analysis of Battery Fire on People in the Battery

	Frec [1/year] or Conditional Prob.	Severity A	Severity B
External Fire Case			
External fire at the Battery location.	1,0E-02		
People present in the battery to activate shut down (external fire is the result of people´s activity; other causes are disregarded in the area of Cerro Dragón)		1	1
Someone activates the Shut Down	9,0E-01		
The operator can´t control the fire and he/she doesn´t drain the separator to the emergency pool. The oil remaining in the separator evaporates, and the PSV opens	5,0E-02		
The operator doesnt´t evacuate the facility.	5,0E-01		
Wind direction from vent point to the battery	2,3E-01	Worst orientation	
The flammable cloud coming from the vent point to the battery finds an ignition point.(the original fire)	1,0E+00		
The increased fire reaches 2 or more people in the battery		5,0E-01	0
The increased fire reaches only one person (added to Sev. B Freq)		3,0E-01	3,0E-01
External Fire		**2,6E-05**	**3,1E-05**

Table 4. Risk Analysis of Battery Fire on People in the Battery (continued)

	Frec [1/year] or Conditional Prob.	Severity A	Severity B
Process Fire Case			
Major release in equipment or Battery line.	4,0E-03		
Inmediate ignition of the battery release	1,7E-01		
People present in the battery to activate shut-down		2,5E-02	1,1E-01
Someone activates the shut-down	9,0E-01		
The operator can´t control the fire and he/she doesn´t drain the separator to the emergency pool. The oil remaining in the separator evaporates, and the PSV opens	1,0E+00		
People don´t evacuate the facility.	5,0E-01		
Wind direction from vent point to the battery	2,3E-01	Worst orientation	
The flammable cloud coming from the vent point to the battery finds an ignition point (the original fire)	1,0E+00		
The increased fire reaches 2 or more people in the battery		5,0E-01	0
The increased fire reaches only one person (added to Sev. B Freq)		3,0E-01	3,0E-01
Process Fire		**8,8E-07**	**2,9E-06**

	Bat orientation from vent	Severity A	Severity B
Battery risk-fire	12 a 7	**2,7E-05**	**3,4E-05**
	8 a 11	**8,1E-06**	**1,0E-05**

Notes:

1.- The indicated frequency values apply to the worst orientation (so the risk would be lower for the other sectors in each group).

2.- Worst orientation between sectors 12 y 7: Sector 9, Wind Prob=23%.
Worst orientation between 8 y 11: Sectors 12 y 2, Wind Prob=7%.

Road Risk – other than Fire

Road fatalities frequency is calculated as the venting frequency multiplied by the following conditional probabilities.

Coverage probability: For the dispersion cloud to reach the road and become dangerous for vehicles, wind must blow towards the road at a wind speed above the transition speed (see above). The geometry of the problem requires one or several sectors from the wind rose to be considered, as shown in Figure 8.

Figure 8. Sectors of wind rose – Roads

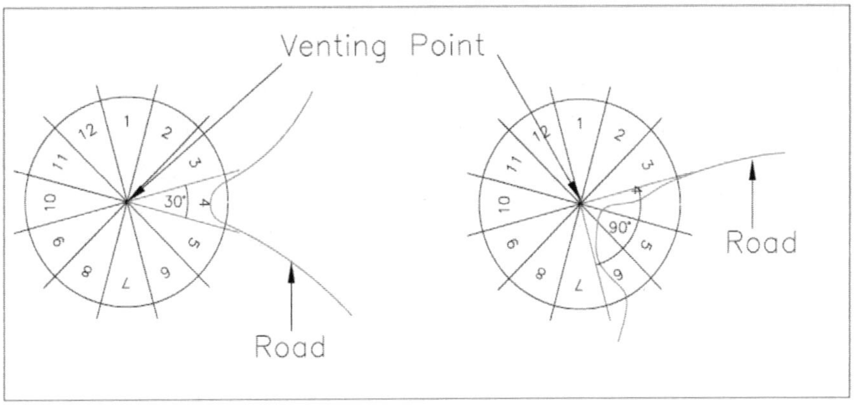

Vent-vehicle coincidence probability: Once the cloud develops and reaches the road, the venting event must take place at the same time the vehicle is passing by. The vent duration and the period of time between successive vehicles are used to calculate the vent-vehicle probability.

Probability of ignition: The passing vehicles are the ignition points. The number of vehicles circulating during the venting event must be considered (each vehicle with an individual ignition probability) in order to calculate the probability that one of them causes the ignition of the dispersion cloud. We have failed to find values of this probability in the literature and so a value of 0.3 was adopted.

Vulnerability: Given the fact that people are located very close to or at the ignition point, a value of 100% vulnerability is used.

Risk Analysis Model

$$F^S = \sum_i \underset{(1)}{f_i} \cdot \underset{(2)}{PS_{i,j}} \cdot \underset{(3)}{P_s} \cdot \underset{(4)}{Pi_{i,j}}$$

i	Nominal flowrates (i=1,...,5)
j	Battery-venting point distances (j=1,...,5)
S	Severity (S = A or B)

Where:

(1) f_i Frequency of venting events of duration i, where i stands for duration period (6 hours, 1 hour or 0.1 hour in this study). Venting events are grouped by their duration, as the vent-vehicle coincidence and ignition probability depends on it.

(2) $PS_{i,j}$ Vent-vehicle coincidence probability for vents of i duration. It depends on two variables: vent duration and period between cars, as follows:

$$Ps_{i,j} = \begin{cases} 0, & \dfrac{1}{T_j} = 0 \\ \dfrac{d_i}{T_j}, & \dfrac{d_i}{T_j} < 1 \\ 1, & \dfrac{d_i}{T_j} \geq 1 \end{cases}$$

where: • d_i Venting event duration • T_j Period between vehicles

The following figure illustrates values for $1/T_j \neq 0$

Figure 9. Vent – Vehicle Coincidence Probability

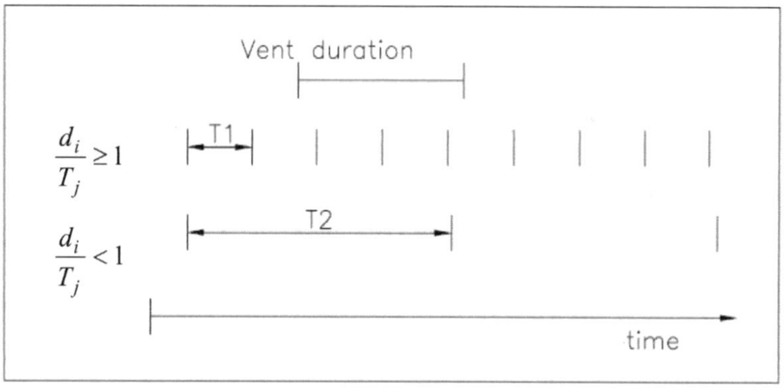

(3) P_c Coverage probability. Similar to the battery case; the only difference in this case being the need to consider 30° to 270° angles in situations where the road surrounds the venting point (see Figure 8 above).

(4) $Pi_{i,j}$ Probability of ignition by vehicles. It is calculated as the complement of the probability that none of the vehicles passing through the explosive cloud during the venting event provokes ignition, considering an individual ignition probability P=0.3, as follows:

$$Pi_{i,j} = 1 - (1 - p)^n$$

where n is calculated as the next integer above d_i / T_j.

Results

Frequencies obtained for lower-than-recommended distances between venting point and roads for any venting scenario except fire are shown in the Table 5.

Table 5. Road Risk other than Fire – Severity A Frequency [1/year]

Severity A: vehicles with 2 o more people

u_t	3	N											
	Segment [°]	1	2	3	4	5	6	7	8	9	10	11	12
High Traffic	30	2.0E-02	3.9E-02	1.7E-02	5.5E-03	1.0E-02	1.0E-02	1.5E-02	5.1E-02	1.2E-01	1.1E-01	7.4E-02	2.7E-02
	90	8.7E-02	7.7E-02	6.2E-02	3.3E-02	2.6E-02	3.6E-02	7.7E-02	1.9E-01	2.9E-01	3.1E-01	2.1E-01	1.2E-01
	150	1.8E-01	1.1E-01	9.3E-02	8.3E-02	5.9E-02	9.2E-02	2.1E-01	3.1E-01	3.7E-01	3.9E-01	3.6E-01	2.7E-01
	210	3.0E-01	1.9E-01	1.3E-01	1.2E-01	1.5E-01	2.3E-01	3.3E-01	4.0E-01	4.1E-01	4.2E-01	4.5E-01	4.1E-01
	270	4.3E-01	3.2E-01	2.2E-01	2.0E-01	2.9E-01	3.8E-01	4.2E-01	4.3E-01	4.4E-01	4.7E-01	4.8E-01	4.7E-01
1/hr	30	1.3E-02	2.6E-02	1.1E-02	3.6E-03	6.7E-03	6.7E-03	9.8E-03	3.3E-02	7.9E-02	7.3E-02	4.8E-02	1.7E-02
	90	5.6E-02	5.0E-02	4.0E-02	2.2E-02	1.7E-02	2.3E-02	5.0E-02	1.2E-01	1.8E-01	2.0E-01	1.4E-01	7.8E-02
	150	1.2E-01	7.1E-02	6.0E-02	5.4E-02	3.8E-02	6.0E-02	1.4E-01	2.0E-01	2.4E-01	2.5E-01	2.3E-01	1.8E-01
	210	1.9E-01	1.3E-01	8.4E-02	7.7E-02	9.7E-02	1.5E-01	2.1E-01	2.6E-01	2.7E-01	2.7E-01	2.9E-01	2.7E-01
	270	2.8E-01	2.1E-01	1.4E-01	1.3E-01	1.9E-01	2.5E-01	2.7E-01	2.8E-01	2.9E-01	3.1E-01	3.1E-01	3.0E-01
1/day	30	2.7E-03	5.3E-03	2.3E-03	7.4E-04	1.4E-03	1.4E-03	2.0E-03	6.8E-03	1.6E-02	1.5E-02	9.9E-03	3.6E-03
	90	1.2E-02	1.0E-02	8.4E-03	4.5E-03	3.5E-03	4.8E-03	1.0E-02	2.5E-02	3.8E-02	4.1E-02	2.9E-02	1.6E-02
	150	2.4E-02	1.5E-02	1.3E-02	1.1E-02	7.9E-03	1.2E-02	2.8E-02	4.2E-02	5.0E-02	5.2E-02	4.8E-02	3.7E-02
	210	4.0E-02	2.6E-02	1.8E-02	1.6E-02	2.0E-02	3.1E-02	4.4E-02	5.3E-02	5.5E-02	5.7E-02	6.0E-02	5.5E-02
	270	5.8E-02	4.3E-02	2.9E-02	2.6E-02	3.9E-02	5.2E-02	5.6E-02	5.7E-02	5.9E-02	6.3E-02	6.4E-02	6.3E-02
1/week	30	3.9E-04	7.6E-04	3.3E-04	1.1E-04	2.0E-04	2.0E-04	2.9E-04	9.8E-04	2.3E-03	2.2E-03	1.4E-03	5.1E-04
	90	1.7E-03	1.5E-03	1.2E-03	6.4E-04	5.0E-04	6.9E-04	1.5E-03	3.6E-03	5.5E-03	5.9E-03	4.1E-03	2.3E-03
	150	3.4E-03	2.1E-03	1.8E-03	1.6E-03	1.1E-03	1.8E-03	4.0E-03	6.0E-03	7.2E-03	7.4E-03	6.8E-03	5.2E-03
	210	5.7E-03	3.7E-03	2.5E-03	2.3E-03	2.9E-03	4.4E-03	6.3E-03	7.6E-03	7.9E-03	8.1E-03	8.6E-03	7.9E-03
	270	8.2E-03	6.1E-03	4.2E-03	3.8E-03	5.6E-03	7.4E-03	8.0E-03	8.2E-03	8.5E-03	9.1E-03	9.2E-03	9.0E-03
1/month	30	9.1E-05	1.8E-04	7.8E-05	2.5E-05	4.7E-05	4.7E-05	6.8E-05	2.3E-04	5.5E-04	5.0E-04	3.3E-04	1.2E-04
	90	3.9E-04	3.5E-04	2.8E-04	1.5E-04	1.2E-04	1.6E-04	3.4E-04	8.4E-04	1.3E-03	1.4E-03	9.6E-04	5.4E-04
	150	8.0E-04	4.9E-04	4.2E-04	3.7E-04	2.6E-04	4.1E-04	9.4E-04	1.4E-03	1.7E-03	1.7E-03	1.6E-03	1.2E-03
	210	1.3E-03	8.7E-04	5.8E-04	5.3E-04	6.7E-04	1.0E-03	1.5E-03	1.8E-03	1.8E-03	1.9E-03	2.0E-03	1.8E-03
	270	1.9E-03	1.4E-03	9.8E-04	8.8E-04	1.3E-03	1.7E-03	1.9E-03	1.9E-03	2.0E-03	2.1E-03	2.1E-03	2.1E-03

A similar table was drawn up for Severity B. However, the results in the table above provide sufficient information to make recommendations.

Battery Fire Risk on Roads

The following table describes the events tree and indicates the conditional probabilities and frequencies for external fire and for process fire.

Road risk evaluates fatalities in vehicles caused by ignition of a flammable cloud reaching the road after battery Shut-Down is activated, applying to cases where the distance between the road and venting point is less than the safe distance calculated by the consequence analysis for this scenario.

Table 6. Risk Analysis of Battery Fire on Roads

A or B Severity (according to the presence of 1, 2 or more people in the vehicles)

External Fire Case

	Freq [1/year] or Conditional. Prob High Traffic	1/hour	1/day	1/week	1/month
Vehicle circulation (A: 2 or more people.; B: Only the Driver). Vent estimated duration: 10 hours. Probability of ignición per vehicle: 0,3.	1.0E+00	9.8E-01	1.3E-01	1.8E-02	4.1E-03
External fire frequency	1.0E-02				
People present in the battery to activate shut down (external fire is the result of people´s activity; other causes are disregarded in the area of Cerro Dragón)	1.0E+00				
Someone activates the Shut Down	9.0E-01				
The operator can´t control the fire and doesn´t drain the separator to the emergency pool. The oil remaining in the separator evaporates, and the PSV opens	5.0E-02				
Wind direction from vent point to the road.	2.3E-01	Worst orientation			
External Fire	**1.0E-04**	**1.0E-04**	**1.3E-05**	**1.8E-06**	**4.2E-07**

Process Fire Case

	High Traffic	1/hour	1/day	1/week	1/month
Vehicle circulation (A: 2 or more people; B: Only the Driver). Vent estimated duration: 10 hours.	1.0E+00	9.8E-01	1.3E-01	1.8E-02	4.1E-03
Major release frequency in standard process units	4.0E-03				
People present in the battery to activate Shut Down	1.4E-01				
Someone activates the shut-down	9.0E-01				
The operator can´t control the fire and doesn´t drain the separator to the emergency pool. The oil remaining in the separator evaporates, and the PSV opens	9.0E-01				
Wind direction from vent point to the road	2.3E-01	Worst orientation			
Process Fire	**1.0E-04**	**9.9E-05**	**1.3E-05**	**1.8E-06**	**4.1E-07**
Road risk fire (External + Processes)	**2.0E-04**	**2.0E-04**	**2.6E-05**	**3.6E-06**	**8.4E-07**

	Road orientation from vent	Severity A	Severity B
Risk Fire on primary and secondary roads	12 a 7	**2.0E-04**	**2.0E-04**
	8 a 11	**6.2E-05**	**6.2E-05**
Risk fire on roads used sporadically	12 a 7	**2.6E-05**	**2.6E-05**
	8 a 11	**7.8E-06**	**7.8E-06**

Notes:

1.- The indicated frequency values apply to the worst orientation (so the risk would be lower for the other sectors in each group).

2.- Worst orientation between sectors 12 and 7: Sector 9, Wind Prob=23%.
Worst orientation between sectos 8 and 11: Sectors 12 y 2, Wind Prob=7%.

RECOMMENDATIONS

Based on the safe distances from dispersion simulations and the risk calculations described in this paper, recommendations were prepared and shown graphically in Figure 10.

Figure 10. Recommendations

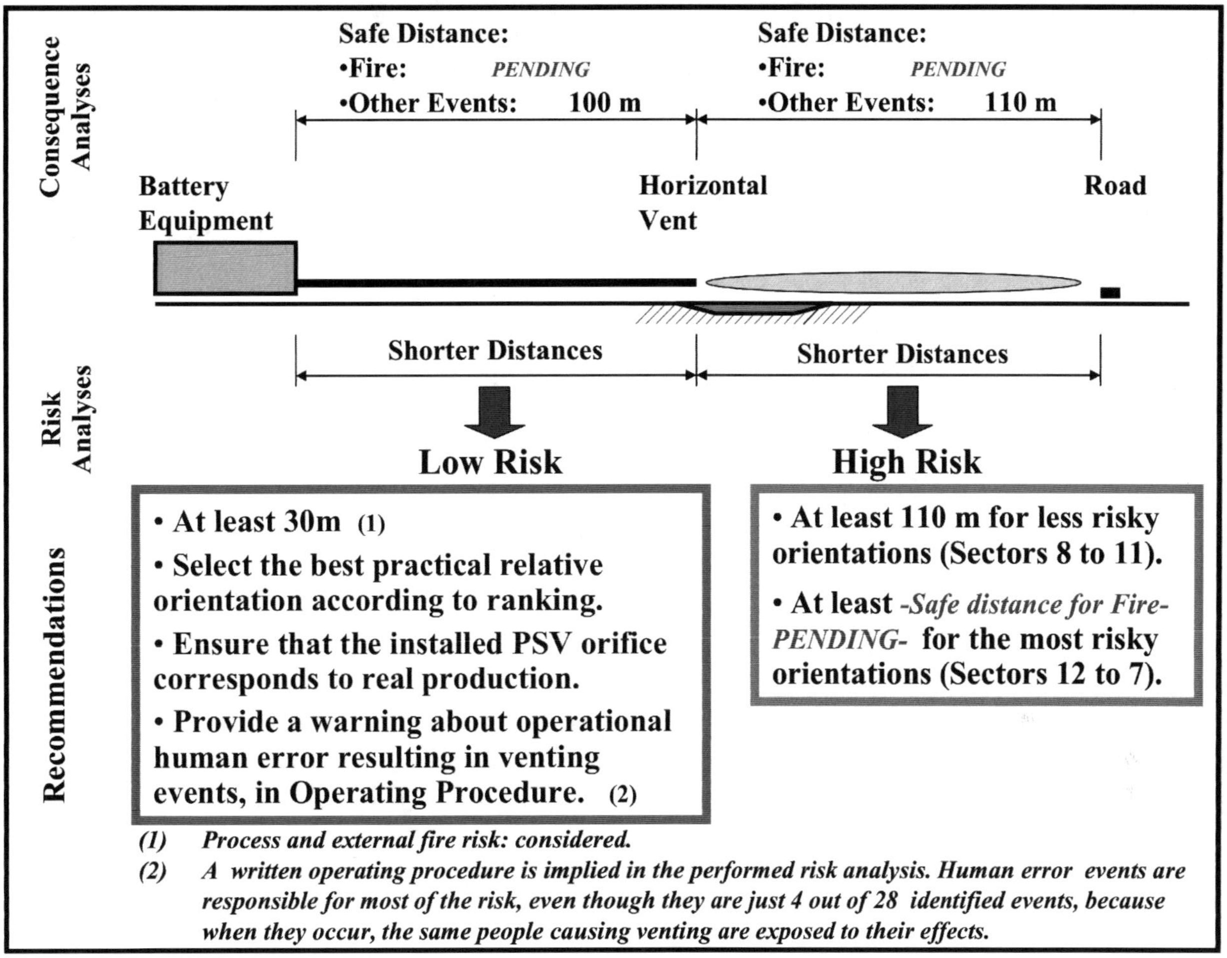

LESSONS LEARNED

Management awareness and commitment to process safety is a key enabler.

Performing the risk analysis in-house has provided significant insight and it will allow other staff to gain PS knowledge.

The analysis may require a higher-than-expected level of detail to obtain a risk assessment of suitable accuracy, and resources must be available to do the job.

Wide-range engineering expertise is required to identify sources of risk.

Involvement of Operations personnel in the analysis is key for a proper description of scenarios.

This kind of studies is a valuable source of PS learning for project engineering and operations staff – the opportunity should not be wasted.

Communication skills, visual aids and minimal jargon are important to convey process safety messages to people other than PS engineers.

BLAST RESPONSE OF SOFT-SKINNED SHELTERS AND OCCUPANT VULNERABILITY

D. V. Ritzel
Dyn-FX Consulting Ltd, 19 Laird Ave North, Amherstburg, Ontario, N9V 2T5 Canada

J. Crocker, D. Whitehouse
Martec Ltd, Suite 400, 1888 Brunswick St., Halifax, Nova Scotia, B3J 3J8 Canada

ABSTRACT

Soft-skinned shelters such as tents offer an expedient means for environmental protection of personnel undertaking construction or maintenance operations at industrial sites, and on occasion these may be deployed within blast hazard zones. Despite their apparent simplicity, the blast response of industrial work tents (soft-skinned, hard-framed) is complex and a rare example of a fully coupled fluid-structure interaction problem in air blast damage analysis. It is not feasible to derive a P-I iso-response diagram by classical methods for example. More importantly, the vulnerability of occupants is not adequately revealed by P-I diagrams since in fact the blast response of a hard-framed tent may be potentially lethal to occupants without having 'failed'. This paper describes the blast response dynamics of certain metal-framed polyester-sheathed tents as well as assessments of occupant vulnerability. Recent analyses of the blast response of fabric-skinned 'air-beam' shelters (soft-skinned, soft-framed) are also presented in comparison. Although some development is required on this latter concept, these systems demonstrate high blast resilience due to their modes of energy absorption and stress dispersal. High-performance air-beam shelters having controlled flexure show good potential for blast protection of occupants as well as mitigating injury risk even at failure.

1. INTRODUCTION

Lightweight relocatable shelters (LRS) include a wide range of both soft and hard-skinned structures such as trailers, huts, tents, as well as certain *ad-hoc* constructions primarily intended for environmental protection of personnel. Such shelters are designed primarily for features of transportability, breakdown size/weight, and ease and speed of setup. For remote deployments, these structures are often the only shelter option where local materials, environment, or required setup speed preclude other constructions. LRS are deployed extensively in expeditionary military camps, stations established by civil authorities during large-scale emergency response operations, as well as industrial/commercial sites under construction or maintenance. Such shelters are used in a great many roles including accommodation, office, workshop, messing, vehicle-bays, and stores; specially adapted units are used for field hospitals or for housing specialized equipment, monitors or sensors.

LRS typically present large surface areas relative to their supporting lightweight, low-rigidity framework and weak ground fixity, and are thereby inherently vulnerable to blast in the event of an accidental explosion or terrorist bombing attack. Of particular interest here is the potential deployment of LRS at certain industrial process operations including petrol-chemical refineries where they serve as office, workshop, rest, or assembly areas including lunch rooms and may be within the regime of blast-hazard zones.

The blast vulnerability of typical hard-framed plywood-sheathed industrial work trailers has been known since the civil defense studies of nuclear blast effects in the 1950s[1] and became tragically highlighted by the casualties of the 2005 Texas City blast[2]. Soft-skinned shelters offer an alternative to hard-sheathed designs and are typically less expensive, easier to store and deploy, are often the only feasible means of enclosing larger field worksites or assembly areas such as mess facilities. However, blast response of soft-skinned shelter systems had been largely unexplored until the late 1990s[3]; the potential for developing an entirely 'soft bunker' concept for protection against moderate blast threats may be credible.

The blast vulnerability of soft-skinned shelters has been the subject of Defense research both in the US and Canada between 1998-2005 with particular regard to terrorist bombing threats such as VBIED (vehicle borne improvised explosive device), or 'truck-bomb', attacks. A comprehensive program of full- and model-scale blast experiments, as well as computational modelling has been conducted to quantify the blast vulnerability of such tents and develop blast hardening upgrades under the Canadian DRDC program[4]. Subsequent to the Defense initiative, industry-sponsored research continued with regard to the blast response of soft-skinned soft-framed (air-beam) shelters[5,6]. Specific results from any studies which link a particular blast threat condition with damage to particular military assets are classified for security reasons. However, in the interests of workplace safety, certain 'sanitized' results (ie, having specific sensitive information deleted) and generalized technical remarks can be presented with regard to deployments of such structures into industrial blast-hazard zones.

The current report will summarize key phenomenology of the blast response of metal-framed soft-skinned tents as well as initial results from investigations of the blast response of soft-framed ('air-beam') soft-skinned shelters. Estimates of the vulnerability of occupants are also discussed. The dynamics of blast interaction with soft-skinned shelters is complex, and the project has required new developments in experimental diagnostics for highly responsive, large-scale structures. High-resolution FSI (fluid-structure interaction) models are required for the computational simulation of this problem which link the response dynamics of the structural components with the compressible gas-dynamics of the blast-wave flow.

2. BLAST RESPONSE OF METAL-FRAMED HEMI-CYLINDRICAL TENTS

Clearly, there are many possible designs, sizes, and material options in the construction of hard-framed fabric-skinned tents. Considering the framework alone, framing elements are typically made of aluminum, steel, or composite tubing, for which there are many options for connectors and ground fixture. Therefore, it is not intended or feasible to draw conclusions regarding the blast resistance of all soft-skinned hard-framed shelters from a single study. However, the evaluation of the stronger currently available tent designs, as indicated by their nominal wind-rating, was expected to provide an upper limit for blast performance of hard-framed tents, as well as provide insight as to the generic response dynamics of this general class of LRS. All tents for which a generic wall segment can be approximated by the simplified representation in Fig. 1, can be expected to exhibit similar blast response behaviour to the robust high-wind designs considered here, although likely to higher extent.

The most robust of the large-scale hard-framed fabric tents are those designed for strong wind conditions such as required for use in arctic regions. Fully framed hemi-cylindrical tents with polyester weather-shell skins as shown in Fig. 2 are a common type of LRS used by armed services at both expeditionary and permanent encampments.

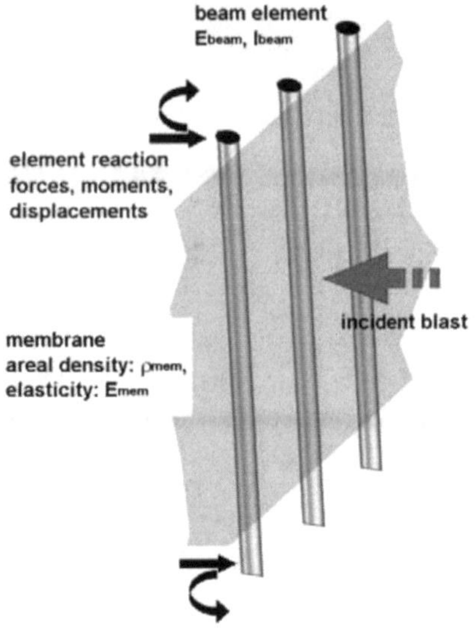

Figure 1. Representation of a generic element from a membrane/beam system typical of hard-framed tents of current study. The weather-shell membrane, typically made of reinforced polyester fabric, is draped over the external surface.

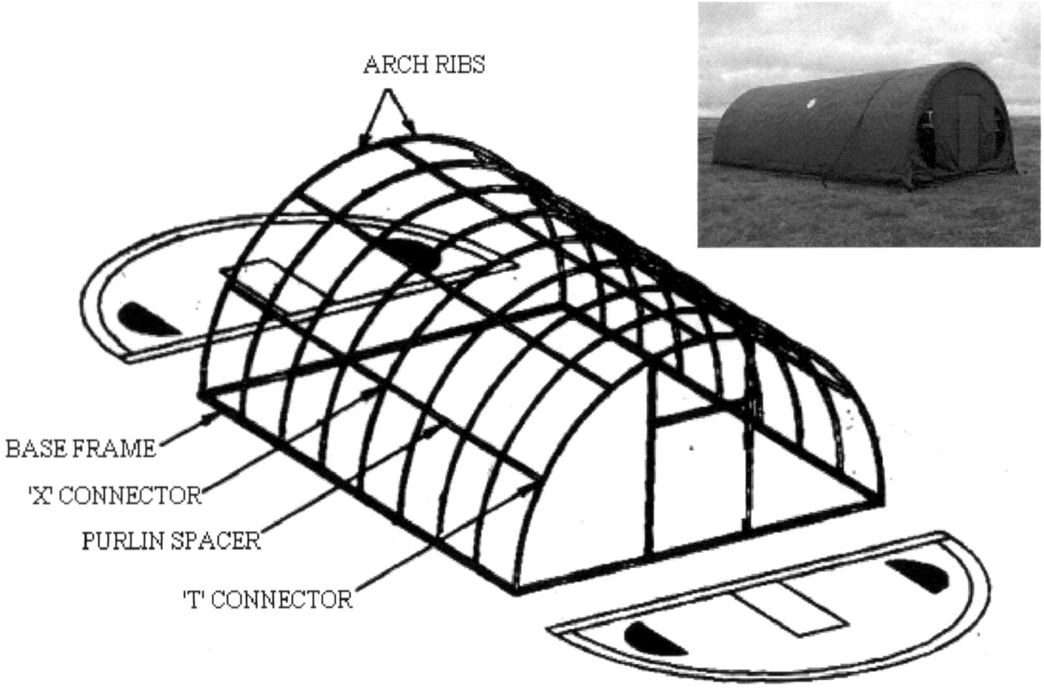

Figure 2. Features of a typical hemi-cylindrical tent for high wind load. Although not having a distinct side 'wall', hard-framed cylindrical and dome tents fall within the category of membrane/frame systems of current interest.

For reasons of intellectual property (IP) and the security of current tent deployments of armed services, specifics of the particular tent manufacturer, tent model names, and details of the blast conditions leading to particular damage outcomes under current study will not be presented. The current paper describes basic phenomenology of the blast response of tents of the type under consideration, advances made in blast diagnostics and experimentation for this problem, as well as general solution concepts.

2.1 Tent Construction

The main structural configuration of the tents under consideration are sketched in Fig. 2. All tents are hemi-cylindrical in shape with a floor area of roughly 5m x 10m with central height of 2.7m (16'x 32' x 9'); the metal superstructure framework consists of arch-ribs and transverse purlin spacers. Although Fig. 2 shows six segments to each arch-rib, typical of larger tents of this style, there were four arch segments for all tent designs studied. Three tent designs were assessed having this basic configuration but with significant differences in features such as the number and stiffness of the arch-ribs. The tent framework is assembled from elemental sections of arch-ribs and purlin spacers which are all joined by connectors. An insulation layer followed by a weather-shell fabric skin are draped over this framework and tethered to the end-frames and base-frame.

Table 1. Summary comparison of tent designs evaluated.

Feature	Design A	Design B	Design C
Frame mat'l	steel, square x-section	aluminum, circular x-section	aluminum, square x-section
No. of Arches	5	9	6
Relative Arch EI	33.7	7.25	13.2
Joint connectors	steel, female, button-pin	steel, male, button or through-pin	steel, male, through-pin
Windows	square, 2 each side at mid-length, 2 on each end-wall	oval segment, 2 on each end-wall	square, 5 each side
Base-Frame	along two sides only	complete rectangle	complete rectangle
End-Frame	optional, secured top only	weakly joined to frame	solidly connected to frame
Door	flap	solid, hinged	flap
Floor	ground sheet	¾" plywood on sheet-metal joist framework	ground sheet
Base-frame fixture	ground staking	bolted/screwed to floor platform	ground staking
Ground tether	optional hold-down strap	optional hold-down strap	four tethers to skin each side

2.2 Full-Scale Blast Field Trials

A total of three large-scale blast field trial events were staged at the DRDC Suffield Experimental Proving Ground over two years spanning September 2004 to September 2006. The tests subjected a total of seven tents of various designs or configurations to the same reference side-on blast conditions from a 2000kg high-explosive charge, roughly representing a large truck bomb. The particular blast-load conditions for the testing had been estimated as likely to cause very severe damage to such tents as currently deployed, yet at a threshold where upgrade measures could be expected to make a difference.

The trials had the primary objective of assessing the relative blast resistance of various tent designs and identify the mechanisms by which the tents failed or demonstrated superior blast resistance. A secondary objective was to assess the injury potential to occupants from the transmitted blast, deflected framework, and projected debris or wall attachments. Instrumentation included internal and external high-speed digital video, internal and external blast pressure gauges, displacement tracking, high-resolution laser-profiling of final deformation, as well as manikin response dynamics. Details of the trial design and instrumentation are described elsewhere[4], and only summary results and conclusions will be presented here for brevity.

Representative data from the field-trial series are shown in Figs, 3-6 showing the response dynamics and blast pressure conditions.

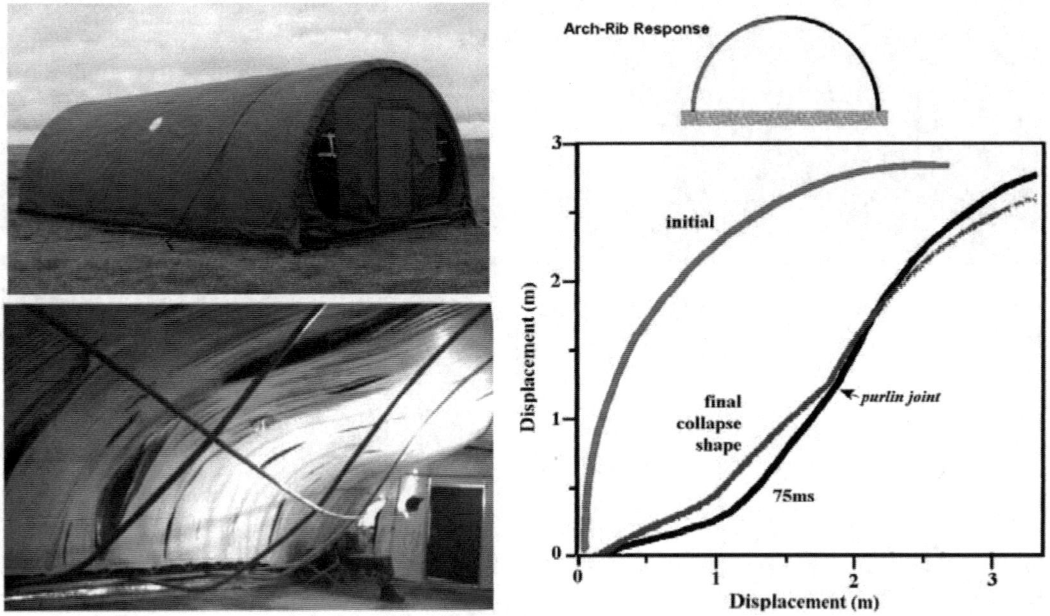

Figure 3. [Upper Left] Tent prior to blast test. [Lower left] Frame from internal high-speed video showing severe deflection of the arch-ribs and detachment at their base; a seated manikin has been struck by a deflecting arch. [Right] Deformation of the front half of central arch-rib comparing initial shape, peak dynamic deflection as resolved by rectified image analysis, and final deformed shape as measured by post-event laser profiling.

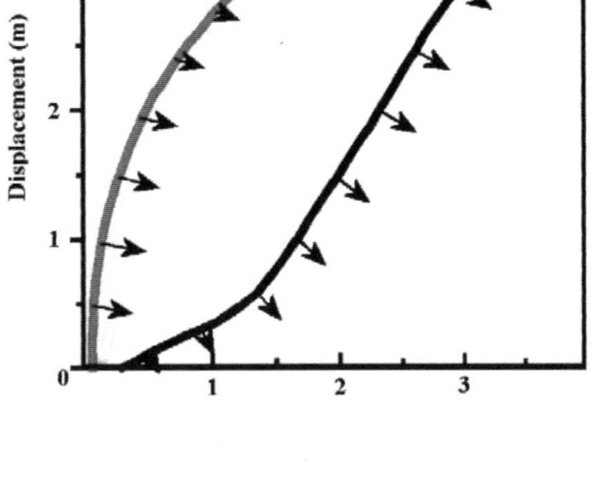

Figure 4. Instantaneous velocity vectors for the deflecting arch-rib shortly after blast arrival (red) and at the end of the positive phase of the blast overpressure (black). Maximum velocities attained by the framework would be sufficient to cause skull fractures, although the blast conditions in the free-field on their own would be well below lethal thresholds from primary blast injury. Membrane forces accelerating the frame to this extent were imparted during the first third of the blast duration.

Figure 5. [Upper sequence] Frames from internal high-speed video showing tent response including rupture of the membrane skin (different tent design to that for Fig. 4). Dynamic deflections reached 1.2m yet the structure rebounded close to its initial condition; the frame deflection would have inflicted skull fracture on occupants in its pathway. [Left] Condition of the framework after the event showing minor frame damage despite the high deflection which would clearly have been hazardous to occupants.

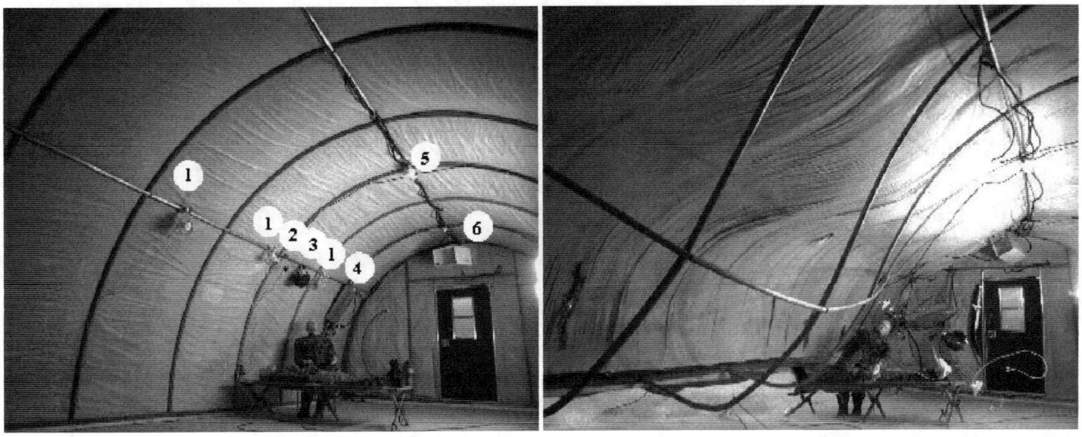

Figure 6. Interior of blast-loaded tent installed with manikins and fittings typical of habited shelters (1: clip-on fan; 2: clip-on lamp; 3: helmet; 4: circuit panel; 5: lamp fixture; 6: 5KW heater. The seated manikin would have suffered skull fracture from the impact of the nearby arch-rib. Detached fittings, gear, debris, and whipping wires are projected throughout much of the internal space.

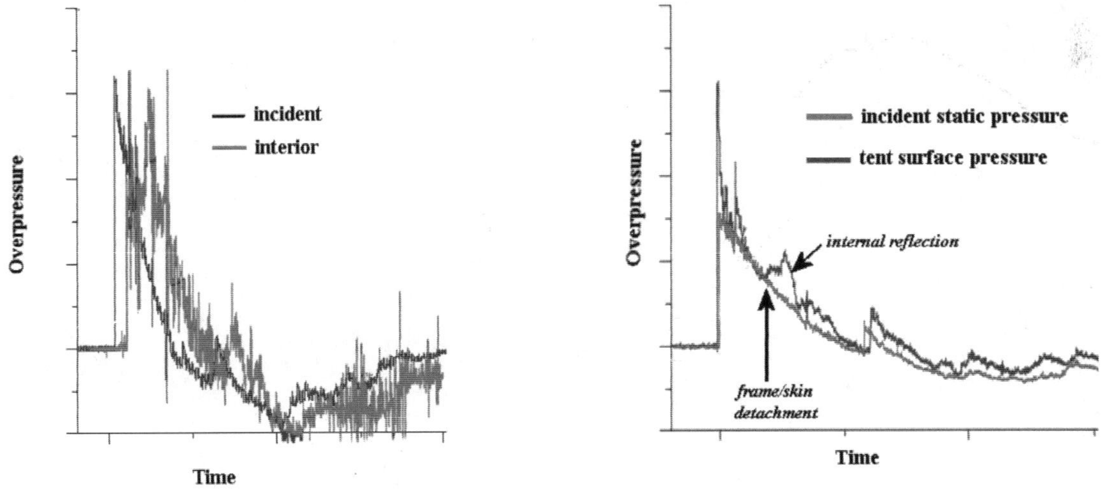

Figure 7. Pressure gauge records comparing external incident blast static pressure, internal pressure at tent center floor level, and the pressure measured on the deflecting fabric surface by a B-H 'Ride-Along' gauge. Note that the surface pressure on the fabric skin does not directly represent the load passed to the framework as

2.3 SCALE-MODEL TESTING AND COMPUTATIONAL MODELING

Despite the consistent and repeatable results for the blast response dynamics demonstrated for a wide range of tent designs in the full-scale field trials, further correlations and validations were conducted by means of both scale-model testing and advanced computational modeling. The scale-model work was required to allow cost-effective parametric investigation of the effects of various upgrades and modifications. The computational modeling was required in order to derive a credible P-I diagram for tents, that is, resolve the response outcomes for a full range of blast conditions beyond the single data-point afforded by the field trials.

Simplified responding model tents were constructed at approximately 1/12th scale to assess the relative importance of factors such as frame stiffness, ground fixity, tethering, and skin-sheathing properties. The DRDC Suffield 1.8m Blast Tube facility was used to generate repeatable and well-defined blast wave profiles. Details of the scale-model testing are described elsewhere[4]. This testing proved to be a highly reliable and cost effective means to conduct the parametric studies required. As shown in Fig. 8, results for elastic deflection and final plastic deformation proved to be remarkably close to those observed from the full-scale field trials, although it was not the intention or expectation to replicate any particular result from the full-scale trials.

Figure 8. [Upper left] The 1.8m Blast Simulator facility at DRDC Suffield; [upper right] 1/12th scale simplified responding model on the test table; [lower left] deformation of the wire-frame structure of the model; [lower right] frame deformation from full-scale trials.

From the trials and preliminary analysis, it was clear that the response of soft-skinned structures is fully coupled with the blast flow. It is not possible to presume, for example, that the loading to the framework was directly and simply due to the pressure on the outer skin surface through time as with standard blast load analyses. When the blast front first encounters the skin there is a brief and weak external reflection; however, depending on the mass/area and flexure of the skin, most of the blast energy becomes propagated into the shelter space as the skin is abruptly accelerated forward. Some energy is also dissipated by visco-elastic processes of the material deformation. Subsequently, the skin motion largely follows the air flow velocity in the blast wave and in fact separates from the framework long before, and unrelated to, the end of the positive phase duration of the static overpressure. The pressure on both sides of the membrane soon approaches the equivalent to the Lagrangian static pressure condition. The loading to the framing is entirely dominated by the early skin membrane action regardless of the pressure history on the outer surface of the skin. Rather than simply apply a designated pressure loading function to the skin surface *a-priori* to undertake P-I response analyses, a fully resolved blast flow with air velocity and density fields must be included in the modeling of the blast/structure interaction.

A high-resolution fully-coupled computational model linking the FE code LS-DYNA with the blast CFD code CHINOOK was implemented with the primary objective to generate P-I diagrams for particular tent designs. Illustrative results from the modeling which show the very good correlation with the experimental field trial results are shown in Figure 9.

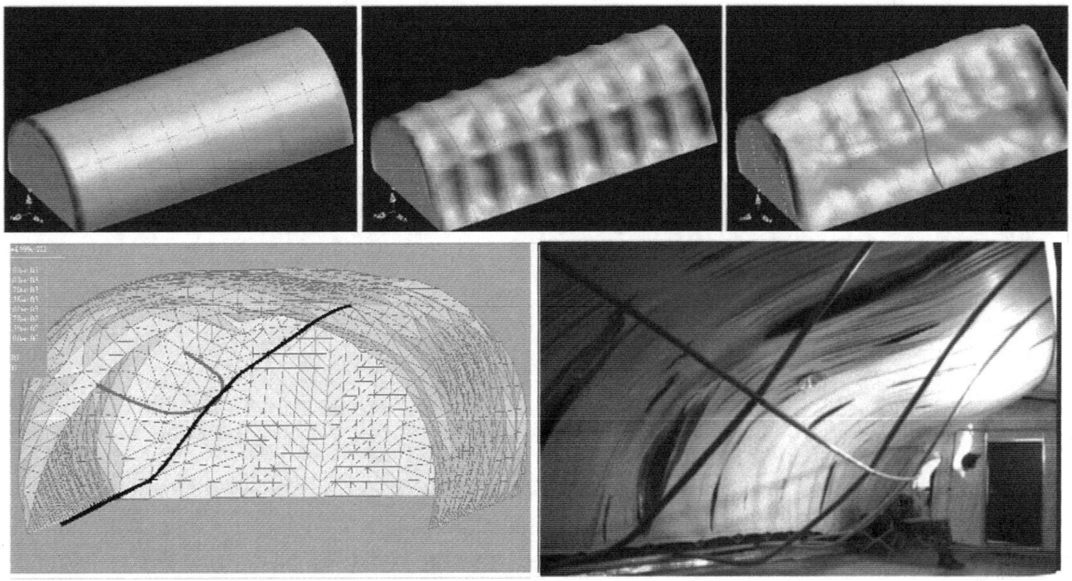

Figure 9. Computational modeling results for the simulation of blast on a hemi-cylindrical metal-framed tent. The model requires the coupling of FE and CFD codes to properly resolve the response; as shown in the lower two figures, the behaviour of the fabric skin is only linked to the framework response for a relatively brief initial period of membrane action whereon the response of the framing is inertial.

2.4 DISCUSSION

Despite the significant differences in the tent designs studied, a consistent picture of the blast response of membrane/hard-frame tents has emerged. The physics of the blast interaction with the tent is not straightforward; the response dynamics of the skin, framework, and air within the tent, all interact as depicted in Fig. 10. On first encounter with the tent skin, the blast wave is partly reflected but mostly transmitted to the tent interior by the shock-induced motion of the skin. The interior pressure records confirm that a sharp-front shockwave with amplitude about ¾ that of the incident blast quickly develops inside the tent even without rupture of the skin. The blast-loaded skin develops strong membrane forces on the frame, and it is this action, rather than the direct effect of the blast wave, that loads and accelerates the framework. However, the tent skin having vast surface area relative to its mass/area will not move faster than the air-flow velocity of the incident blast wave, and soon detaches from the frame having accelerated it forward in a 'sling-shot' effect. This detachment of the skin from the frame occurs well before the negative phase of the free-field blast, and in fact the skin billows backwards towards GZ before arrival of the negative phase due to blast-wave reflections within the tent.

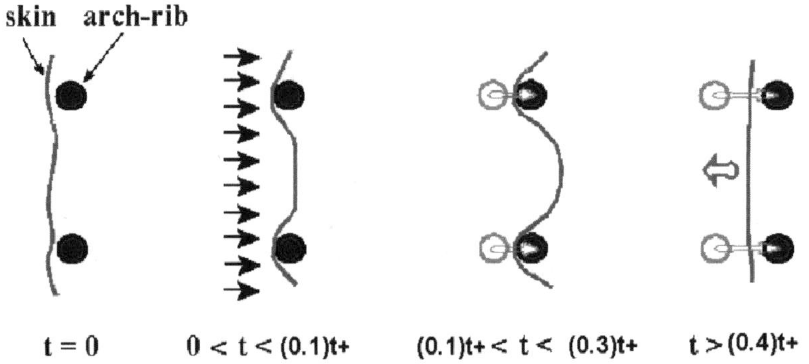

Figure 10. Simplified depiction of blast interaction with a typical membrane/beam system of hard-framed tents. The membrane action of the skin throws the framing ahead in a 'sling-shot' effect; the time record of the membrane action on the framing is largely unrelated to the surface pressure on the skin at any particular time. The time scales shown for the process of accelerating the framework are with respect to the positive duration of the blast wave static overpressure, t+; values shown are particular to the tents and loading

Therefore, the critical blast loading and acceleration of the metal framework due to the skin membrane action occurs over a much shorter period than the positive overpressure duration would suggest. However, high loading and momentum are imparted to the frame during this initial membrane action. The lower segment of the arch-rib is typically bent inwards around its mid-section as shown in Fig. 9, and the arch is globally collapsed away from the blast. This bending and collapsing action of the arches twists the base frame along its length thereby prying at any pins or bolts holding the base-frame to the ground or flooring; the base-frame may become detached if that base fixture is weak.

There are clearly implications for injury risk to personnel occupying hard-framed tents even at exposure to low-level blast. The fabric membrane skin of typical metal-framed tents serves to 'capture' the blast load over the entire projected area exposed to the blast and propel the framework dangerously into the occupied space. Even at low threshold blasts from which the tents may rebound, the elastic deflection of the metal framework presents high risk to any personnel within the vicinity of the deflection. Furthermore, any internal attachments to the membrane or framework can become flung into the habitable space at high velocity creating further risk of injury. Electrical lines and junction boxes are likely to be disrupted with consequent risks of fires or exposure to live wires.

3. BLAST RESPONSE OF SOFT-SKINNED AIR-BEAM SHELTERS

It would be desirable to keep the advantages of soft-skinned shelters as compared to other LRS options with regard to versatility, cost, weight, ease of storage/transport/deployment, yet deal with the blast risk factors described in the previous section. At first consideration, the most straightforward blast upgrade for hard-framed tents would be to pad and/or further restrain the beam elements from deflection. In fact, such upgrades can be achieved without great cost or re-design effort. However, at best such modifications ultimately extend the blast survivability limit by tens of percent, and do not change the fundamental injury risk.

Air-inflatable temporary structures have been commercially available for decades although very few designs are sufficiently robust to be considered for evaluation against blast threat conditions even to 10kPa overpressure levels. Using the prior pre-screening selection requirement of being rated for industrial use in potentially severe wind conditions, an air-beam shelter of proven performance in arctic-type wind conditions was identified. A unique tethering and restraint system for the primary air-beam support structures afforded this performance, although the shelters had not been designed explicitly for blast resilience. The air-beam shelters have no hard framing or paneling in their construction and are self-supporting by means of large pressurized-air columns of polyester-fabric tubing formed into arches as shown in Fig. 11. The supporting air-beam columns are covered by a polyester fabric skin, also having an aramid (Kevlar®) fabric option for ballistic protection.

Two commercially available models were evaluated for blast resilience as shown in Fig. 11, being different in scale by a factor of nearly two, designated here by X10 and X19 (not the manufacturer's model identifiers) to indicate their respective nominal spans of 10m (33ft) and 18.9m (62ft) respectively. The larger of the shelters has a different profile to the air-beam support arch, including a vertical side-wall segment of 2.4m (8ft). The air-beam arches are 1.3m in diameter at 1.8m spacing in comparison to 1m diameter at 1.55m spacing for the smaller X10. Another critical difference in the X19 is the inclusion of the lightweight non-structural inner liner, essentially attached as a drape to give a wall appearance to the inside of the structure. This feature will be shown to have a significant effect on the blast response.

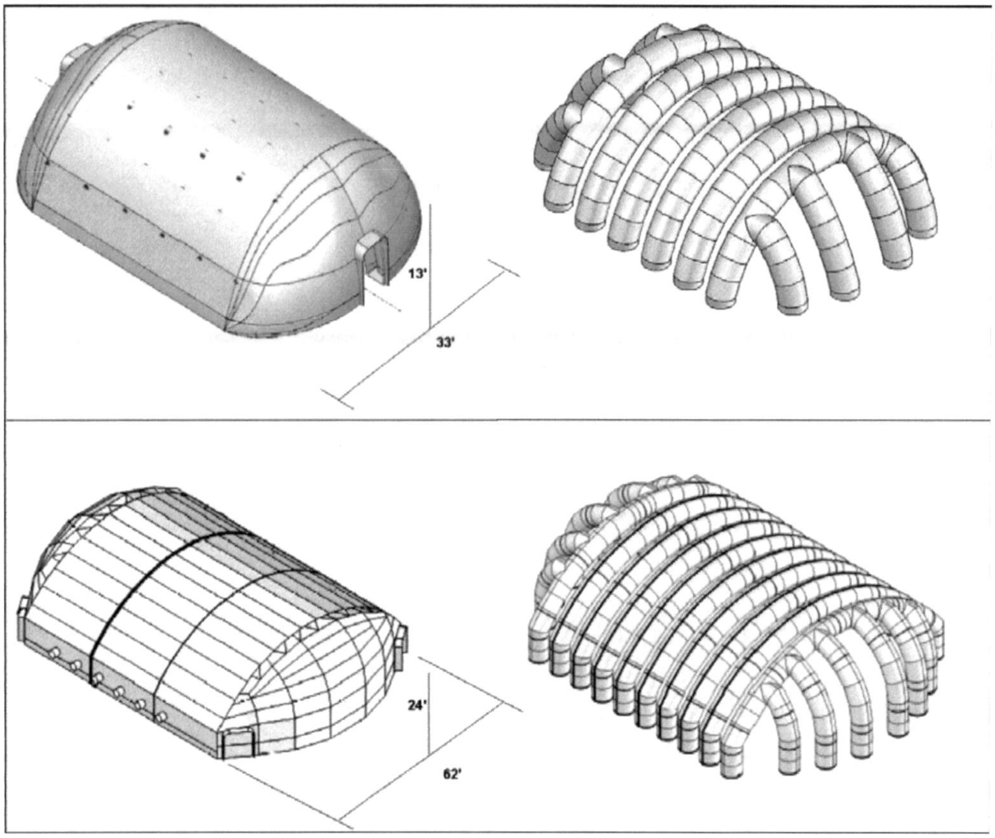

Figure 11. Comparison of the general configuration of the X10 (above) and X19 (below); the underlying air-beam arches supporting the structures are shown to the right in both cases.

The two shelters were evaluated for blast resistance principally by means of the FSI modeling technique developed and validated previously for the hard-framed tents; the X10 was also tested in a large-scale field trial to blast levels slightly higher than previous tent trials.

The FSI blast response model has been simplified to that of a single air-beam arch as shown in Fig. 12 as if this were a segment from an infinitely long structure. The modeling approach chosen for this analysis saves substantial computational effort and yields very conservative results for the following reasons:

- loads are presumed from the most vulnerable direction, lateral to the long axis;
- blast profiles were representative of fuel-air explosive events having typically 50% higher impulse than comparable high-explosive blast source for given peak pressure and duration;
- global stiffness given by the 'buttress' ends of the true structure is neglected; and
- load diminishment from blast diffraction around the ends of the true structure and spherical expansion across the computational domain are not included.

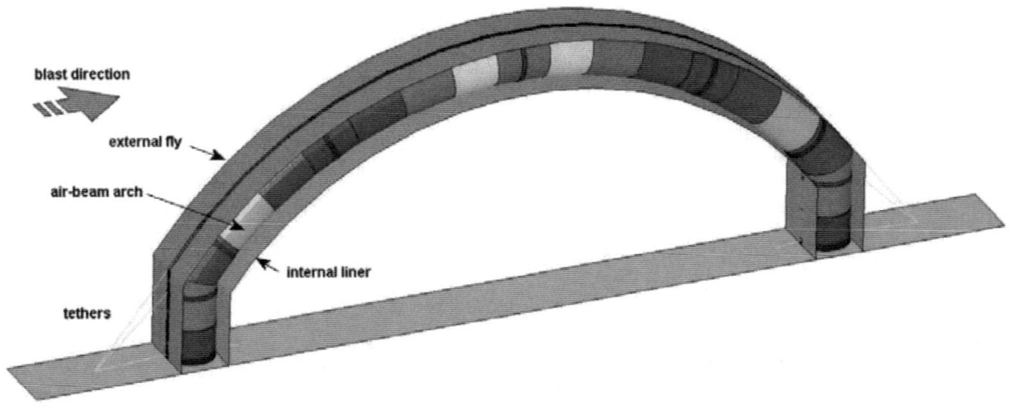

Figure 12. LS-DYNA finite element model of single arch-beam segment for the X19 showing the air-beam, outer fly, external tethers, and inner liner. Air is treated as a compressible material within the air-beam. This LS-DYNA FE model is two-way coupled with the blast CFD code Chinook since accounting for the FSI dynamics is critical to the response.

It is important to define a 'measure of effectiveness' or response criterion relevant to the critical issue of concern which in this case is injury risk to occupants. In this case, the deflection of the air-beam walls would present some degree of injury risk to personnel within the shelter and also relates to the strength of the transmitted blast to the interior, and this has been taken as the critical response criterion. A person standing in a zone into which the air-beams deflect is likely to be knocked to the ground although the initial impact itself is unlikely to be injurious. Since in all cases relevant here the ultimate response of the structure is to rebound, the maximum elastic incursion of the deflecting walls into the *primary habitable space (PHS)* was considered to be the critical measure of response. The PHS for the X19 was defined as the central rectangular cross-sectional zone starting 0.5m (20in) from either wall and 2.0m (6.6ft) high defining a cross-section area of 30.6m^2.

Figure 13 shows the results for the maximum wall deflection for three blast conditions of similar peak overpressure near 20kPa (3psi), but increasing duration and hence increasing impulse. It can be seen that the first blast condition of 38ms duration does not cause the walls to deflect into the PHS. The 102ms blast causes the air-beam to intrude into a corner of this zone having an 'intruded area' of 0.83m^2. As expected, the severe load case of 177ms causes the most deflection of about 1.5m (5ft), an area intrusion into the PHS of about 0.98 m^2, or about 3.3%. Since the structure rebounds, the velocity of the wall at these maximum deflections approaches zero and presents the greatest injury risk as it enters the PHS.

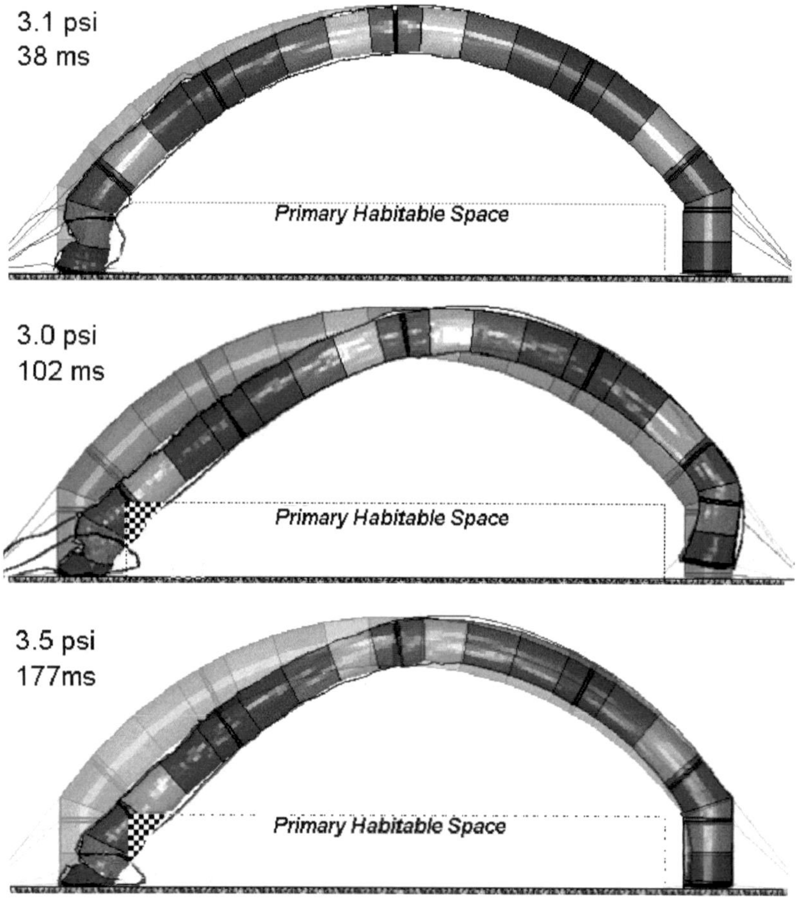

Figure 13. Maximum deflection of the air-beam arches for blast conditions of about 20 kPa (3 psi) and increasing duration or impulse. The initial shape of the air-beam profile is shown as a ghosted image.

It is typical for larger structures presenting a greater projected area to the incident blast to have greater blast vulnerability, ie, it is more difficult to protect a large or high area using the standard 'hard bunker' design approach due to the greater spans and moments to be resisted. The air-beam shelter exhibited somewhat superior blast resistance with size due to the fact the blast essentially 'ripples through' the air-beam system.

An unexpected outcome from the X19 evaluation relates to the particular mode of response of the vertical wall segment as shown in Figs. 14 and 15, for which the combined effect of the inner liner, arch geometry, and tethering are very significant. The blast wave transmitted through the air-beam becomes reflected and confined by the inner liner causing development of a pressurized air pocket between the liner and the deflecting air beam. The back-pressure from this zone of compressed air stabilizes the inner surface of the air-beam and inhibits its inward motion. The combined effect of this restraining pressure, the weight of the structure on the buckling wall, and the action of the tethers, causes the air-beam wall to collapse downwards and 'roll-under' rather than deflect inward. The restraint given by the uppermost tethers on the vertical wall segment has an important role in this process. The entire volume of air between the outer fly and inner liner becomes part of the response system ultimately restrained by the external tethers. The adiabatic compression of trapped air volumes as well as the inherent visco-elastic behavior of most of the structural materials suggest some portion of the blast energy is being absorbed or dissipated by the structure. These and other energy dissipation processes could be exploited in future designs such as by use of 'dashpot' damper components in the tethering system.

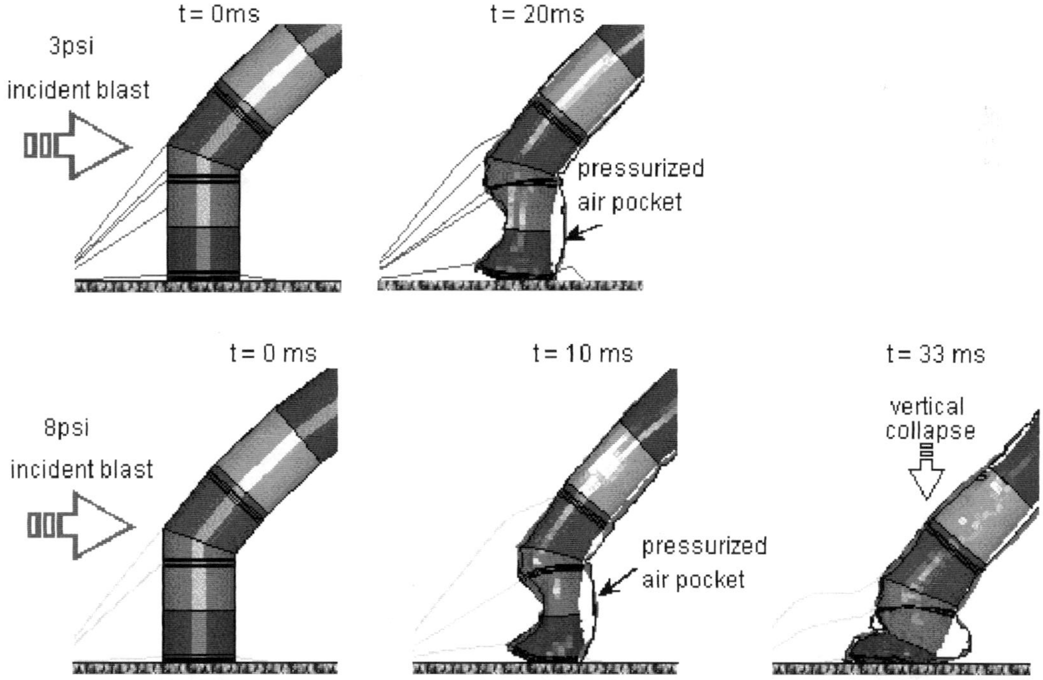

Figure 14. Effect of the inner liner in stabilizing the wall deflection.

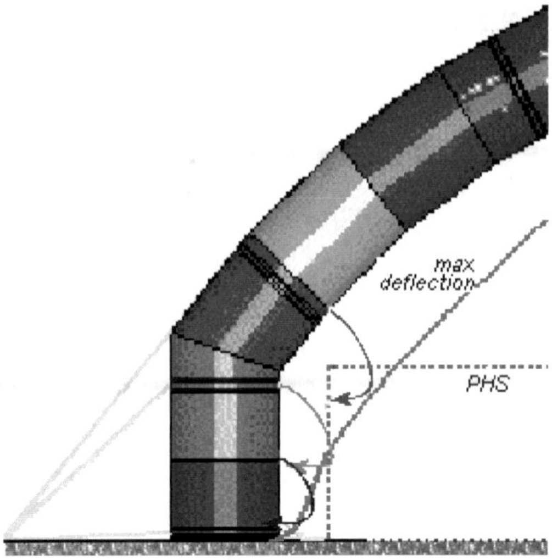

Figure 15. Trajectories of three distinct points on the air-beam arch show that the air-beam deflection into the PHS has a predominantly downward and 'rolling-under' motion having a relatively low horizontal velocity component.

The smaller X10 airbeam structure was evaluated for blast response both through FSI computational modeling and a full-scale blast field trial. The blast field test is described in detail in Ref. x, and only summary results will be presented here. The blast exposure was 30.6kPa at 45ms duration as shown in Fig. 16.

Figure 16. Blast testing of the X10 shelter showing the incident blast conditions.

Results from the internal high-speed video tracking the internal deflection of the air-beam columns under blast load are shown in Fig. 17. As previously noted, the air-beam structures had not been specifically upgraded or fortified for the blast test, but were tested as-supplied for their standard industrial field deployments. Several of the key ground tethers broke early in the response, hence deflections were higher than expected from the modeling. Furthermore, one of the airbeam columns was torn open due to the protrusion of a particular internal ground stake. Despite these localized structural failures which would be easily rectified in an upgraded design, the air-beam deflection was both less severe and would be less injurious to occupants.

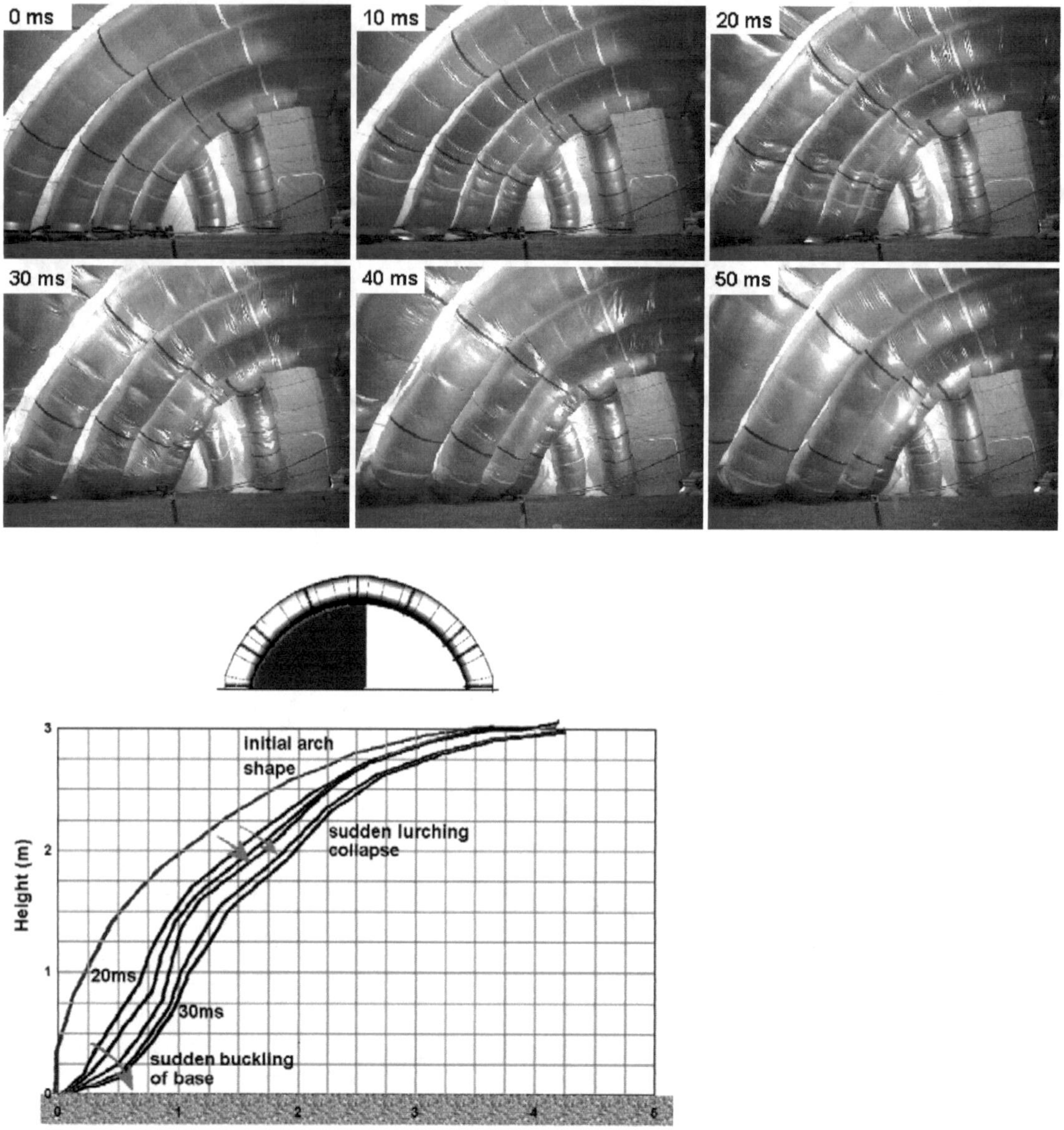

Figure 17. Blast deflection of the X10 shelter, exaggerated in this case due to failure of the base tethering restraints early in the loading which caused a sudden buckling collapse at the base of the air-beam column. Strengthening and some reconfiguration of the tethering system were seen as straightforward upgrades for a blast-upgraded version of the shelter.

Due to the flexing nature of the shelter response, the propagation of a pressure wave into the interior habitable space is of some concern. Measurements of the internal pressurization were taken in the trial and also analyzed by means of the FSI computational modeling. The test results confirmed that a diminished and somewhat distorted pressure wave is propagated into the shelter space as shown in Fig. 18, comparable to that observed in the blast testing of the hard-framed tents. In all cases of relevance here the incident blast is well below the threshold for lethality from primary blast injury for humans exposed in the open (nominally, 100kPa). Since the transmitted pressure wave has been further attenuated and dispersed somewhat into compressions as opposed to shocks, the risk of primary blast injury to occupants is negligible for the test case and always lower than blast exposure in the open.

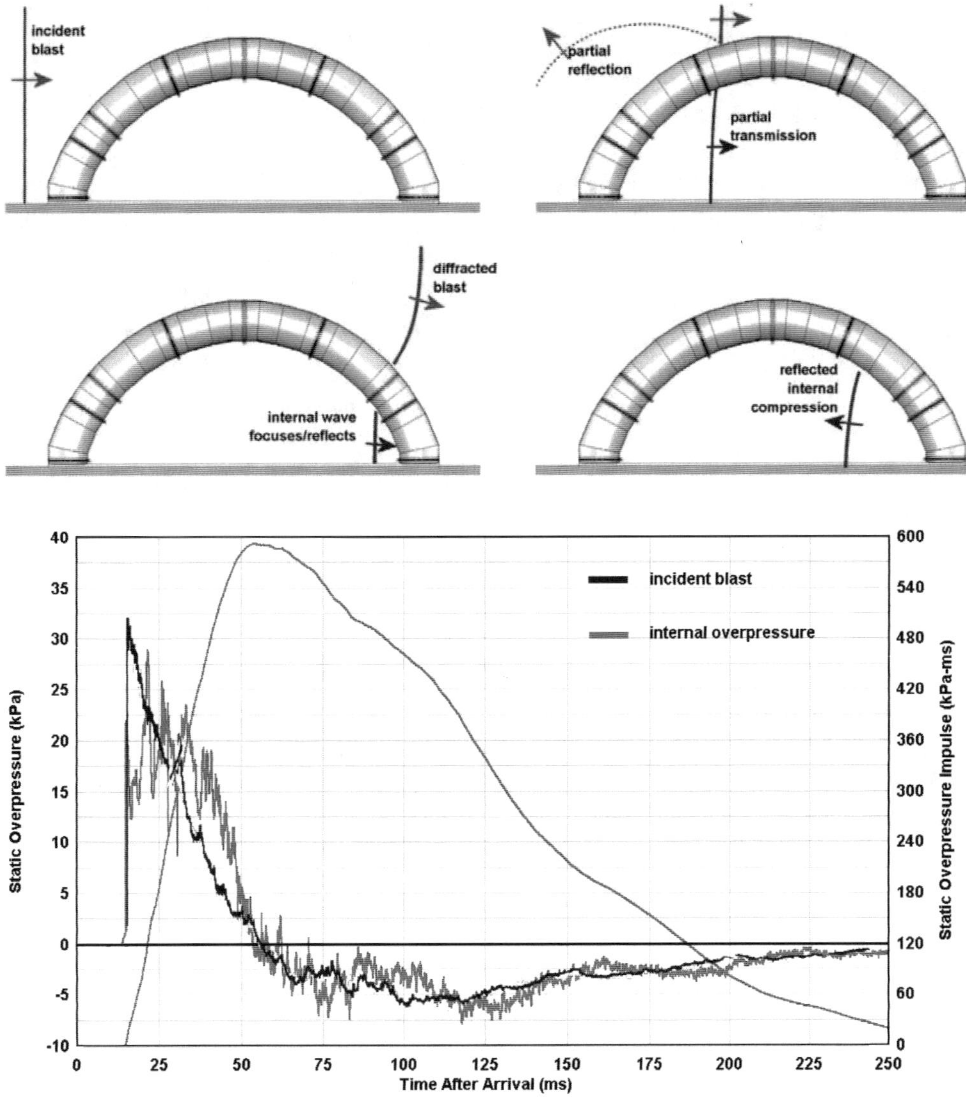

Figure 18. Depiction of the pressure wave dynamics of the blast interaction with the X10 air-beam shelter and comparison of measurements of the external incident static over- pressure with the internal overpressure record as obtained in the X10 blast field trial.

A P-I diagram summarizing the blast response of the air-beam shelters using the criterion of 'maximum intruded area' expressed as a percentage of the primary habitable space, is given in Fig. 19.

3. 2 DISCUSSION

Blast essentially 'ripples over' the air-beam structures evaluated in this study causing some deflection from which the shelter rebounds. The supporting air-beam and tether system does not transmit bending moments or shear as with hard framing, and forces are taken by membrane and tensile stresses for which the materials are inherently strong and energy-absorbing. The particular wrap-around restraint or 'hug-strap' for the air-beam column designed for performance under high wind loads is also critical to its good blast resilience. Furthermore, the visco-elastic flexing of the fabric surfaces and compression of trapped volumes of air partially absorbs the incident blast. The particular downward buckling response of the air-beam shelter having a vertical side-wall segment could be further exploited to diminish the more dangerous modes of inward wall deflection. For wall deflections into the primary habitable space there is some risk of 'knock-down' injury from the impact of the air-beams. Methods to evaluate this injury risk using computational simulations validated for car-crash air-bag deployments can be applied if detailed assessments are required.

Strengthening of certain critical tethers and attachments is recommended and straightforward. In fact a revamping of the external ground tethering system could greatly extend the blast resilience of the design. However, it was noteworthy that even with failures of the tethers as currently designed, considerable energy is absorbed and injury risk is greatly mitigated in comparison to the response dynamics of hard-framed structures.

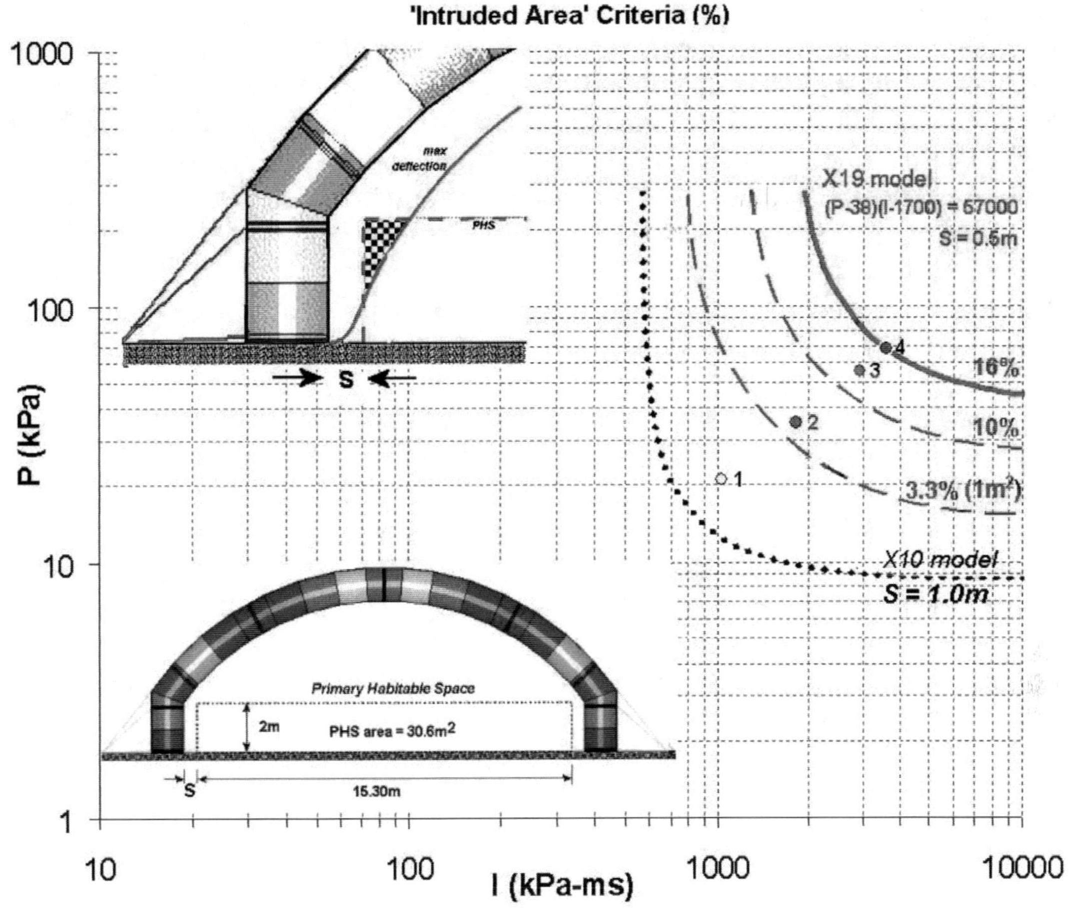

Case Study	Peak Overpressure ΔP		Positive Phase Impulse ΔI		Duration Δt+
	psi	kPa	psi-ms	kPa-ms	ms
1	3	20.9	149	1027	102
2	5	34.7	264	1820	102
3	8	55.6	425	2929	104
4	10	68.5	518	3570	103

Figure 19. P-I diagram showing iso-response contours for air-beam shelters X10 and X19.

4. CONCLUSION

The injury risk to occupants of hard-sheathed hard-framed temporary shelters such as typical industrial work trailers exposed to blast is now well known. Soft-skinned shelters such as tents offer an alternative to hard-sheathed designs and are typically less expensive, easier to store and deploy, are often the only feasible means of enclosing larger field worksites or assembly areas such as mess facilities. However, soft-skinned shelters may also be deployed to sites having blast hazards and have not been evaluated for blast resilience and injury risk until relatively recently.

The current report summarizes progress in the analysis of the blast response of soft-skinned shelters. Due to the highly responsive nature of the fabric surfaces, the blast response dynamics of soft-skinned shelters is in fact quite complex and an example of a fluid-structure interaction problem unlike conventional air-blast response analyses for structures. It has been shown that typical hard-framed tents are highly responsive to blast, and framework can deflect dangerously into the habitable space even without having structural failures. The impact of the deflected framework or components, as well as the possible projection of wall attachments or electrical wiring, are shown to present significant risk of injury to occupants and present a comparable or higher threat than in the case of hard-sheathed shelters.

However, specially designed 'soft-skinned, soft-framed' air-beam shelters have demonstrated remarkable potential for blast resilience and mitigation of injury risk to occupants. The supporting air-beam and tether system does not transmit complex bending moments or shear stresses as with hard framing. Such a structural system flexes when subjected to blast and ultimately takes shock loads as membrane and tensile stresses for which the materials are inherently strong and energy absorbing. The degree and mode of deflection can be largely controlled by tethering, and consequences of air-beam impact clearly less injurious than hard-framed structures. Although unconventional and requiring some further development, testing and computational analyses have demonstrated that carefully designed air-beam shelters may offer a 'soft bunker' solution for blast protection within certain threat limits.

5. REFERENCES

1. Glasstone S., and Dolan, P.J., (Editors), "The Effects of Nuclear Weapons", *Chapter V: Structural Damage from Air Blast*, Third Edition, US Dept. of Defense and the Energy Research and Development Administration, 1977.
2. "Trailer Blast Damage Information, BP Texas City Accident", News Release, US Chemical Safety Board, 30 June 2006.
3. Stevens, D., Marchand, K., Young, L.A., Moriarty, R., Cropsey, L., "Evaluation of Structural Response and Human Injury in Shock-Loaded Expeditionary and Temporary Shelters", Proceedings of the 11th International Symposium on the Interaction of the Effects of Munitions with Structures, 5-9 May 2003, Mannheim Germany. *Releasable to NATO Military Authorities and Government Agencies of Military Aligned NATO Member Nations (US Authority: Defense Threat Reduction Agency, 8725 John J. Kingman Rd., Ft. Belvoir, VA 22060-6201)*
4. Ritzel, D.V., Crocker, J., "Blast Response of Hemi-Cylindrical Tents", 19th Int'l Symposium on Military Aspects of Blast and Shock (MABS19), Banff, Alberta, Canada, 1-5 Oct 2006.
5. Ritzel, D.V., "Blast Response of the DSI-10 Air-Beam Shelter", Dyn-FX Contract Report CR040801, Dyn-FX Consulting Ltd, May 2008. COMMERCIAL-IN-CONFIDENCE.

6. Ritzel, D.V., “Blast Response of the DSI-19 Air-Beam Shelter”, Dyn-FX Contract Report CR100801, Dyn-FX Consulting Ltd, Nov 2008. COMMERCIAL-IN-CONFIDENCE.

ACKNOWLEDGMENTS AND DISCLAIMER

Much of the material presented here was derived from published research by Defence R&D Canada under the program “Force Protection Against Enhanced Blast”. Mr. Kevin Scherbatiuk (contract scientific authority); and Dr. Stephen Murray (Program Scientific Authority). DRDC neither endorses nor refutes the analyses or interpretations presented which are entirely those of the authors. Mr. Simon Yang. Ms. Sara Baftechi, and Mr. Tim Dunbar at Martec Ltd conducted the FSI modeling. Additional support for the modelling effort was provided by Dr. Rick Link, Mr. Ken MacKay, Dr. Lars Voxen Hansen and Mr. Eric Li.

Identifying Key Safety Roles during Organisational Change

P.A. Davidson; Unilever plc., Colworth Science Park, Sharnbrook, Bedford, MK44 1LQ, UK. paul.davidson@unilever.com
S.D. Mooney; Unilever UK Ltd., Leeds Sourcing Unit, Coal Road, Seacroft, Leeds, LS14 2AR, UK. dominic.mooney@unilever.com

1. ABSTRACT

As businesses move to lean manufacture to minimise costs this inherently leads to reduced staffing levels and the removal of layers of management. During these organisational changes it is important that safety is not compromised by the removal of key staff. In order to prevent major accident events occurring, a number of layers of controls and defences are put in place to both prevent an event and to mitigate the consequences if an event occurs. There are many competing techniques for assessing the impact of organisational change, each of which has its pros and cons.

This paper discusses a technique based on the use of the Bow Tie methodology to analyse major accident scenarios for a high hazard site, presenting examples from a real life case study. Controls identified in the analysis then have a responsible person assigned to them and a summary of these roles is used to define the impact of removing a role. Keys roles responsible for safety controls will be maintained and any eliminated roles can have their responsibilities reassigned. After the change, final role descriptions will specifically include descriptions of key safety control responsibilities.

2. THE MANUFACTURING SITE

The manufacturing site at which the work described in this paper was carried out is one of the largest personal care aerosol manufacturing sites in the world. Producing over 400 million aerosols per year, the site is located in the North of England. The site operates 24 hours per day, 7 days per week and employed 465 staff prior to restructuring. In the case of the changes considered in this paper, a significant delayering exercise was planned with a role reduction of 15% (65 people), all of whom were at supervisory / managerial level or provided technical support.

The manufacturing site's management procedures are certified under ISO9001, ISO14001 and BS EN 18001:2007 and referred to as the 'Integrated Management System' (IMS).

Personal Care aerosols use a flammable LPG propellant (a blend of propane, butane and isobutane) as well as a flammable base of either ethanol or volatile siloxane. The site has bulk storage of 150 tonnes of LPG, 120 tonnes of ethanol and 100 tonnes of volatile siloxane, as well as up to 40 tonnes of LPG in packed stock awaiting shipment offsite.

3. REGULATORY ENVIRONMENT

3.1. Seveso II Directive

In 1976, the "Seveso" accident happened at a chemical plant in Seveso, Italy that manufactured pesticides and herbicides. A dense vapour cloud containing tetrachlorodibenzoparadioxin (commonly known as dioxin) was released from a reactor, used for the production of trichlorophenol, due to an uncontrolled exothermic reaction. Although no immediate fatalities were reported, kilogramme quantities of the substance lethal to man even in microgramme doses were widely dispersed. This resulted in an immediate contamination of some ten square miles of land and vegetation. More than 600 people had to be evacuated from their homes and as many as 2000 were treated for dioxin poisoning.

Following the accident, the European Union adopted the first Seveso Directive[1] in 1982. Following several other major chemical accidents, including Bhopal and the Sandoz Warehouse fire, the original directive was amended and, finally, at the end of 1996, the so called Seveso II directive[2] was adopted. Several further notable accidents then occurred (Toulouse, Baia Mare and Enschede) and the directive was further extended in 2003. Whereas the original Seveso directive focussed on technical measures for controlling hazards, the modified Seveso II Directive focuses more on the management systems needed to ensure the technical measures are in place.

3.2. UK Regulations

As an EU Directive, the Seveso II Directive is not directly applicable across the whole of Europe; it must be incorporated into national law in order to be enforceable. Within the UK, the Statutory Instrument used to incorporate the directive are the Control of Major Accident Hazards (COMAH) Regulations[3] which was amended in 2005[4] to incorporate the 2003 directive extensions.

3.3. Requirements

The COMAH Regulations apply to sites which store or process quantities of hazardous materials in excess of quantities detailed in the Regulations. There are two levels contained in the Regulations - a lower threshold and a higher threshold. Sites which fall below the lower tier threshold for their inventory of hazardous chemicals are not subject to the regulations. Those sites that exceed the lower tier threshold but are below the top tier, must follow the requirements for a lower tier site, while those that exceed the top tier threshold must follow the full requirements of the Regulations.

3.3.1. Lower Tier

All sites that qualify at least as lower tier must notify the competent authority of the intention to operate a major hazard site.

Sites must take all necessary measures to prevent major accidents. They must carry out a risk analysis to understand and predict:

- How a major accident could happen on the site
- The possible consequences of a major accident

Sites must prepare a major accident prevention policy (MAPP) document signed off by the senior management on site. This must include details of the measures used to control major accident hazards on the site. Sites must also prepare a safety management system document showing the organisational structure, responsibilities, procedures and resources for implementing the MAPP. Documented evidence must be provided to the Competent Authority that precautions have been taken to prevent major accidents, including plans, systems and procedures.

As a result of the quantities of hazardous materials handled on the manufacturing site discussed in this paper, it operates as a lower tier COMAH site.

3.3.2. Top Tier

In addition to the requirements for lower tier sites, top tier sites must prepare a safety report for the site to be submitted to the Competent Authority. This safety report details the MAPP and the safety management system required to ensure its delivery. It also incorporates significant detail describing the potential consequence of an incident involving the major accident hazards and a technical description of the control measures deployed against the hazards. The safety report must be reviewed at least every five years or when any significant change occurs on the site.

Top tier sites must prepare an on-site emergency plan to deal with any foreseeable major accident and provide information to local planners so that an off-site plan can be prepared. Additionally, top tier sites must inform the schools, hospitals, residents and businesses which may be affected by a major accident of the fact that the site is a major hazard site subject to operation under the COMAH regulations, as well as the hazards on the site and how major accidents will be handled and communicated.

3.3.3. Major Accident Prevention Policy

Foreseeable events

In order to deliver a MAPP, the first step is to identify the foreseeable major accidents that could occur. The foreseeable events are a combination of the fundamental hazards, the system design and how it is operated. There are many techniques for identifying hazards ranging from simple brainstorming techniques through checklists to more formal methods such as FMEA or HAZOP. During the procedure of inspection, the Competent Authority will challenge the completeness of the foreseeable event matrix and so a robust identification of hazards is essential to the process.

Control Measures

As already mentioned, the MAPP takes the foreseeable events that could result in a major hazard accident on the site and details how these are to be prevented or mitigated. The MAPP should not only include the hardware controls and defences in place but also any procedural controls that are in place. The management systems in place to ensure that the hardware and procedural controls remain operational also need to form part of the MAPP.

Linking the controls to the parts of the management system that ensure they remain operational is a difficult challenge. For a large site, the number of individual controls in place can run into thousands and either the management system definitions in the MAPP are very top level with significant clustering assumed, or they are detailed and become difficult to manage.

3.3.4. Organisational Change

When significant changes are made to the management and technical structure on a site, the safety management system can be adversely impacted. The loss of jobs can have an effect on morale, as well as the fact that the elimination of roles can leave holes in the controls in place to prevent safety system failure.

As part of the regulatory framework an organisational change risk assessment is needed whenever significant changes are made to the management structure. There have been a number of techniques developed for assessing organisational change – including the so called Entec method [5] – each of which has its own limitations. Most methods are very time and resource demanding as they require a detailed understanding of the hardware, tasks and procedures on the plant.

The site in question is a highly automated plant for which the Entec method is known to have weaknesses; it was decided to try something new and revisit the foreseeable event matrix with the Bow Tie technique to define the controls and defences in place. During the exercise, the responsibility for these controls and defences would be identified by role so that the impact of the organisational change could be identified.

4. BOW TIE

The Bow Tie methodology can partly trace its roots back to the "Swiss Cheese" model of Reason [6] and was probably first used by ICI in the 1970's. Its current form was primarily developed by Shell in the 1990's in response to the 1988 Piper Alpha disaster [7] to reinvigorate their safety management systems. The Bow Tie approach is primarily aimed at hazard identification and risk communication and takes its name from the shape of the final diagram. There are many software packages currently on the market that incorporate the Bow Tie technique: the work described in this paper was undertaken using the software package BowTiePro [8].

4.1. Method

The Bow Tie technique involves identifying a hazard and then defining an event that can occur as a result of the hazard being realised. For example, for a storage tank of LPG, the hazard is the LPG in the tank and the event could be ignition of leaking LPG gas. Once an event is defined, this forms the knot centre in the Bow Tie. To the left of the knot a fault tree is developed:

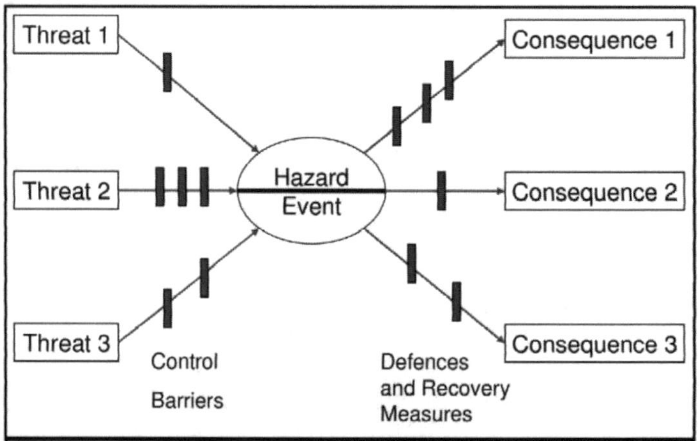

Figure 1 - Generic Bowtie Diagram

threats that could cause the event to happen are connected to the knot with barriers or controls identified to prevent the event. On the right of the knot, an event tree is developed with the possible consequences if the event happens. The event is connected to the consequences with defences or recovery measures acting as barriers. Figure 1 summarises this graphically.

As each control or defence identified is vulnerable to failure, threats to each control can be identified with additional controls in place to prevent that threat being realised and causing the barrier to fail. Figure 2 shows a typical control, in this case the procedural control of a Management of Change System. This control has two threats, "failure to follow procedure" and "failure to assess promptly" that can potentially make it fail. Additional controls are put in place to prevent each of these threats propagating.

Each control has a controlling role, a control type and control code associated with it. In figure 2, three control codes are shown: OR – Organisational, PR – Procedural and TR – Training. The control types (the shaded squares above the controls) represent types such as company mandatory, legal requirement or site specific; normally these are colour coded and easier to distinguish. The responsible role for the control is noted at the base of control box.

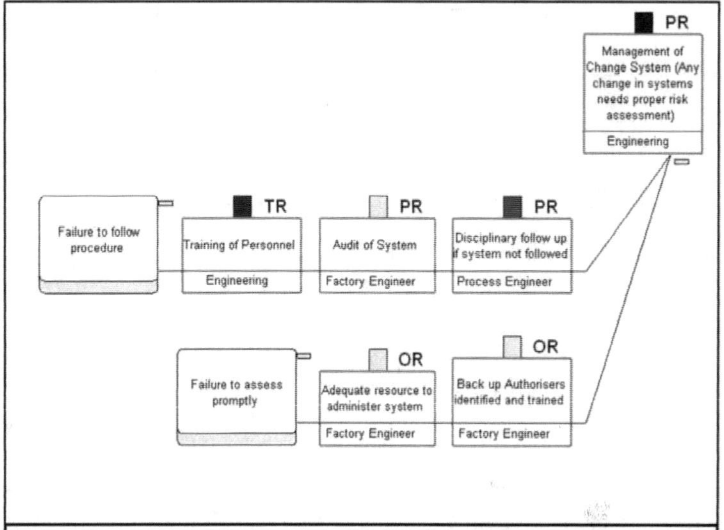

Figure 2 - Typical Control with Threats and Controls

4.2. Team Make Up

The proposed reduction of staff in a major reorganisation can have a major impact on morale and staff management relationships. This also has an impact on the means of conducting an organisational change risk assessment. At the initial stages of the reorganisation the scope of the changes is fluid and can be impacted by the risk assessment. Therefore the consultation between management and unions is confidential and work on the restructuring is handled by a restricted team. Formal consultation must also be completed within a fixed timeframe as laid out in law. A restricted timescale and team has an impact on not only the technique used for the risk assessment but also the complexity that can be considered.

The activities described in this paper were carried out during the formal consultation process and the development of the Bow Ties was done by teams consisting of site safety management, engineering and the Trade Union representatives from the consultation forum. The Trade Union representatives were mainly line operators and so had a good knowledge of the on the ground controls but were not generally trained in formal risk assessment methods such as HAZOP or FMEA.

An initial half day training session for the teams was held where the concept was introduced and a first partial Bow Tie was developed. Subsequent sessions to complete the exercise consisted of teams between 4 and 8 depending on shifts and staff availability. Each session was led either by an Engineering or Safety Manager to ensure consistency in approach.

5. BOW TIE PROCESS

5.1. Identifying Events

In carrying out a Bow Tie assessment, the key element is to ensure a meaningful event is identified. The event relates to a specific hazard, but should not be a final consequence. The technique will work with any event, but by choosing an event that is more of a consequence this can result in a distorted bow tie with no event tree. For example, taking the hazard "Flammable gas" it would be possible to work with the event "Tank BLEVE", but this is more a consequence and there would be several events to cover the range for this hazard. A more sensible event would be "ignition of released gas" as this will have several possible consequences one of which may be Tank BLEVE (Boiling Liquid Expanding Vapour Explosion). There is no right or wrong in this choice as both will work, it is a team decision. Occasionally part way through developing the Bow Tie, the team may decide the event is not the best one and they can then restructure the Bow Tie by changing the event perhaps to make the Bow Tie easier to understand.

As the site already had a foreseeable event matrix in its existing MAPP, this formed the basis of the analysis. However, the foreseeable event matrix in the MAPP is a summary of the consequence from the bow ties. The bowtie knot events are therefore not immediately obvious and some work is required to frame the correct question when identifying the events to be analysed. The work described in this paper identified 21 events to be analysed via 21 Bow Tie diagrams for the site being considered compared to the original 40 foreseeable events in the existing MAPP. Not all of the Bow Ties consequences are strictly major hazard events with respect to the definition in COMAH, but they are all significant for the site and are therefore included in the site plans.

5.2. Identifying Threats

Threats are the primary things that could result in the central event in the Bow Tie knot. For example, if we consider the event of a car accident, the threats would include vehicle failure, driver error and other vehicles. All of these threats could cause the central event of an accident.

5.3. Identifying Consequences

The consequences of an event are the escalations that could result from the event. Taking the car accident example, possible consequences could be: vehicle damage, driver injury, pedestrian injury, Police Action or death.

Where consequences can be escalations of other consequences then a linked Bow Tie is advisable, here the consequences of one or more Bow Ties become the threats to another event. For example the consequence of "jet fire" from LPG leak ignition will be a threat to the event "escalation of LPG fire to tank BLEVE".

Consequences are risk assessed to identify their severity and likelihood (without any preventative or mitigating barriers). The risk from the combination of severity and likelihood can be used to rank the consequences and to decide on the number of primary defences that must be put in place. For each consequence, the risk to people, plant, environment and reputation will be different and therefore, separate risk ratings are done for each

5.4. Identifying Controls and Defences

In this paper we use the term control on the threat side of the Bow Tie and defence on the consequence side.

When we consider the threats that could result in the central event we consider what barriers have been put in place to prevent the threat occurring. These will be both primary hardware controls as well as secondary procedural controls. For each control it is possible that they could fail to operate and therefore each control can be assigned threats which in turn may have controls to ensure the operation of the top level control. Looking at figure 2 above, the control of management of change system could fail if the procedures in the system were not followed or if the procedure took too long to assess a change. For these two threats to the control, additional controls are provided.

On the consequence side, the defences are barriers that are present to prevent the escalation of the central event to the consequences. Again each defence can have threats which may have additional controls to prevent the defence barrier failing.

Each control or defence is assigned a responsible person (role), these can later be used to identify key roles during the organisational change. For each any applicable procedures are listed and any deficiencies can be noted for subsequent correction.

6. ESTABLISHING THE CURRENT ORGANISATION POSITION

6.1. Foreseeable Events and MAPP

The foreseeable event matrix required under COMAH to develop the MAPP can simply be taken from the consequences identified during the Bow Tie exercise. The foreseeable events are ranked by their risk rating from the severity likelihood assessment. The ranking of risks can be done by looking at an individual risk types, e.g. rank by risk to people or alternatively, a combined total of all types can be used. The priority is then given to the consequences with the highest summed risk. Within this work, 40 foreseeable events that could be significant to the site were identified by the use of 21 Bow Tie diagrams.

The description for the MAPP of the safety management systems in place encompasses the totality of the Bow Tie diagrams. These form a visible description of the hardware and management systems in place to prevent the foreseeable events occurring. Where consequences are major hazard events these are cross referenced in the site emergency plan.

6.2. Identifying Key Safety Roles

As discussed earlier, each control is assigned to an individual role and therefore by looking at controls assigned to individuals, the key roles with respect to the safety on the plant can be identified. In assigning the role to a control, the primary responsibility is assigned. For procedural controls other roles may actually execute the control, but their failure will be picked up as threats to the original control. As part of the assessment,

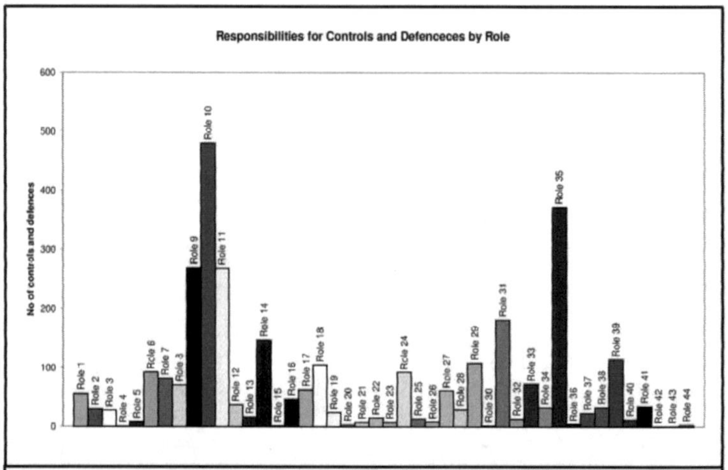

Figure 3 - Responsibilities for Controls and Defences by Role

overloaded roles or controls without responsibility may be identified, these can be addressed during the review process. During the exercise, over 3000 controls and defences were identified across 44 different role types. Each role type may have more than one person carrying out the role, for example shift electrician.

Figure 3 shows a graphical summary of the controls and defences assigned to each role. If as a result of the change; one of the significant safety related roles such as Role 9, 10, 11 or 35 were eliminated, there could be a significant impact on the operational robustness of the MAPP. The significance of the controls plays a part in the importance of the role for safety as well as the sheer number of controls. This fact needs to be also considered in the change review.

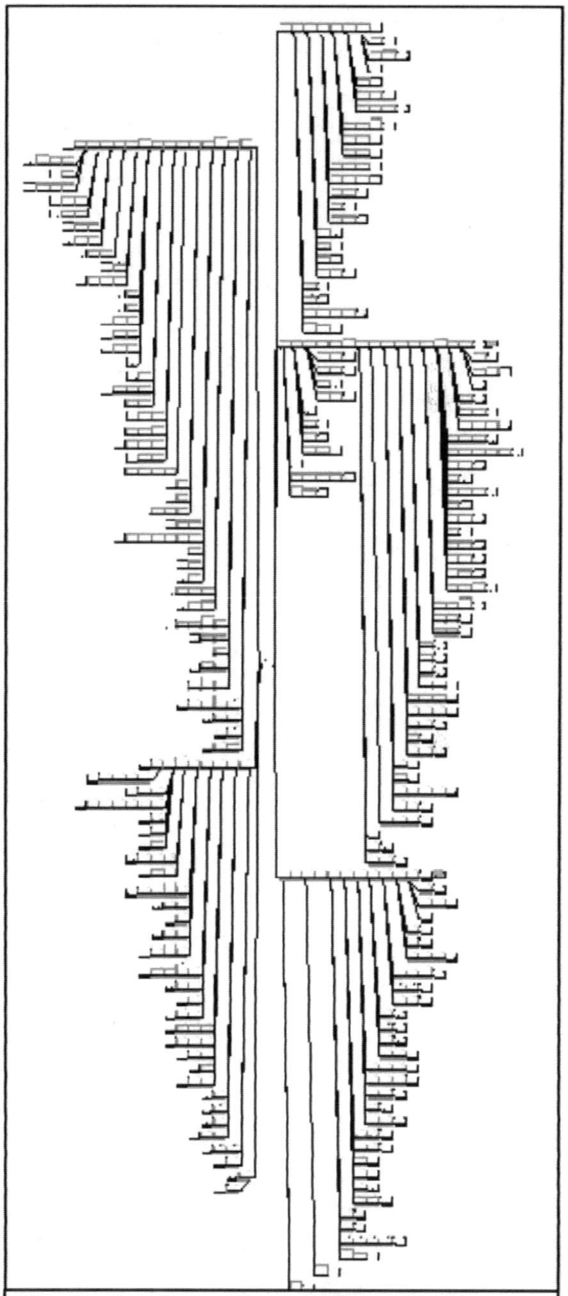

Figure 4 – Example Full Bow Tie Diagram

7. REVIEW OF IMPACT

Once the Bow Tie diagrams have been constructed for the site, assessment of organisational change can be undertaken. The change assessment comprises two separate activities; a review of the impact upon procedures and a review of the impact upon the controls and defences identified in the Bow Tie diagrams.

7.1. Identify Scope of Organisational Change

The scope of the organisational change is defined by the business case which needs to be met and is usually prepared by the site management team. This provides initial identification of the roles which will be affected with changes to responsibility, changes to numbers in the role and/or roles which will become redundant.

As previously noted, under UK legislation, significant changes such as those seen at the manufacturing site, require a formal consultation to be undertaken between management, staff and Trades Unions. Part of the consultation for this change included the assessment of the impact of the changes to safety.

7.2. Review Key Safety Role matrix for impact on Management Systems

Prior to the organisational change assessment being carried out, the site safety team carried out a review of roles identified within the various procedures and work instructions that form the ISO9001, ISO14001 and BS EN 18001:2007 management system. These roles were recorded in a matrix that identified the key safety roles against the document in which they had responsibilities detailed. The matrix included the quality and environmental processes as well as safety, but for the purpose of this paper only safety is being considered.

The key safety roles which fell into the scope of the organisational change were all referenced in the matrix, and the individual documents reviewed by the management team and safety professionals to assess the impact of the change. In some cases, this identified gaps where key safety tasks would have no owner once the change was implemented.

Once a gap was identified, a proposal was made for change to the procedure to address any vulnerability. The proposals moved through a cycle of consultation, with an opportunity for staff and Trades Union input to come to an effective solution. Once proposals were agreed to close all the gaps identified, the documentation was updated prior to the change and training material prepared.

It is important to carry out this review prior to the review of the Bow Ties described below because these proposals potentially change responsibilities and employee numbers, modifying the scope of the organisational change.

7.3. Review vs. Bow Tie for Gaps

Each Bow Tie diagram was reviewed by the management team members associated with the areas included in the diagram. Where tasks involved significant technical skills members of the engineering function were also involved. The assessment was also available to the staff and Trades Unions to comment upon.

The review was carried out using a record sheet and a large scale print out of the Bow Tie diagram. The reason for the large scale print out is shown in the complexity of just one of the 21 Bow Ties shown in figure 4.

A deliberate decision was taken to keep the process as simple and visual as possible to facilitate understanding amongst the team who may rarely be involved in such assessments. The Bow Tie diagrams were marked up by hand and the record sheets were filled in to explain any changes or decisions made.

The review had five steps:

1) Validate that the Bow Tie was accurate (in case of changes or errors).
2) Review the post-change organisation against the existing organisation shown on the Bow Tie diagram.
3) Mark up any required changes to the Bow Tie diagram or procedures and detail them on the record sheet.
4) Sign off by local management.
5) Sign off by safety management.

The review of the existing organisation against the post-change organisation considered staff workload (i.e. impact of numbers reduction), gaps (tasks with no owner after the change), and competency (do new task holders have appropriate training and skills). A key consideration was the question 'Is any control or defence removed by this change?' as such a change required documentation to demonstrate that it had been adequately considered.

Sign off by local management was included as they are the responsible persons for the activities detailed under the Bow Tie diagram. The safety management sign off was a final 'sanity check' of the assessment by a competent professional.

7.4. Amend Bow Tie, Procedures or Organisational Change to Close Gaps

The processes for review of the Key Safety Role matrix and the Bow Tie diagrams are inherently iterative and may require a number of repetitions to reach an adequate and acceptable solution.

Once the review is completed, the Bow Tie diagrams need to be updated to reflect any changes, and final changes to procedures need to be documented. As a result of this, the scope of the organisational change may also be changed.

In the case of the assessment for the manufacturing site described, some aspects of the organisational change were modified as a result of this assessment process, particularly those related to cover and emergency arrangements. The Bow Ties and Procedures were also updated to reflect the changes.

8. CONCLUSIONS

Within this paper, we have discussed the application of the Bow Tie technique to Organisational Change in a sourcing unit during a 15% staff reduction. The technique has shown itself to be easily applied with little training needed for the team carrying out the study. The Bow Tie technique itself is very visual and has shown itself to be a good method for capturing the safety management procedures on a site, being understandable by both technical and non-technical staff and delivering a visual representation of the safety structure. When this is coupled to the gap analysis and review, it can clearly be shown that the management systems and the major accident prevention policy have not been compromised by these significant changes. The approach used was discussed with the Competent Authority and the Organisational Change Risk Assessment was accepted as robust. As a method for determining foreseeable events and producing a site MAPP, the technique is easily followed, leading to identification of gaps in the management systems.

The technique is currently being rolled out across other lower tier COMAH Sites run by the Company in the UK with the full agreement of the Competent Authority, as a method for reviewing their MAPP documents.

9. REFERENCES

1. Council Directive 82/501/EEC on the Major-accident Hazards of Certain Industrial Activities , European Communities / Union (EUR-OP/OOPEC/OPOCE), 1990, ISBN: 9 28 261456 5

2. Council Directive 96/82/EC of 9 December 1996 on the control of major-accident hazards involving dangerous substances, European Communities / Union (EUR-OP/OOPEC/OPOCE), 1996, web link referenced 2 January 2009. http://eur-lex.europa.eu/LexUriServ/LexUriServ.do?uri=CELEX:31996L0082:EN:HTML

3. The Control of Major Accident Hazards Regulations 1999 , HMSO, 1999, ISBN 0 11 082192 0.

4. The Control of Major Accident Hazards (Amendment) Regulations 2005, HMSO, 2005, ISBN 0 11 072766 5.

5. Assessing the safety of staffing arrangements for process operations in the chemical and allied industries, Contract Research Report 348, HSE Books, 2001, ISBN 0 71 762044 1.

6. Human Error, James Reason, Cambridge University Press, 1990, ISBN 0 52 131419 4.

7. The Public Enquiry into the Piper Alpha Disaster, Lord Cullen, HMSO, 1990, ISBN 0 10 11310 2.

8. BowTiepro – web link referenced 21 January 2009. http://www.BowTiepro.com.

Organizational Change Risk Assessment
One Company's Approach

John C. Wincek, CSP
Croda, Inc.
8 Croda Way
Mill Hall, PA 17751
j.wincek@crodausa.com

ABSTRACT

Organizations can undergo change in many forms, including job transfers, promotions, resignations, and position elimination/additions. When these changes occur, it is imperative that the duties and responsibilities (including formal and informal) of the "old" organization be accounted for in the "new" organization.

In every organization, each person fulfills many roles and responsibilities. These duties may be listed in a job brief, position description, or other document. There also frequently exist tasks, responsibilities, or areas of expertise that are not identified in a job brief, if one exists at all.

This paper presents one company's methodology for assessing the risks associated with organizational change. These changes may take the form of resignations, promotions, reorganizations, and the creation of new positions. The paper describes a team-based approach, which includes methods of identifying what the "old" organization did, and how it will be accomplished in the "new" organization. Gaps in assigning responsibilities in the "new" organization are identified, risks assessed, and corrective actions agreed upon and tracked to completion.

Keywords: Organizational Change, Risk Assessment, Change Management

INTRODUCTION

Organizations routinely undergo changes such as hiring, resignation, staff reductions, reorganizations and promotions. Every individual who leaves a position takes with them the knowledge of what functions they performed. While many of these functions may be documented in a job brief or description, there are often tasks, duties and functions performed that few others know about, or that wouldn't be traditionally included in the role the person filled.

Departing individuals take with them the experiences they gained during their tenure. CCPS, in Guidelines for Technical Management of Process Safety, describes the individual who has worked at the site since startup. This individual knows where underground piping runs, why equipment is operated in a certain manner, what major accidents have occurred, and many other valuable pieces of information that may never have been documented.[1] Obviously, the loss of this person would greatly decrease organizational memory.

Changes affecting the working environment can also impact Process Safety. Changing from a 5 days per week to a 7 days per week work operation, changing the shift length from eight to twelve hours, and downsizing the maintenance crew during night and weekend shifts can all impact process safety, and must be evaluated to determine what risks have been created or exacerbated.

Guidelines for Management of Change for Process Safety[2] identifies "Organizational and Staffing Changes" as a category of changes that should be managed or could increase risk. Examples of such organizational changes include:

- Relocating the site technical group to a remote central corporate location without considering the impact on their ability to provide support to the facility.
- Changing from an 8-hour shift to a 12-hour shift without evaluating the potential effect of greater fatigue associated with longer shifts.
- Replacing an operations unit manager without considering the training needs for the new unit manager.
- Deciding not to replace a retiring corporate loss prevention expert who previously reviewed all relief system designs, or replacing the expert with an inexperienced engineer.
- Realigning the corporate PSM auditing function to place primary auditing responsibility at the site level, without considering the possible reduced expertise or independence of local auditors.

PROCESS CHANGES VS. ORGANIZATIONAL CHANGES

In today's world, many organizations have management systems to evaluate process changes. These changes may include piping changes, equipment upgrades, reconfigurations, procedure changes, and many others. From one company to another, many of the same questions are asked when evaluating these changes. These evaluations are typically described in a Management of Change policy or procedure, which requires review of the change by individuals from several different perspectives and areas of expertise.

Organizational changes can have as big an impact on process safety as process changes can. The loss of key expertise or experience can cost an organization the experience of past incidents. Increasing the number of responsibilities an individual has might mean there is less time to complete inspections, or track learnings from their own or other organizations. Evaluation of organizational changes requires that questions be asked that are quite different from those asked for process changes.

For this reason, it can be difficult to devise a Management of Change process that can accommodate both types of changes. Croda felt that developing such a comprehensive Management of Change process would result in large portions of the process being applicable to only one type (organizational or process) of change or the other. It was thought that this would result in confusion, wasted time evaluating what parts of the assessment would apply to a particular change, and potential "pencil-whipping" the parts of the process that only apply to one type of change or the other.

Croda determined that it would be best to develop a process for evaluating organizational changes that is separate from the process-related Management of Change process.

CHANGES REQUIRING RISK ASSESSMENT

Croda's Organizational Change Risk Assessment procedure applies to proposed and imposed organizational change (roles, responsibilities, terms and conditions of employment), including position elimination/addition, personnel replacement, changes to hours and working conditions.

Croda has defined a Relevant Change as one that could potentially have an impact on the organization's ability to control accident hazards, or one that could have a significant impact on the organization's ability to manufacture and deliver high quality materials to its customers. As a guideline, all of the following would be examples of a Relevant Change:

- An established position is removed from the organization
- A new position is added to the organization
- Individuals are required to take on new responsibilities demanding skills and competencies unconnected with those previously required
- Individuals relinquish responsibilities for tasks without those tasks being reallocated.
- Changes to terms and conditions of employment such as hours, shifts, location, etc.

RISK ASSESSMENT INITIATION

An Organizational Change Risk Assessment (OCRA) is initiated by the member of the Site Management Team (SMT) within whose organization the change will occur. This individual will ultimately approve the outcome of the Risk Assessment before submitting it for approval at the Vice-President level of the company.

The SMT member begins the process by completing a Section I of the OCRA form. This provides basic information about the proposed change, including:

- A description of the proposed change.
- The reason for the proposed change.
- The name of the person proposing the change.
- Who will serve on the team that completes the OCRA.

RISK ASSESSMENT TEAM

Make-up of the OCRA team is critical to the success of the process. Like a Process Hazard Analysis team, if the members do not have the right expertise, process (organizational) understanding, etc., the outcome is likely to miss key contributors to risk. Croda requires that the team members be selected from the following employees:

- Manager of the department undergoing change
- Customers of the position/organization undergoing change
- The employee(s) involved in the change
- Coworkers of the employee/positions undergoing change
- SHE/Regulatory Affairs Manager
- Process Safety Manager

In selecting the team, the following mix of attributes must be achieved, as appropriate to the change under consideration:

- Adequate seniority/experience
- Adequate understanding of Human Factors
- Adequate technical understanding of, and experience with, the issues involved
- Adequate understanding of organizational and business issues involved in the change
- Adequate understanding of the concept of risk

As many team members as necessary should be selected to ensure identification of duties relevant to the safe operation of the plant, and the production of quality product.

CONDUCTING THE RISK ASSESSMENT

The main function of the team is to identify all of the functions of the current organization, and to document how they will be performed in the new organization. The team uses its collective knowledge and experience to identify relevant duties performed by personnel/positions undergoing change. It is very likely that the SMT member or department manager already has a list of functions/duties from the current organization, and ideas on how they will be accomplished in the new organization.

Although this information is often not thorough (because it was likely completed by an individual instead of a team), it serves as an excellent starting point for the OCRA team.

Like hazard identification is critical to the success of a hazard study, identification of all of the functions/duties of the current organization is critical to the Risk Assessment. In a hazard study, unidentified hazards cannot be managed. Similarly, every overlooked function can leave a gap in the new organization, presenting some level of risk. Croda instructs its OCRA teams to include the following, at a minimum:

- Duties identified in the position Job Brief(s)
- Duties identified by the relevant manager
- Duties identified by the position's customers
- Duties identified by employees who held the positions involved in the change
- SHE, Responsible Care, Regulatory and Process Safety-related duties

Ancillary duties, while not identified in a job brief or position description, may be vital to the functioning of the organization. For example, an existing project engineer, who previously held a process engineering position, may have continued to provide valuable process engineering expertise after transfer to the new position. This may have occurred despite having been transferred to the project engineering position, without being identified in the project engineer's job brief. Should this employee later leave the company, the replacement hired to fill this position may not possess the skills necessary to provide process engineering expertise, resulting in a loss of technical expertise to the organization.

Wherever required, the team should consult as widely as necessary to allow them to come to an informed decision on the risks associated with the change.

This phase of the Risk Assessment is documented on the Identification and Transfer of Duties section of the OCRA form. For each function of the element of the organization undergoing change, a corresponding entry must be made describing how the function will be accomplished in the new organization.

How duties will be performed in the new organization will often be determined by the manager of the department undergoing change. In fact, most managers will have already identified this, at least for the functions he/she identified on his/her own. While other team members may advise on the effectiveness of the proposals, it is typically the responsibility of the manager to decide exactly how and to whom responsibilities will be assigned.

This method has also been used to evaluate the risks associated with changes to the work environment. Due to the current economic climate, Croda recently began operating on a five days per week schedule instead of seven. During this assessment, the "duties" identified were more generic, such as:

- Fire detection and response
- Response to alarms
- Leak detection and response

Additionally, hazardous storage systems were evaluated to determine appropriate ways the shut down, such as storage tank isolation, to minimize the potential for, or severity of, containment losses.

Mitigations for these risks included items such as:

- Increasing the frequency of rounds conducted by security.
- Developing a checklist of pressures and temperatures, along with acceptable ranges for each, to be monitored by security.
- Development of a DCS screen showing all parameters to be monitored by security.
- Developing an on-call list, ensuring a knowledgeable employee is always available for the security officer to contact.
- Developing a list of circumstances which require contacting the on-call employee.
- Communicating with local emergency responders, ensuring they are aware of the plant status, and methods to contact the facility's emergency response team.

The assessment also recommended that several other assessments be completed, including assessing employee fatigue during the new shift schedule, and re-assessing Layer of Protection Analyses to ensure their validity during the absence of plant employees.

IDENTIFICATION OF RISKS

Risks exist where duties, functions or capabilities of the current organization are not adequately accounted for in the new organization. Activities critical to process safety, such as inspections, design reviews, monitoring, even spare parts ordering, can go unnoticed until it is too late. Overlooking these functions now can leave their absence undiscovered until it is too late.

It is sometimes the case that the capabilities of the new organization are not adequate to accomplish some of the functions of the old organization. If this is the case, the new organization may be required to acquire additional knowledge or expertise. This may require additional personnel, training or experience, which can take time to acquire. If this is the case, plans must be made for accomplishing various functions until the new organization's capabilities can be improved.

RISK MITIGATION

Initially, it is the responsibility of the OCRA team to propose solutions to the risks they discover. Depending on the makeup of the team, they may be able to propose groups, or even individuals, within the organization with the capability to accomplish a specific function. Although the department manager will ultimately make the final decisions, the team can recommend efficient or even synergistic ways of accomplishing organizational and/or specific functions.

At this point in the Risk Assessment, the functions of the current organization have been identified, along with either how those functions will be performed in the new organization, or recommendations on how to accomplish them. It is now the responsibility of the SMT member, and often the department manager, to decide how best to mitigate any risks identified by the OCRA team. They may, of course, simply accept the recommendations made by the team. Alternatively, they may decide on alternative risk mitigation strategies. Whatever actions are decided upon, they must have someone assigned the responsibility of completing them, and a target date for completion.

Once the risk mitigation actions have been decided, documented and assigned, the SMT member will sign the plan and submit it to the relevant Vice President for review and approval. After making changes requested by the VP, and receiving final approval, the plan may be implemented.

It is typically the responsibility of the SMT member to track the risk mitigation actions to completion. In practice, this is often assigned to relevant department managers, who periodically report the status of individual items to the SMT member.

RISK ASSESSMENT OF SENSITIVE CHANGES

Often, due to the sensitive nature of changes, employees who would be best suited to serve on the Risk Assessment team cannot be informed of the changes until they are formally announced. In this situation, the complete assessment as described cannot be conducted until after the change has been announced or implemented. This can also make the assessment more difficult, as former members of the organization may not be available to discuss the functions that they performed.

These situations must be managed carefully by the relevant department manager and/or SMT member. Those "in-the-know" can and should complete as much of the assessment as possible before the change is implemented. After the change has been implemented, an assessment by a full assessment team must be completed. The results of the preliminary assessment are given to the OCRA team to further develop and complete. In these cases, Croda requires an assessment by a full OCRA team be completed within a reasonable period (<30 days) from implementation of the change.

Similar to the concept that "bolt-on" safety improvements are more expensive than including safety features in the original design, recommendations made by an assessment team after a change has been announced or implemented can be more "expensive" or difficult to implement. Management must be willing to change course where necessary to ensure the organization can continue to effectively manage Process Safety.

CONCLUSION

Organizational changes, like process changes, can negatively and severely affect process safety. They should be reviewed by a multidisciplinary team, which includes members with different roles and from different levels of the organization. The review process must be guided by a policy and procedure, and the outcome documented. Any corrective actions generated by the review must be tracked to completion. If these ingredients are included in an Organizational Change Management program, many of the risks that would otherwise go unnoticed can be successfully managed.

---

[1] American Institute of Chemical Engineers; Technical Management of Chemical Process Safety; 1989; pp.77-78

[2] American Institute of Chemical Engineers. Guidelines for Management of Change for Process Safety, 2008, Table 1.3

AN OVERVIEW OF THE CCPS GUIDELINE ON PROCESS SAFETY CONSIDERATION DURING THE MERGER AND ACQUISITION PROCESS

By Georges A. Melhem, Ph.D., Henry Ozog, Gary Kenney, George Groves, and Ashok Kalelkar, Ph.D. – ioMosaic Corporation 93 Stiles Road / Salem, New Hampshire 03079

ABSTRACT

The Center for Chemical Process Safety (CCPS) will publish a new book titled "Guidelines for Acquisition Evaluation and Post Merger Integration", with a strong focus on risks associated with process safety and its management. This publication is very timely since there is a plethora of mergers and acquisitions (M&A) underway as chemical companies seek M&A as one avenue of growth and strategic consolidation in a highly mature industry.

It is very important to note that CCPS has chosen to not only address the pre-merger *process safety due diligence* but also the salient and far more important issues of post merger management of safety risk related issues that are often overlooked to the detriment of the acquiring company. A typical due diligence process may not correctly identify process safety gaps which can lead to increased integration cost and risk for the acquiring company.

Among other things, this book will contain detailed check lists identifying the "must do" items on a pre-merger basis and a "must do and how to manage" check list and directions on a post merger, smooth and seamless integration basis. In any organization the responsibility, authority and accountability (RAA) chain is clearly delineated, but the specific responsibility for process safety management issues may vary between the acquiring corporation and the acquired entity. During post merger integration it is vital that a unified RAA chain be articulated and be clearly understood and be operational on day one after the merger. We know from real life experience that post merger process safety hiccups tend to occur with greater frequency in the acquired entity rather than the acquiring entity. This book's content is cognizant of the salient issues inherent in mergers and acquisitions and clearly lays forth the path to smooth integration in the aftermath of a pre merger due diligence audit for process safety.

INTRODUCTION AND BACKGROUND

The world of mergers and acquisitions has become standard headline news in the financial markets. As a result, the chemical and petroleum industry's business portfolios are ever evolving. Organizations are now merging, acquiring, divesting, splitting, and even spinning off new companies. Many employees can speak to working for an employer that has been bought out, merged or spun off under a new name. Quite often this happens in a few years. The profile of corporations and the core businesses within the corporations is an extremely dynamic process in this current world market.

The European Community remains the largest producer of chemical products followed by the USA and Japan. The traditional dominance of chemical production by these countries is being challenged by changes in feedstock availability and price, labor cost, energy cost, differential rates of economic growth and environmental pressures. Instrumental in the changing structure of the global chemical industry has been the growth in countries such as China, India, and Korea among many others. Not only are stereotypical chemical and refinery industries acquiring and merging in the world market to make new organizations, but nonchemical manufacturers are also venturing into the chemical industry as it has recently proven to be a profitable investment. This new world market and diversity within corporate profiles, includes multiple business strategies. These diversified strategies have led to mergers and acquisitions at an ever increasing rate. Early 2003 was the turning point.

Chemical companies with an improved balance sheet became more active in M&A deals. The chemical industry's interests in M&A are rising because they need a way to increase scale. After years of holding off and being internally focused, chemical companies have started to look externally to find ways of improving their balance sheets and creating value"[1]. In addition the investment community and many corporations, now make it their business to drive mergers and acquisitions. The chemical industry attracts private equity firms. "The traditional model is for private equity firms to either strip costs out of their investments or bundle them with similar or complementary businesses before selling them. The average life of an investment up to the point of sale is three and a half to five years."[1]

Statistics from Bloomberg News on merger and acquisition activity in the chemical industry (see Table 1) indicate that over the three year period of 2003, '04 and '05 the value of acquisitions just within what is designated the 'chemical industry' increased from 28.3 to 55.6 billion dollars.

Table 1. Reported Merger and Acquisition Activity in the Global Chemical Industry (Bloomberg News)

Year	Number	Value US$
2003	62	28.3 Bil
2004	88	47.6 Bil
2005	95	55.6 Bil

A key element that needs to be managed through every one of these mergers, splits, acquisitions and spin offs is "process safety". CCPS developed the M&A Guideline with a clear focus on "process safety" issues that should be considered when an organization is contemplating a possible merger and/or acquisition.

The CCPS M&A Guideline will serve as a working tool rather than a textbook on the merger and acquisition process. This Guideline can be used by companies that are acquiring chemical, petrochemical and petroleum process facilities through mergers and acquisitions. The environmental, safety and risk considerations related to companies or facilities that handle highly hazardous compounds add a layer of complexity to merger and acquisition considerations that are not usually a concern in the nonchemical industry business sectors.

GUIDELINE OUTLINE AND BRIEF OVERVIEW

The CCPS M&A Guideline consists of seven chapters and several appendices including a master checklist.

Chapter I: An Overview of Process Safety

A short overview of process safety requirements is provided to ground potential users who are unfamiliar with process safety. This chapter incorporates major hazard concepts as used in the US, the European Union, and other international countries with a clear distinction between process safety and occupational safety. The text addresses process safety as a concept and not just as US Process Safety Management or European Union Major Hazard regulations. Chapter I also highlights the need for creating awareness in and among upper management of the issues they need to 'have in their minds' with respect to process safety during the merger & acquisition process.

Chapter II: An Overview of the M&A Process

The M&A process is fairly simple, but the details can have a major impact on the outcome of the process. Surprises late in the negotiations can derail the negotiations and/or lengthen the sales process.

M&A deals can range from the very large mega-merger of two huge multinational companies to the small, such as, a sale of a service station. No matter what the size of the deal, the process is similar but the complexity and the size of the army will be different. The use of a checklist, such as the one included in the CCPS Guideline, allows for multiple teams to work concurrently in different locations and deliver their findings in a consistent manner.

Note that M&A is a term that is commonly used to define various business transactions that involve the transferring of assets and personnel from one entity to another and/or certain contractual agreements between parties. These include: mergers, acquisitions, joint venture formations, divestments, decapitalizations, Initial Public Offerings (IPOs), asset swaps, etc., and contractual agreements, such as, toll processing agreements, etc.

A process safety professional will likely be a member of a specific M&A team, usually Health, Safety and Environmental (HSE) or perhaps an engineering team. The checklist included with the CCPS Guideline can be readily incorporated into the overall checklists of either organization.

The overall objectives of the HSE or engineering M&A teams are similar: capture value and manage risks through knowledge. The team should raise potential issues for special management to the business team, provide the financial team with the correct valuation of HSE costs with expenditure timeline for the financial model and provide the legal team with sale agreement language, mechanisms and concepts.

Chapter III: Screening Potential Candidates

There are sources of information on the internet that can provide a jumpstart to a due diligence while sitting at your desk. These include the target company's web site; federal, state and local agency web sites; newspaper and television media sites local to the plant and aerial photos from a site like Google Earth. In a couple of hours of research a tremendous amount of background information on a plant and plant site can be obtained by starting with only the company name and site location. But note that this *background* information should be confirmed by actual data room information or during the site visit. Newspaper articles might not be completely factual, so use them as red flags and for guidance about an issue, but not as facts to be included in the due diligence report. A good place to start is the target company's own web site. Here information on the safety culture of the company can be found which will provide insight into what could be found at the plant. Many companies have their HSE management system on the site, which can be reviewed to see if process safety is included. Look for information regarding process safety metrics.

Federal, state and local agency internet sites are also a good source of information. These give information on permits, but also many sites contain permit applications and environmental impact assessments for a number of projects that can be viewed for background information plus some process safety information. These sites are sometimes more difficult to search through, but jewels can be found.

Some states have their own regulations with process safety requirements that go beyond those required by OSHA and EPA. For example, New Jersey under their Toxic Catastrophe Prevention Act Program has requirements for reactive chemicals management and inherently safer technology analyses. Other states with significant additional process safety regulations include California, Nevada and Delaware.

Companies that are active members in process safety industry associations such as CCPS will generally have better process safety programs than their peers that do not participate. When checking if a company is a member of any of these industry groups, the number of years they have been a member also gives some insight into the company culture. For example, being one of the founding companies of CCPS is more significant than a company that joined last year. A more detailed search may include representation on various process safety committees by company personnel. A company where individuals actively participate on committees and present papers at conferences is more likely to have superior process safety programs.

When conducting these searches, if a major problem is found the team leaders of the project should be informed without delay. A recent news article about a local town council voting to shut down the target facility should be addressed quickly. Other red flags may show up in the preliminary searches, which the project leaders should consider in their decisions to continue the project. M&A projects are people resource intensive and can be quite expensive, so the team leader will want to shut down bad projects early in the process if possible. Some process safety issues that may derail an M&A project at this stage might include:

- Significant PSM issues that may have been identified as part of an OSHA inspection or a recent National Emphasis Program (NEP) inspection
- Significant facility siting issues requiring major building relocation or modification
- History of incidents, especially those impacting offsite

- Proximity to sensitive populations such as schools, hospitals and recreational areas

When working a divestment, the same type of search can be conducted on the site that is being divesting. If news articles about the plant site are found by conducting a Google search, they will need to be addressed in the data room and vendor due diligence report as the information will be requested in a question and answer (Q&A) session. There may be a significant amount of information on a company's internal web site relative to a facility. This information can be obtained in a confidential manner and used to populate a data room.

For a Joint Venture or Merger, this type of search can be conducted on the partner organization and their assets. The HSE Policy and management systems can be found on the partner company's web site. There will likely be other indicators of their standards and corporate direction. One added step can be to Google each of the key members of the target company's board of directors (which can be found in the annual report) to determine if there are process safety issues surrounding them that could create negative impact.

There are other web services that are usually paid subscriptions that the legal teams use to gather information on lawsuits, enforcement actions and the like. Though these might produce some good information, for this level of review it might be best to just ask the Legal team if they have run the search and request the output.

Chapter IV: The Due Diligence Phase

The skills and experience of the process safety members of the due diligence team must be clearly outlined before the due diligence process starts.

While a preliminary due diligence might begin with an internet search, the actual due diligence starts with a checklist. Whether you are working on a divestment or an acquisition, the uses of the checklist are the same, but the order of events is different. The divestment process (see Figure 1) the site visit and document review are conducted to generate the Vendor Due Diligence (VDD) report and populate the Data Room with pertinent documents, whereas in an acquisition (see Figure 2) the VDD and Data Room are completed first followed by the site visit and preparation of the due diligence report.

Figure 1. Typical M&A Divestments Activities Covered in Chapter IV

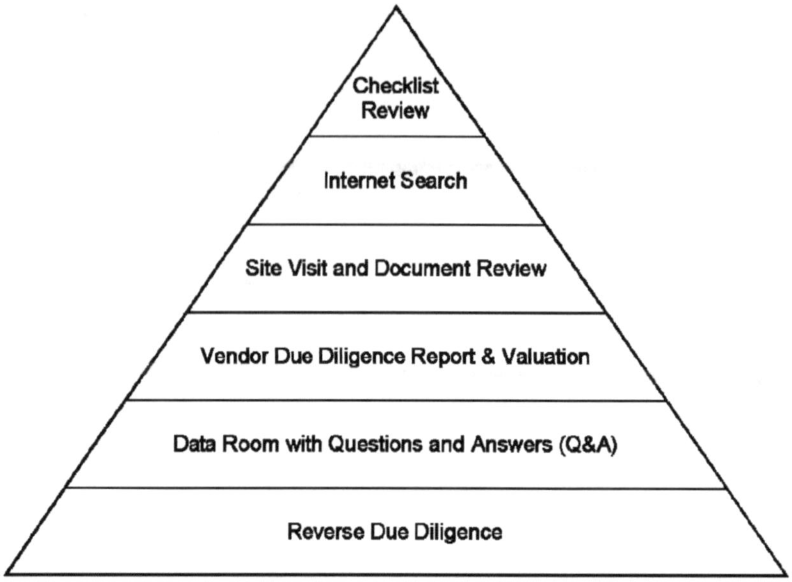

Both the divestment and acquisition processes start with the Checklist. Checklists should be living documents that are modified as regulations or issues get more complex. Review the checklist to assure that new issues haven't arisen that should be considered, or contact CCPS to see if a newer version of the checklist is available.

Figure 2. Typical M&A Acquisition Activities Covered in Chapter IV

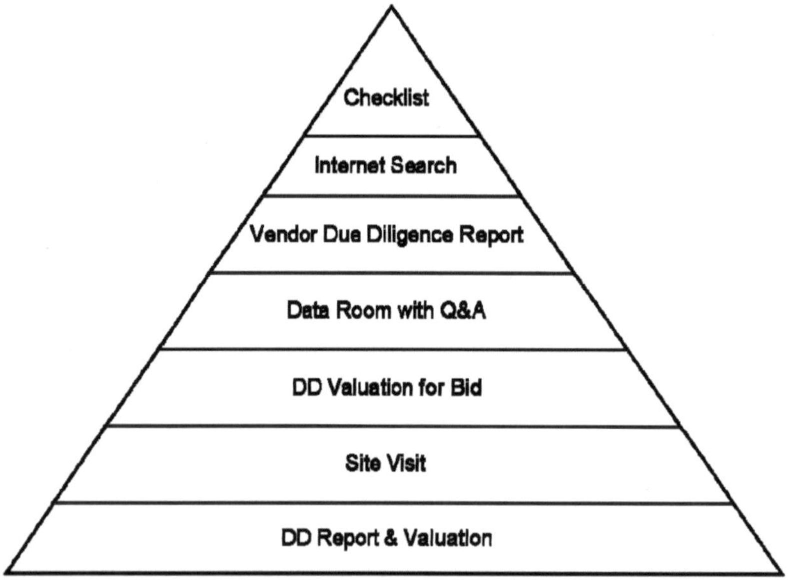

Chapter IV discusses all the details of information to be made available in the Data Room as well as all the steps outlined in Figures 1 and 2. This chapter represents a wealth of information and is a must read.

Chapter V: Developing the Integration Plan

Poor post deal integration was identified as the major factor or primary reason that, in as high as 43 percent of mergers, the merged company or operations did not outperform competitors. It is well established that poor planning is often a contributory cause to poor performance. Once the deal is closed there is a need to set aside sufficient time and resources to establish a well-founded strategy and plan.

Figure 3 outlines the basic process for developing a comprehensive process safety integration plan. It starts with collecting and gaining an understanding and mutual agreement with senior and executive management on the boundaries that will surround the future operation of the newly acquired and combined businesses (i.e., the overall integration strategy). From there it continues on through a process of review(s) to identify gaps and the extent of those gaps and culminates in the development of a resourced plan for addressing or closing all such gaps. In many organizations the first time any process safety expertise is brought into the picture is after the deal has closed. In those cases, the process safety integration will have to start at 'ground zero' and undertake a series of reviews of the new business and operation to establish whether any potential gaps exist and the resources required to address any such gaps.

Developing the integration plan starts with developing the boundaries for the overall strategy that will be selected for integrating or merging the process safety aspects of the newly acquired business into the new combined business or organization. One should consider:

- The 'extent' to which the newly acquired business or facilities are to be integrated,
- The 'height of the bar' senior management is setting for the operation of the new business or facility with respect to process safety, and
- The resources and budget that will be committed to achieve integration.

Figure 3. Overview of the Integration Process

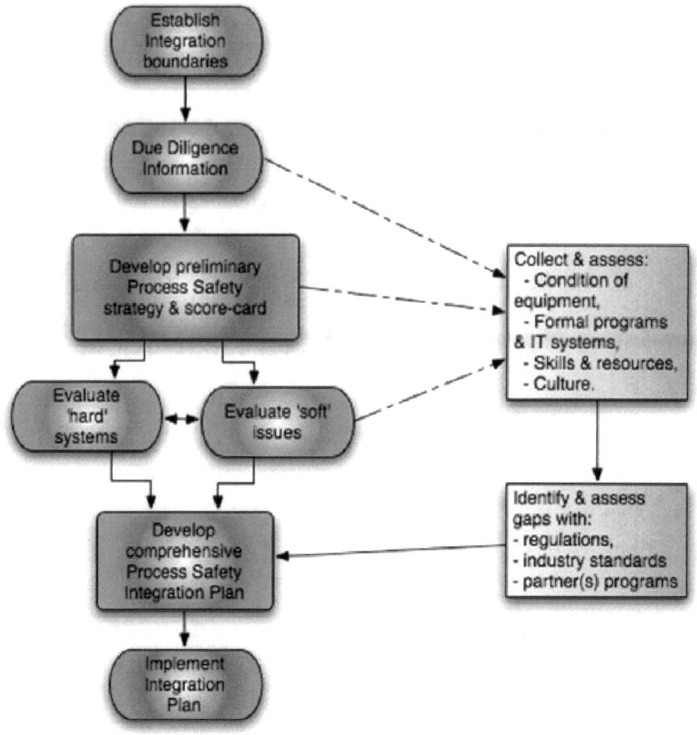

Next, one should develop a set of measurable expectations in each of the four main elements within the boundaries established for the integration strategy of:

- The physical plant, equipment and operating processes,
- The human resources required to drive and support an effective process safety program,
- The desired state of the various management systems, programs and procedures that constitute an approach to process safety,
- The culture required to assure these requirements are effectively carried through to day-to-day operations.

It is important to identify and form a capable process safety integration team. The team(s) will need to have technical skill and experience in each of the four main elements of integration (i.e., the state of the physical plant or equipment, the management systems, the human resources and cultural issues). During the first phase of the integration process or that of establishing the gaps that exist in each of the four elements the team will need to have skill and experience in auditing and reviewing such systems and programs. During the second phase of 'plugging the gaps' the team members will need to have skill or experience in developing and implementing systems, programs and procedures. Be prepared and plan for team members to be rotated or replaced whether as a result of burnout or potential advancements in their careers by taking other positions.

To effectively assess the gap between the current approach and expectations, consider the following actions:

- Establish the expectations for the newly acquired business.

- Use data and information from the due diligence phase to establish this baseline, or if necessary develop a protocol of issues to be examined,

- Form an interdisciplinary and inter-company team to undertake a series of reviews of each of the sites or facilities.

- Identify and quantify potential gaps between the current state and expectations:

    - The general condition of the physical plant, equipment and process operations,
    - The process safety related management systems or programs in place,
    - The technical skills and resources that are employed to drive or support the process safety programs,
    - The culture towards process safety of the management, technical and supervisory staff and the operators and maintenance staff,

- Develop a gap analysis or report between the current and desired states,

- Complete this review in a period of preferably three months and not more than nine months.

Finally, develop an action plan that will address the following:

- Development of a program to address and put into place temporary controls for all risks deemed 'intolerable' within a period of 90 to 100 days,

- Development of a capital works program to address deficiencies or gaps found in the physical plant, equipment and processes. Responsibility for completing these works is likely to rest with groups or departments other than the process safety group.

- Development of a plan to rationalize resources and organizational responsibility for process safety related activities. As part of this plan address the need to establish the desired cultural changes required to assure process safety is imbedded into the day-to-day activities of the newly acquired business.

- Development of a plan to address gaps in the policies, procedures, and systems that govern the desired approach that senior management wish the new organization to achieve.

    - Where there are gaps in the state of current information on the physical plant, equipment and processes the plan should address and 'plug' these gaps first,
    - Once these gaps are addressed, undertake the necessary changes to address and modify the process safety governing processes and systems.

- Development of a consolidated plan to manage the overall integration process, track progress towards goals, and implement corrective measures as soon as variances are identified.

- Development of a Management of Organizational Change plan that assesses whether the planned changes will still control risks to acceptable levels and identifies the controls necessary to assure such risks continue to be controlled.

Chapter VI: Implementing the Integration Plan

An integration process is a change process. For integration to be successful, the feelings or emotions of those who will be impacted by the changes need to be engaged, more so than just presenting a series of 'logical arguments' of the need for the change.

To successfully complete a process safety integration task requires building momentum in and among various managers, teams and individuals of the newly acquired business to accept change. Without the desire to change, the probability of revising the management systems, modifying equipment or processes, etc. in line with senior management's expectations, is remote. Anyone that has just been assigned the responsibility to integrate, or merge, the two approaches to process safety of a newly acquired business and the parent company's together should take the time to sit down and review Chapter V and at least pull together the skeleton of such an integration plan. If there is sufficient time to sit down and develop an integration plan as outlined in Chapter V, implementation will entail the mechanics of working through that plan. However, having a good plan does not automatically guarantee success. A study by KPMG of change programs found 80% of all such programs failed to achieve the stated objectives for those programs.

Much of the integration process involves modifying equipment or changes to the formal written management systems. However, all of that work will be done by and through the efforts of various individuals. As so much of the change then involves people, it is vital any integration process addresses the need to shift culture and the behaviors of people. And this change must occur not only in the newly acquired business, but also in the parent company as well, to be successful in achieving the expectations of senior management.

Figure 4. Seven Step Process for Implementing Change (Taken from Kotter)

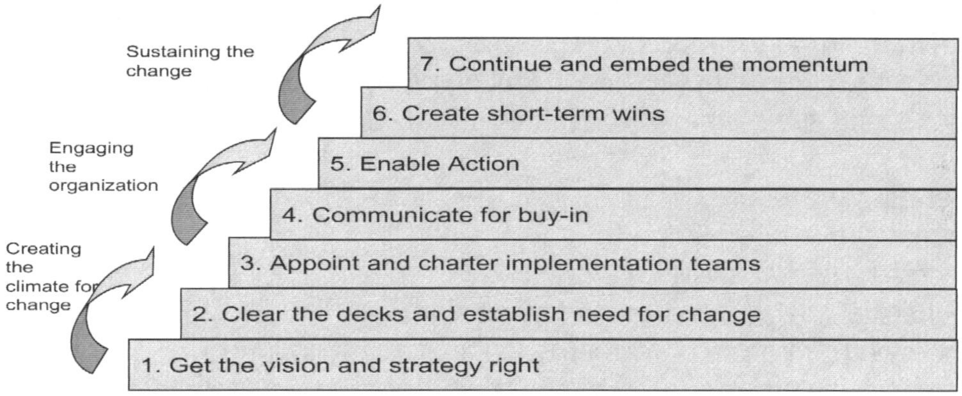

Chapter VI provides detailed recommendations on how to effectively implement each step of the process outlined in Figure 4.

Chapter VII: M&A in the Future

In early 2009, M&A activity worldwide has slowed to a snail's pace. The planned $15.3 billion merger between Dow Chemical and Rohm & Haas has been delayed by at least five months because Dow cannot get the financing that it needs without ruining its credit rating. Rohm and Haas then sued Dow for not closing the deal on January 27, 2009. [2]

If an M&A activity faces financial issues the results can be catastrophic. The US operations of the recent merger between Basell and Lyondell (LyondellBasell) filed for bankruptcy on January 6, 2009 under the weight of a massive debt load and slumping demand for its products.[3] The merger created the world's the third largest petrochemical company.

In tough financial times, the resources to conduct adequate due diligence of process safety issues during M&A activity may be limited and could compromise the final entity. Even in a good financial environment, poor merger integration of process safety could lead to catastrophic consequences.

An invaluable resource for the M&A team is the master checklist of M&A Process Safety Related Issues provided with the CCPS Guideline. The vision of the CCPS M&A Subcommittee is that all future M&A activities will use this checklist and provide feedback to CCPS on any changes or improvements that are suggested based on its use.

In the US, regulatory emphasis on process safety has increased significantly with the launch of OSHA's Petroleum Refinery Process Safety Management National Emphasis Program (NEP) on June 7, 2007. This program identified significant process safety issues during their 2008 inspections resulting in over 200 citations for the top seven PSM elements and sub-elements listed below:

1. Operating procedures
2. Process safety information for equipment
3. Process hazard analysis criteria
4. Mechanical integrity testing and inspection
5. Process hazard analysis recommendations and follow up
6. Mechanical integrity related to operating deficient equipment
7. Management of change implementation

OSHA's inspections have validated that process safety is still not being adequately implemented despite being a regulation since 1992. As the refinery NEP nears completion, OSHA has indicated a desire to initiate a similar program for the chemical industry. This increased regulatory emphasis will also increase the need for M&A teams to devote greater resources to evaluating process safety.

Eventually the current worldwide economic slowdown will end and the strongest surviving companies will accelerate their M&A activities as there will be many good companies to select from. Many companies may choose to expand globally to access more economical raw materials and labor or gain access to the local markets. For example, in 2008 Dow Chemical tried unsuccessfully to form a joint venture with Petrochemical Industries Co (PCI) a unit of Kuwait Petroleum Corp (KPC) to gain access to petrochemical raw materials. In 2007 Saudi Basic Industries Corp (SABIC) bought GE Plastics to gain a foothold in the North American market. Any M&A activity that involves companies from different countries can create significant barriers to successful integration. Language and local culture are two of the most significant barriers that would need to be overcome in addition to the barriers related to process safety language (i.e., local regulations and industry standards) and culture (i.e., rules based verses risk based approach to process safety).

Many regulations and industry standards have been developed as a result of a major accident. The creation of CCPS resulted from the toxic gas release in Bhopal followed by other accidents that caused significant onsite and offsite impact. The BP accident in Texas City will probably be viewed as one of the events that led to the writing of this book. The vision of CCPS is that all companies involved in M&A activities will learn from the mistakes of the past and use the information in this book and the included checklist to improve their evaluation of process safety issues. In addition to the methods provided in this book for evaluating process safety at each stage of the M&A process, going forward companies will have a set of process safety metrics to more quantitatively evaluate performance of companies involved in M&A activity. In fact poor process safety metrics may become a significant deal breaker when screening potential candidates for M&A. These metrics are described in detail in an upcoming CCPS publication.[4] As companies use these tools and recognize their value, evaluation of process safety in the M&A process will become a mandatory requirement and not an afterthought. We do not envision that the overall M&A process will change significantly in the future, however we do believe that process safety will become a "core" issue to evaluate during M&A activities and that process safety metrics will become a major factor in the decision making process.

APPENDICES:

The Appendices include a master checklist and examples of various flowcharts to assist with the planning of the various activities and/or steps in the M&A process.

Appendix A. The Master Checklist

Due diligence starts with a detailed checklist. The CCPS Guideline checklist was compiled based on input from several CCPS member companies. The checklist will be provided in electronic format with the Guideline or made available for download from the CCPS website. The checklist (see Figure 5) addresses the following topics:

(a) Commercial Evaluation,
(b) The M&A Team,
(c) The Data Room,
(d) Planning Visits,
(e) Site Visits,
(f) Process Safety,
(g) Assessing Major Risks,
(h) Process Safety Management and Culture,
(i) Process Safety Staffing,
(j) Hazard Identification,
(k) Management of Change,
(l) Process Safety Information,
(m) Mechanical Integrity,
(n) Process Safety and Operating Procedures, and
(o) Auditing.

The use of detailed checklist such as the one included in the CCPS Guideline allows for multiple teams to work concurrently in different locations and to deliver their findings in a consistent manner.

Figure 5: M&A Master Checklist Worksheet Sample Element (for illustration only)

A1 Information or Data Item

	A	B	C	D	E	F
31	Is there an obvious under-spend?					
32	Request a list of major equipment with age and general condition with reference to maintenance schedules. Major equipment would typically include:					
33	Compressors,					
34	Large pumps (e.g. >50Hp),					
35	Pressure vessels,					
36	Reactors,					
37	Heat exchangers,					
38	Storage tanks,					
39	Fired heaters, etc.					
40	Request a list of all identified critical equipment with, at a minimum, any legally required information, for example					
41	Materials of construction,					
42	Engineering drawings (i.e. P&ID's),					
43	Electrical classification,					
44	Relief system design and design basis,					
45	Design codes and standards used,					
46	Safety interlocks.					
47	Request process safety information, including:					
48	Physiochemical and toxicological properties,					
49	Flammability,					
50	Stability,					
51	Reactivity,					
52	Corrosivity.					
53	Request information on the integrity and maintenance of process safety systems, including:					
54	Electrical distribution preventive maintenance program(s),					
55	Mechanical integrity program(s),					
56	A selected sample of integrity test and inspection records and inspector qualifications.					
57	Request process safety related metrics, including:					
58	Leaks, breaks and spills data,					
	List of incidents in the last 'X'					

Commercial Evaluation | The M&A Team | **Data Room** | Planning Visits | Site Visits | Process Safety | Assessing Maj Risks

ACKNOWLEDGMENTS

The author would like to acknowledge the contributions of the CCPS leadership and Committee as well as the ioMosaic book chapter authors.

REFERENCES:

1. Kamakura, Yasuhiko "Corporate structural change and social dialogue in the chemical industry", International Labour Office, CH-1211 Geneva 22, Switzerland.

2. Wall Street Journal, "Dow Chemical: We Don't Feel Like Closing the Merger. Is That a Problem?", January 27, 2009.

3. Forbes.com, "LyondellBasell U.S. units file for bankruptcy", January 6, 2009.

4. CCPS "Guidelines for Process Safety Metrics", American Institute of Chemical Engineers, New York, New York, expected publication in 2009.

Choosing Appropriate Quantitative Safety Risk Criteria
Applications from the New CCPS Guidelines

Walt Frank
Frank Risk Solutions, Inc.
1110 Shallcross Avenue
Wilmington, DE 19806
wlf@FrankRisk.com
302-521-7588

Dave Jones
Chevron Energy Technology Company
Richmond, CA 94804
DWJones@chevron.com
510-242-9258

Abstract

In some situations, risk can best be understood, and therefore managed, through risk quantification. Establishing quantitative risk criteria can be a key component in informed risk decision making. Currently many companies use quantitative risk criteria, either formally or informally. Some companies base their criteria on regulatory precedents (e.g., Netherlands, UK HSE, etc.) while others have developed "home grown" criteria, involving company-specific judgments. This paper will provide an overview of the new American Institute of Chemical Engineers, CCPS *Guidelines for Developing Quantitative Safety Risk Criteria,* which is intended to provide assistance to companies seeking to develop such criteria. The paper will provide an early view of the book contents, and highlight challenges to be addressed by any company attempting to establish risk criteria.

"To measure is to know... If you can not measure it, you can not improve it."

Lord Kelvin (1824 - 1907)

1. Introduction

This paper offers a necessarily brief summary of the new CCPS *Guidelines for Developing Safety Risk Criteria (RCGuidelines).* All of material presented below is extracted from the *RCGuidelines*, and represents only a few of the major themes discussed in the book. The reader should refer to the full *RCGuidelines* for additional information. The book provides: (1) a discussion of the importance of risk criteria, (2) a review of the types and application of risk criteria, (3) risk criteria precedents, and (4) guidance for companies to consider in establishing risk criteria.

Risk, risk assessment and risk management have been discussed extensively in other CCPS publications [References 1, 2 and 3]. The *RCGuidelines* do not repeat that discussion, except for those aspects that pertain to the selection and use of quantitative safety risk criteria. Of special importance is the definition of risk itself [Reference 2]:

> *Risk is a measure of human injury, environmental damage, or economic loss in terms of both the incident likelihood and the magnitude of the loss or injury.*

The concept that risk is a combination of the consequences and the likelihood of an event (or events) first appeared in the mid-1600s and has remained unchanged for 3½ centuries, much like some of the fundamental laws of physics, which were created about the same time.

The consequences described above can cover an extensive range of events, from personal injury to environmental and property damage. While all consequences are important, the *RCGuidelines* specifically focus on fatalities from episodic events associated with process hazards, such as fires, explosions or hazardous material releases. Fatality is the consequence of concern most frequently addressed by applicable regulatory and government precedents, and most often used in establishing safety risk criteria.

The *RCGuidelines* reinforce several other important concepts that are integral to the use of risk criteria: Specifically:

- There is no such thing as zero risk. Residual risk always remains, no matter how great the resources applied to reduce it. An extreme example to illustrate this fact would be staying in bed to avoid all risk. While seemingly harmless, the health consequences as a result of a sedentary life style would expose the individual to other significant risks.
- Risk management requires resources. Significant step changes in risk reduction require increasingly greater expenditures of resources. The most effective risk management strategy applies available resources such that greatest overall reduction is achieved, taking into account all risks potentially faced by an organization.
- Successful operating history is no guarantee of future success. Operating a hazardous material processing facility for a period of time without a major incident does not prove that an incident will not happen.
- Risk tolerance is not the same as risk acceptance. The *RCGuidelines* address what are often termed risk tolerance criteria and not risk acceptance criteria, reflecting the belief that no incident harming people should be considered acceptable. However, based upon the fact that there is no such thing as zero risk, society tolerates some level of risk in return for the benefits derived from the activity posing the risk. Air travel is a good example of this.

In addition, the *RCGuidelines* offer an observation on the apparent contradiction between establishing risk criteria (which assumes some level of residual risk) and promoting a goal, as many companies have, of "zero incidents." Risk assessment, and setting risk criteria to help drive risk reduction, represent just one component of an overall process safety management system, each part of which contributes to reducing the likelihood of incidents.

Finally, and most importantly, the *RCGuidelines* do not propose specific risk criteria for the process industries. Such a task is beyond the charter of the CCPS, and involves a complex mixture of regulatory, social, legal and business considerations, each of which includes company-specific aspects.

2. The Case for Quantitative Safety Risk Criteria

The need for quantitative safety risk criteria is based upon an evolving need for improved risk assessment tools and more effective risk decision-making. The *RCGuidelines* trace the history of risk assessment and risk management from its early origins in the maritime insurance industry through the aerospace and nuclear industries to current applications in the processing industries. Risk assessment methods and models have changed with the times, reflecting the increasingly complex risk issues being addressed, as well changing science and technology.

The *RCGuidelines* compare current applications of qualitative and quantitative risk assessment (QRA) methodologies. Process safety professionals and many operating facility personnel are acquainted with the use of hazard and operability studies (HAZOPs) and What If Studies to gain a qualitative appreciation of process risks. In contrast to complex QRAs, these qualitative assessments can be easily applied to a wide range of operations, require fewer personnel with high-end technical expertise, and can drive immediate actions that reduce risk. On the negative side, such qualitative assessments are highly dependent on the experience levels and abilities of the teams conducting the studies. In addition to this subjectivity, and regardless of the qualifications of the personnel, qualitative assessments are not the best mechanism for evaluating high consequence, low frequency incident events. This is because high consequence events occur, statistically, on the order of once every several thousand years for a given facility. Even the most knowledgeable team is challenged when trying to assess the likelihood of two-thousand year events, based upon their 10, 20 or 30 years of work-related experience. Also, the scale of catastrophic incident consequences (e.g. major explosion, fire or hazardous material release) is beyond the experience of most people. Few people have the misfortune to witness firsthand the outcome of a major process plant incident.

QRAs have shortcomings in that they require, as a minimum, highly specialized expertise and reliable failure data. However, once these requirements are met, QRAs are more effective than qualitative assessments in determining the risks associated with infrequent incident events. Accordingly, as the *RCGuidelines* note, companies are increasingly turning to QRA to better understand their major risks. On a parallel path, worldwide many governments and regulatory agencies are mandating the use of QRA for (1) planning approval for new chemical or petroleum facilities, (2) planning approval for new off-site developments in the vicinity of existing chemical or petroleum facilities, (3) safety cases in support of the renewal of operating licenses, and (4) assessment of proposed hazardous material transportation routes.

The above trend has driven the need for quantitative safety risk criteria. While QRA can be used to make many risk-based decisions that do not require risk criteria (for performing risk comparisons or for identifying major risk contributors) the *RCGuidelines* observe that, in some situations, risk management objectives require the outcome of the QRA be compared to a risk criteria. These include:

- <u>Making hard decisions about the tolerability of risk</u> – QRA can form the basis for absolute decisions about required actions, or even about continuing operations.
- <u>Achieving regulatory compliance</u> – Some regulatory authorities require QRA to justify the initiation or continuation of certain hazardous activities.

In all applications of QRA, it is important to recognize that the inherent uncertainties in the assumptions and the models used demand caution in making any risk-based decisions. This is particularly important when applied to absolute decisions about the tolerability of risks.

3. Key Concepts

Worldwide regulatory drivers have had a significant impact on the development and use of risk criteria. These are evaluated in detail in the *RCGuidelines* and briefly summarized in the next section. Before discussing regulatory precedents, it is important to review a few important risk concepts often appearing in literature addressing risk criteria. The definitions below are taken from the *RCGuidelines*:

- Individual Risk – Individual risk expresses the risk to a single person exposed to a hazard or hazards. In other words, individual risk is a measure of all relevant risks superimposed on any given individual. Versions of individual risk include *maximum individual risk,* usually applied to the most exposed individual, and *average individual risk,* applied to different groups of exposed populations The scale of any incident, in terms of the number of people impacted by a single event, does not affect individual risk. Individual risk is often expressed as a single number, usually the frequency per year of a fatal incident involving the individual.
- Societal Risk – Some major incidents have the potential to affect large numbers of people; *societal risk* expresses the cumulative risk to groups of people who might be affected by such events. In other words, *societal risk* measures the potential for impacts to a group of people located in the effect zone of an incident or set of incidents. Thus, societal risk estimates include a measure of incident scale in terms of the number of people impacted. Societal risk is often expressed in the form of cumulative frequency graph, called an F-N curve, with F referring to the cumulative frequency of N or more fatalities.
- As Low As Reasonably Practicable/Achievable (ALARP/A) – The is the concept that efforts to reduce risk should continue until the incremental resource requirement (in terms of time, effort, cost, or other expenditure of resources) is grossly disproportionate to the value of the incremental risk reduction achieved.
- Risk Tolerance – The *RCGuidelines* cite the United Kingdom Health and Safety Executive (UKHSE) in discussing risk tolerance [Reference 4]:

  > "..... It refers...to a willingness by society as a whole to live with a risk so as to secure certain benefits in the confidence that the risk is one that is worth taking and that it is being properly controlled. However, it does not imply that ... everyone would agree without reservation to take this risk or have it imposed on them."

  Central to the definition of risk tolerance is the concept that the public provides permission to operate and represents the ultimate arbiter of risk tolerability.

- Voluntary versus involuntary risk – This refers to the issue of whether the risk is taken on voluntarily or whether the risk is imposed, with people in the workplace generally assumed to have taken the risk voluntarily in exchange for the benefits of employment. As the *RCGuidelines* point out "This statement is particularly valid when the employer has taken reasonable and prudent measures to (1) control workplace hazards, (2) inform workers of the hazards, and (3) train workers in prevention and control of the hazards, in necessary personnel protection, and in emergency response actions." On the other hand, risks to members of the public are involuntary, even for those persons electing to live near a chemical processing facility.

4. Regulatory Precedents

QRA was pioneered in the United States, but other countries have taken the lead in setting standards for quantitative safety risk criteria. More specifically, two countries, the United Kingdom and the Netherlands, have devoted a combined total of over 100 years of effort to addressing risk criteria. They continue to monitor and update their publications and requirements to reflect new information. In this short paper it is impossible to do justice to the extensive body of work generated; however the *RCGuidelines* analyze the efforts of the UK and the Netherlands in detail.

Table 1 and Figure 1 provide an overview comparison of the UK and Netherlands individual and societal risk criteria, respectively.

Table 1 – Comparison of UK/Netherlands Individual Risk Criteria

Individual Risk Criteria, fatality/year	UK	Netherlands
Risk To 'Vulnerable Objects'		
Maximum tolerable risk to workers	10^{-3}	Dutch risk criteria do not address risks to workers
Maximum tolerable risk to the public	10^{-4}	New establishments: 10^{-6}
		Existing establishments: Beginning October 2007: 10^{-5} Beginning January 2010: 10^{-6}
Broadly acceptable (or negligible) risk to workers and public	10^{-6}	Risk criterion for negligible risk has been abandoned
Risk To 'Less Vulnerable Objects'		
Maximum tolerable risk to workers	UK does not make this distinction	The risk criteria above also apply to less vulnerable objects. However, the risk criteria are administered as target values rather than mandatory limits.
Maximum tolerable risk to the public		
Broadly acceptable (or negligible) risk to workers and public		

Figure 1 – Comparison of UK/Netherlands Societal Risk Criteria

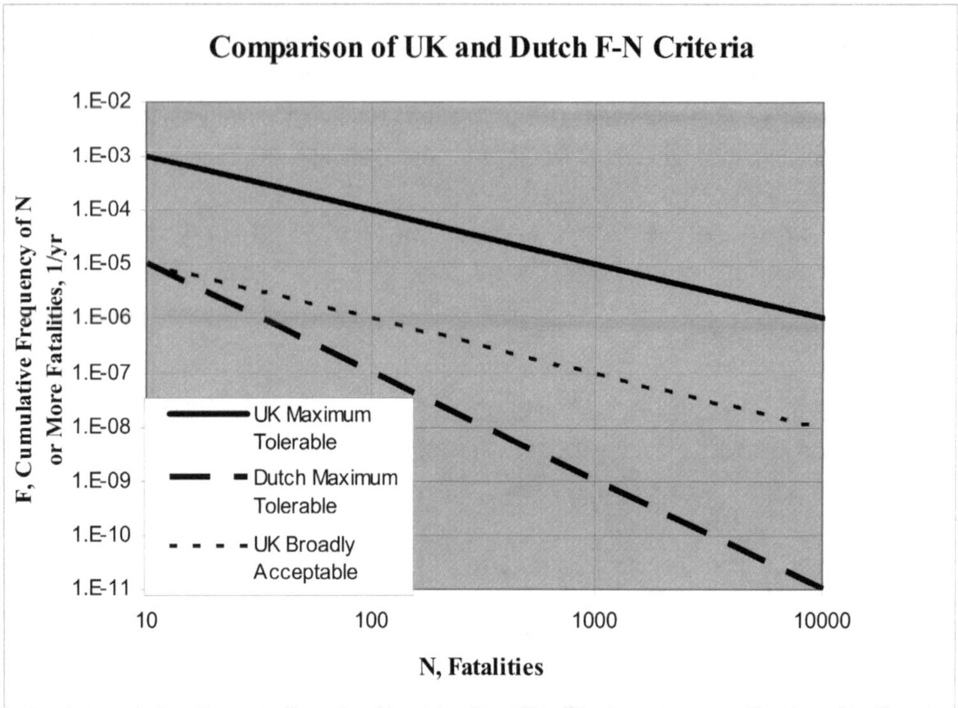

The *RCGuidelines* analyze the differences between the two sets of risk criteria – both for individual and societal risk. In the case of individual risk, these include the UK distinction for workplace risks and the Netherlands distinction between new and existing facilities, as well as the Netherlands concept of vulnerable and less vulnerable "objects".

While the development history, and differing scope and application, can help explain the disparity in individual risk criteria, the gap between the UK and Netherlands societal risk criteria appears, on initial observation, to be so large as to defy explanation – with the Netherlands criteria seemingly several orders of magnitude more conservative. The *RCGuidelines* describe a number of factors for why the two criteria are so different. However, in application, the two criteria may produce rather similar results. Some factors that tend to force this convergence are shown in Figure 2 and the *RCGuidelines* provide a comprehensive assessment of this issue. Some of the factors include:

- Difficulties in satisfying the more stringent Netherlands risk criteria – Application experience reveals challenges in complying with the stringent Netherlands criteria, especially with respect to societal risk. Multiple case studies show the criteria waived because it could not be practicably achieved.
- Risk criteria scaling – Societal risk curves are highly dependent upon the size of the system or facility to which they are applied. A small, less complex operation like a terminal would pose much different societal risks than an

operation on the scale of an entire refinery. The UK criteria are based upon major facilities, such as a refinery, while the Netherlands were initially developed for smaller facilities.

- Differences in application of the ALARP principle – The premise behind the UK approach is that risks are always being driven down, to the extent practicable. This tends to push facility risk lower, in the direction of Netherlands criteria. There is reported to be less emphasis on continuous risk reduction in the application of the Netherlands criteria.

The above discussion, and the more comprehensive analysis in the *RCGuidelines*, suggests that actual risk levels achieved from the application of the UK and Netherlands risk criteria may be similar, despite the apparent disparity between the two, and differences in starting points. While the UK and Netherlands are central to the worldwide development of risk criteria, the *RCGuidelines* also include a wide-ranging evaluation of all other available government/regulatory publications.

5. Considerations in Selecting Quantitative Risk Criteria

The *RCGuidelines* offer the following considerations in establishing risk criteria:

> "An organization seeking to establish risk criteria faces at least two challenges. The first is to ensure appropriate technical rigor and practical applicability; in other words, the risk criteria must be suitable for achieving the goal of effectively guiding risk management decision making. The second challenge is to ensure that the risk criteria can be viewed as being credible and equitable – noting that the organization may be working in an environment where the public, and perhaps employees, are skeptical of the levels of risk they perceive being imposed on them."

More specifically, the guidelines suggest the risk criteria should be:

- Consistent with societal expectations
- Technically sound
- Scalable, reflecting variations in company operations
- Understandable to users and consumers of risk information alike
- Effective in helping businesses evaluate and manage the risk of their operations
- Consistently applicable worldwide, for multinational companies
-

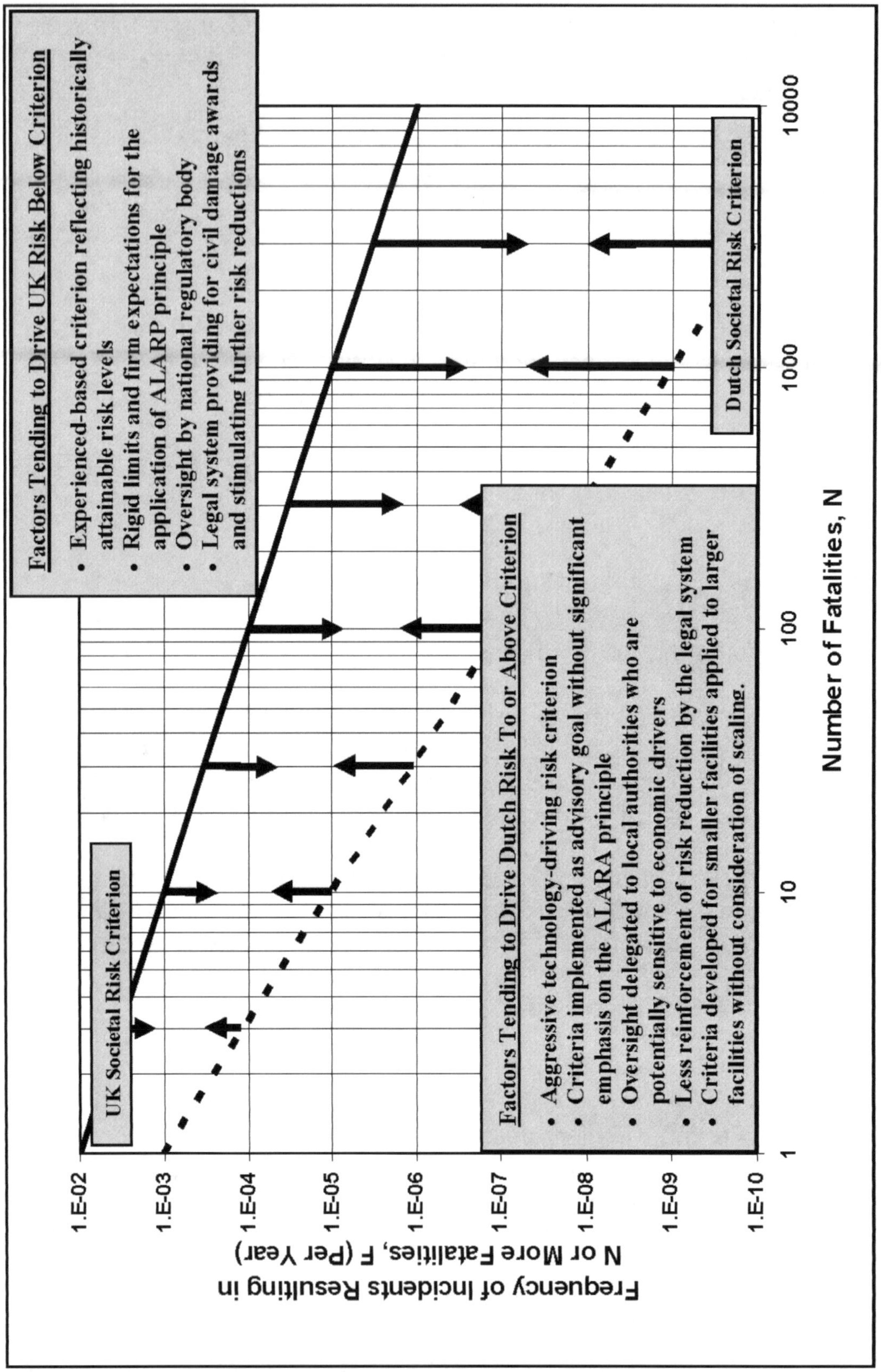

Figure 2 – Convergence of the UK and Netherlands Societal Risk Criteria

Figure 3 shows some steps to consider in developing risk criteria. This paper highlights a few key points:

- Determine the need to establish risk criteria – Before deciding whether or not to establish risk criteria, it is very important for the organization to determine how criteria will be used – what decisions will be made. As discussed above, many risk based decisions (e.g. performing risk comparisons, identifying major risk contributors) do not require risk criteria. Only if an organization needs to meet regulatory requirements, or make absolute judgments about risk does it need risk criteria.
- Determine what risk criteria to develop – Individual and societal risk criteria measure different aspects of risk. The guidelines note that "while the organization has the option of determining which types of risk it will address, industry is becoming increasingly aware of the benefits of a balanced risk perspective that considers both individual and societal risk."
- Determine philosophy for continuing risk reduction – One of more difficult risk management decisions to be made is "how much risk reduction is enough?" In other words, at what point are the resources required for further risk reduction grossly disproportionate to the amount of risk reduction achieved. The procedures for making this determination need to be established for any risk decision-making, qualitative or quantitative, and independent of the quantitative risk criteria that are used. When quantitative risk criteria are used, the ALARP concept fundamentally drives continuing risk reduction. The *RCGuidelines* state: "without ALARP, the maximum tolerable risk criterion may be perceived as the final goal of risk reduction efforts; in other words, efforts could cease once the risk level is reduced to less than the criterion."

6. Conclusions

The new CCPS *Guidelines for Developing Quantitative Safety Risk Criteria* provide a basis and framework for companies to (1) understand basic concepts of risk, risk assessment and risk management, (2) understand regulatory precedents that need to be considered when establishing risk criteria, (3) understand the types of risk criteria to be considered and steps to be taken to develop criteria (4) understand important concepts that need to be incorporated into the development of risk criteria.

This paper discussed, at a high level, a limited number of topics that are addressed in depth in the *RCGuidelines*. The reader is encouraged to reference the book for detailed analysis.

Figure 3 – Risk Criteria Development Flow Chart

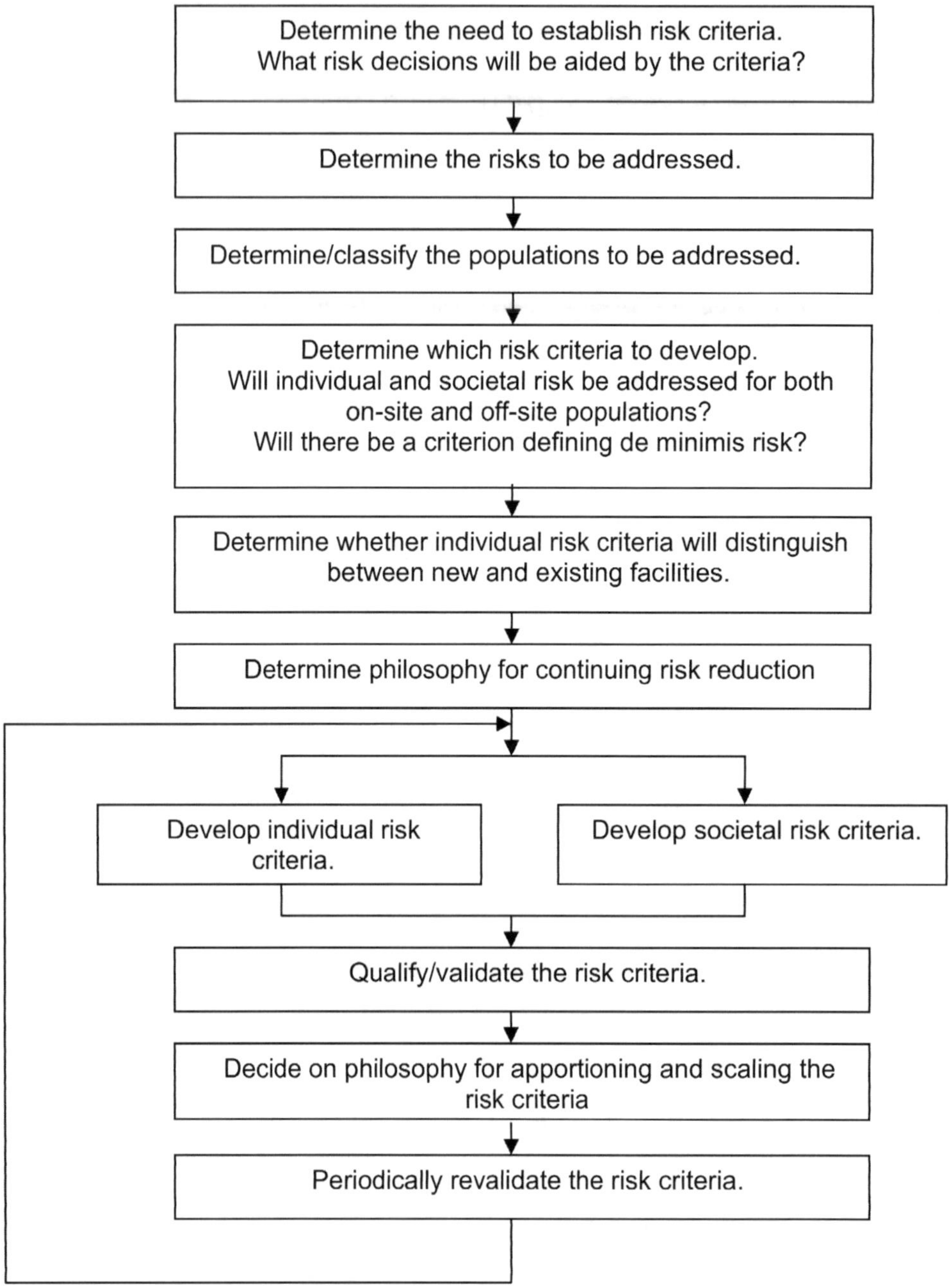

References:

1. *Tools for Making Acute Risk Decisions with Chemical Process Applications,* Center for Chemical Process Safety, New York, 1995.
2. *Guidelines for Chemical Process Quantitative Risk Analysis, Second Edition,* Center for Chemical Process Safety, New York, 2000.
3. *Evaluating Process Safety in the Chemical Industry – A User's Guide to Quantitative Risk Analysis,* Center for Chemical Process Safety, New York, 2000.
4. *Reducing Risks, Protecting People – HSE's Decision-making Process,* Her Majesty's Stationery Office, London, 2001.

94b Single Scenario Risk Criteria: Why would I want them and how do I get them?

J. Wayne Chastain
Eastman Chemical Company
Kingsport, TN 37662
chastain@eastman.com

Abstract:

Traditional criteria for risk evaluation have been based on individual risk and societal risk. Both types of risk criteria are useful in techniques such as Quantitative Risk Assessment. Individual and societal risk criteria are also values which have been specified by several governmental authorities outside the United States. In recent history, the Layer of Protection Analysis (LOPA) technique has gained prominence as a popular risk analysis method particularly for the development of Safety Integrity Levels (SILs) for Safety Instrumented Systems (SIS). Some companies are using this technique as a method to evaluate cumulative risk and comparing this calculated risk to individual or societal risk criteria. However, since LOPA is a method for evaluating a single cause – consequence scenario, it lends itself to the application of risk criteria that are calibrated for addressing the risk from individual scenarios. This paper discusses reasons for the development of scenario based risk criteria and presents a methodology of developing risk criteria for a single cause – consequence scenario starting with individual and societal risk criteria supported by one governmental authority.

The Application of Peak Risk Concepts within a Corporate Risk Program

Fred Henselwood, NOVA Chemicals, Calgary, Alberta, Canada
henselfw@novachem.com

ABSTRACT

Typically corporate risk criteria are designed for the evaluation of risks associated with ongoing activities, and are normally expressed in units consistent with an annual basis, such as events per year. This approach generally aligns well with routine activities and normal operating conditions. However, for one-time events or periodic events, this approach can be misapplied in that the short time period (exposure time or time at risk) for the occurrence of the event, may skew the results. More concerning is that this skewing of results can create an environment where high risk, short-term activities are viewed as being acceptable. For example, if an event occurs only once per year with a duration of one week (approximated as $1/50^{th}$ of a year) then the inclusion of this 1 in 50 factor within a risk assessment may result in the risk associated with this activity being deemed acceptable even if the peak risk associated with this time period is as much as 50 times greater than that which would normally be accepted. As such a hypothetical facility that produces a different product each week of the year could have an aggregate risk level that is 50 times higher than that which would be allowed if the facility were to be involved in the production of only one product over the course of the entire year. Not including the exposure time/time at risk factor within the risk assessment process can effectively solve this problem, however, this will result in a situation where considerable risk resources may be directed at one-time events. As such this approach likely does not deploy corporate capital in an efficient manner. It is therefore being proposed that organizations construct a second risk criterion to address peak risk. This second criterion attempts to balance the need to protect the organization from high risk activities while recognizing that the deployment of capital towards one-time events also has an impact on the business. This paper will discuss the development of a peak risk criterion and its application to short-term and one-time activities.

INTRODUCTION

Often, process safety risk tolerance criteria are expressed in terms of some allotted threshold value representing the likelihood of an adverse event occurrence on a per year basis[1]. For example, the Layer of Protection Analysis (LOPA) book in Appendix E[2] lists a number of fatalities per year values representing various industry risk tolerance criteria, such as 1 fatality per 10,000 years. In assessing short-term or one-time events a problem arises with respect to how you treat the time element, particularly for events which will last far shorter than one year. If one is assessing an event which is expected to last 1 month out of a year, two options can be readily identified for determining if this risk represents an acceptable risk. The first option is to look at the annual risk tolerance criterion and divide this value by 12 so as to generate a monthly risk tolerance criterion (preserve the criteria). The second option would be to consider the potential adverse event frequencies for the one month activity and then divide these frequencies by 12 to account for the fact that this event will only occur for one month out of the year (preserve the frequency). Essentially the first option treats this one month activity as though it were to occur on an ongoing basis while the second option treats this one month activity as though it were to occur followed by 11 months in which this activity is not expected to occur. The net result is that depending on which approach is utilized the risk taken on by an organization will differ.

Compelling arguments can be made to defend either approach, however, the two approaches do not agree with each other, with the second option resulting in the application of a risk threshold which is 12 times higher than that utilized in the first option. Both paths represent a cost to the business and as such how this situation is managed is of importance. Take on too much risk and the business may fail; spend too much on mitigating risk and the business may fail.

In determining which path is appropriate it is important to step back and assess what it is that the application of risk criteria is attempting to accomplish. When establishing risk criteria for managing specific systems the overall goal is to protect the business from harm. Often the intent is to look at a reasonable risk level for an organization or facility and then step this risk level down to a level that is appropriate to apply at an individual system or activity level. For example, a risk criterion that is utilized when conducting a LOPA assessment on a specific piece of equipment is likely not appropriate to apply against the aggregate risk represented by all pieces of equipment forming a particular facility. As the intent of the risk criterion is to manage the overall risk to the business the allowance of short-term activities to be associated with significantly higher risk thresholds suggests that the use of the 1 month out of 12 frequency is likely inappropriate. One could envision a risk assessment strategy where all scenarios were allotted to a day of the month and as such the acceptable aggregate risk to the organization could be shifted a factor of 30 times higher than that which was intended.

The second goal when setting risk criteria is to ensure the efficient use of capital through only mitigating those scenarios which are identified as requiring attention. As such applying the same risk threshold to one-time activities as would be applied to ongoing activities likely inefficiently directs capital towards these short-term activities. Through spending on short-term activities it is less likely that the desired return on capital would be the same as could be achieved by spending on ongoing activities. As such not accounting for the short-term nature of these activities also has a potential to harm the business.

PEAK RISK

> Definition of Peak Risk: the level of risk associated with an activity while that activity is occurring.

Recognizing the need to deploy capital efficiently and recognizing the need to manage aggregate risk the approach being proposed in this paper is a compromise between the two options. Clearly, allowing short-term activities which are associated with high risk levels is a concern as is the deployment of capital to one-time and short-term activities. The resulting proposal is that the time at risk (i.e. the 1 month of 12) term should be considered when evaluating these short-term activities, however, a minimum time at risk term of 10% is suggested as a limit on this parameter. In this way momentary activities should never be associated with a peak risk which is more than 10 times the established annual risk threshold. At the same time these short-term or one-time activities are not held to the same risk mitigation capital requirements as those which are applied to ongoing or long-term activities.

The following equation shows the risk evaluation methodology being proposed:

> Risk = Product of (Frequency of the Event, Event Severity, and Time at Risk)
>
> With a lower limit of 10% on the "Time at Risk" term.

Originally this approach was designed when addressing individual risk as it was found that in some cases one could construct situations where an individual could be exposed to an extremely high risk for a short period of time while not exceeding their annual risk allotment. Further, one could create situations whereby through alternating which individual did a particular high risk job the individual risk could be made acceptable to each individual while if only one employee was assigned to that particular high risk job the risk would not be acceptable. It was recognized that while the goal behind ensuring that individuals are at acceptable risk levels involves protecting both the individual and the organization, this approach, although protecting the individuals, was failing to protect the organization and therefore needed change. At the same time there was a recognition that individuals were exposed to different levels of risk throughout their day and as such some fluctuation in risk level was likely appropriate while the total risk to the individual needed to be managed. The concept of Peak Risk was originally developed to address these situations.

Additionally this approach allowed us to readily link geographic risk to peak individual risk in that if an area was associated with a geographic risk which was equal to or greater than the established peak individual risk level, access to this area should then be restricted. The benefit to the peak risk approach is that the time period the individual may be within the high risk area (as typically related to the evaluation of individual risk) was not required for the calculation. This linkage between peak individual risk (as applied to members of the public) and geographic risk also allowed for the establishment of appropriate geographic risk levels for the fence lines associated with our facilities.

Although peak risk concepts were originally developed for addressing individual risk scenarios the methodology has also been found to be extremely helpful when applying corporate risk criteria to addressing equipment risk. One of the areas we have found this methodology to be of greatest value is when assessing risks associated with taking safety systems out of service for activities such as maintenance or testing. As an example through Process Safety Life Management activities, as linked to tools such as LOPA, it is possible to establish the expected operational availability of various independent protection layers. However, a question arises when these protection layers are to be removed from service if the elevated risk level during this period is appropriate or if alternative protection measures need to be considered. The approach we are suggesting for these periods is to return to the original LOPA assessments and see if the protection layer in question provides greater than a factor of 10 protection (based on changes in both severity and consequence as related to the change in estimated risk). If the change in risk is less than a factor of 10 and the time period in question is short, the system in question can be removed temporarily as required without bringing in alternate protection layers. If the change in risk is greater than a factor of 10, it is likely that temporary protection measures need to be utilized during this period where the system in question will not be in service. (In some cases the original LOPA study may indicate that the risk related to the system is acceptable by a wide margin and as such the removal of the protection layer, despite having an effect greater than one order of magnitude, may not be sufficient to exceed the defined peak risk criterion.)

This methodology suggests that if a system is to be taken out of service temporary alternative protective measures should be utilized if this system provides more than an order of magnitude reduction in risk. In this way these temporary impairments can be quickly assessed and the appropriate actions identified without the need for a detailed study to be completed. As an example, if one were to consider a fire sprinkler as being critical with respect to meeting the appropriate risk criteria this methodology allows a determination of the appropriate course of action to be taken if this sprinkler system needed to be taken out of service for a short period of time. If a risk reduction factor of 100 had been credited for the sprinkler system within the appropriate risk assessments it becomes clear that during the period of time when this sprinkler system is impaired the peak risk will rise by a factor of 100. As a factor of 100 is greater than the peak risk limit of 10, this approach would suggest that while this sprinkler system is impaired other safety activities should be undertaken as the short-term risk is greater than that allowed by the company. If, as an example, it is determined that posting a fire watch will reduce the risk associated with the scenarios of concern by a factor of 10 this action would then allow for the risk associated with this system to be brought to a level which would be appropriate during this short period of time when the primary protection system (sprinkler system) is unavailable.

This linkage between Peak Risk and annual risk criteria is also of value when assessing one-time activities. As an example, if an out of service vessel were to be physically removed from a facility the allowable risk for this action could be assessed by applying the Peak Risk criteria. In essence this approach allows you to assess situations where you only have probability terms and no frequency term in that you essentially assume a frequency of once in ten years. If the likelihood of a crane dropping this vessel while removing it from the site is once in 1,000 times and this event would result in a 1 in 100 chance of fatality you now have enough information to assess if the risk is acceptable (i.e. a fatality probability of 1 in 100,000 attempts equates to a fatality frequency of once every 1,000,000 years which can then be assessed against existing risk criteria).

Another area where we found this technique to be useful was in a case where a new product was being proposed to be prepared on existing equipment. In this case we first assessed the risk based on what the risk from this process would be if the new product were to become a full time activity. Where the peak risk methodology then became of value is that it allowed us to consider what the appropriate risk level would be for the initial trial run. Using the peak risk concept an upper limit on the allowable risk during this trial run could be readily established and the need for additional safeguards could be determined. Depending on the resulting risk values this approach then gives guidance as to what is appropriate to spend on safeguards for the trial run and what is appropriate to spend on safeguards for ongoing use.

CONCLUSION

Through the development of peak risk criteria it was found to be possible to balance the need to protect the organization from short-term, high-risk activities against deploying excess capital to address risks of a relatively short duration. In addition to addressing these corporate concerns this approach was found to readily add value to situations involving individual risk and in particular allows for a methodology whereby geographic and individual risk criteria may be aligned. Further, this methodology was also found to be of value in assessing the need for alternate risk mitigation strategies when reviewing situations where required safety systems are to be temporarily impaired. In addition to addressing temporary impairment situations this methodology was also of value in assessing the need for safeguards for trial periods versus the need for safeguards for ongoing operations.

It should also be noted that the mathematics associated with this approach are relatively straight forward and in most cases this technique can be added to existing risk assessment programs without significantly increasing the complexities associated with these programs. Further, when applied to tools such as LOPA the assessment of risk for short-term activities (usually 1 month or less in duration) typically reduces the need for safety instrument systems by one relative to the requirements for ongoing activities.

RFERENCES

1. Center for Chemical Process Safety, *Guidelines for Chemical Process Quantitative Risk Analysis, 2nd Edition,* ISBN 0-8169-0720-x, American Institute of Chemical Engineers, New York, NY, 2000.

2. Center for Chemical Process Safety, *Layer Of Protection Analysis: Simplified Process Risk Assessment,* ISBN 0-8169-0811-7, American Institute of Chemical Engineers, New York, NY, 2001.

AUTHOR INDEX

American Institute of Chemical Engineers
3 Park Avenue
New York, New York 10016-5991

CURRAN ASSOCIATES INC
proceedings.com